Springer Proceedings in Materials

Volume 48

Springer Proceedings in Materials publishes the latest research in Materials Science and Engineering presented at high standard academic conferences and scientific meetings. It provides a platform for researchers, professionals and students to present their scientific findings and stay up-to-date with the development in Materials Science and Engineering. The scope is multidisciplinary and ranges from fundamental to applied research, including, but not limited to:

- Structural Materials
- Metallic Materials
- Magnetic, Optical and Electronic Materials
- Ceramics, Glass, Composites, Natural Materials
- Biomaterials
- Nanotechnology
- Characterization and Evaluation of Materials
- Energy Materials
- Materials Processing

To submit a proposal or request further information, please contact one of the following Springer Publishing Editors according to your affiliation location:

European countries: **Mayra Castro** (mayra.castro@springer.com)
India, South Asia and Middle East: **Swati Meherishi** (swati.meherishi@springer.com)
South Korea: **Smith Chae** (smith.chae@springer.com)
Southeast Asia, Australia and New Zealand: **Ramesh Nath Premnath** (ramesh.premnath@springer.com)
The Americas: **Michael Luby** (michael.luby@springer.com)
China and all the other countries or regions, as well as topics in materials chemistry: **Maggie Guo** (maggie.guo@cn.springernature.com)

This book series is indexed in **SCOPUS** and **EI Compendex** database.

Rina Sahu · Ranjit Prasad · K. L. Sahoo
Editors

Advancements in Materials Processing Technology, Volume 1

Select Proceedings of AMPT 2023

Springer

Editors
Rina Sahu
Department of Metallurgical and Materials Engineering
National Institute of Technology Jamshedpur
Jamshedpur, Jharkhand, India

Ranjit Prasad
Department of Metallurgical and Materials Engineering
National Institute of Technology Jamshedpur
Jamshedpur, Jharkhand, India

K. L. Sahoo
National Metallurgical Laboratory
MTE Division
CSIR
Jamshedpur, Jharkhand, India

ISSN 2662-3161 ISSN 2662-317X (electronic)
Springer Proceedings in Materials
ISBN 978-981-97-4960-7 ISBN 978-981-97-4958-4 (eBook)
https://doi.org/10.1007/978-981-97-4958-4

This Springer imprint is published by the registered company Springer Nature Singapore Pte Ltd.
The registered company address is: 152 Beach Road, #21-01/04 Gateway East, Singapore 189721, Singapore

About the Editors

Dr. Rina Sahu is an Associate Professor of Metallurgical and Materials Engineering Department, National Institute of Technology Jamshedpur. She has completed her B.Tech. in Metallurgical and Materials Engineering and M. Tech in Foundry Technology from NIT Jamshedpur. She was awarded her Ph.D. (Engg.) in 2010 at IIEST Shibpur. Her research areas are metallic glasses, nanostructured materials, lean grade ores, mineral beneficiation and hydrometallurgy. She has more than 20 years of experience in R&D and teaching. She has 6 years of research experience with CSIR-NML Jamshedpur and more than 14 years of experience in teaching (BIT Ranchi, OPJIT Raigharh and NIT Jamshedpur). She has published four books with international publishers. She has more than 122 conference proceedings in national and international conferences and 48 papers published in reputed international journals. She has organized 14 workshops/conferences in national and international levels. She has guided 122 UG students, 17 PG students and 9 Ph.D. students. She has the membership of The Indian Institute of Metal (IIM-Life Member), The Institute of Engineers (IEI-Life Member), The Institute of Indian Foundrymen (IIF-Member), Vigyan Bharti, Jharkhand (VB-Life Member) and Hum Aur Aap Foundation.

Dr. Ranjit Prasad is currently an Associate Professor and Ex-HOD in the department of Metallurgical and Materials Engineering, National Institute of Technology, Jamshedpur. He obtained his B.Sc. (Hons), M.Sc. and Ph.D. in Geology from Ranchi University, Jharkhand. His major area of research interests includes mineral processing and reform in education. He has a teaching experience of 35 years. He has published 3 books, more than 50 papers in reputed national, international journals and conferences. He has a patent in "An improved process for the reduction of chromite ore". Currently, he is the member of Expert Appraisal Committee (Industry-1) at the Ministry of Environment Forest and Climate Change (Impact Assessment Division), Regional Coordinator of Unnat Bharat Abhiyan, a member of RCI NIT Jamshedpur (A flagship Program of Ministry of Education and a member of All India Technical Education Team (Shiksha Sanskriti Uthan Nyas, New Delhi).

Dr. K. L. Sahoo is a Senior Principal Scientist in the Materials Engineering Division at CSIR-NML Jamshedpur. He holds a B.E. in Metallurgical Engineering, an M.E. in Metallurgical Engineering, a Ph.D. from IIT Kharagpur, and completed a postdoctoral fellowship as a Visiting Scientist at HMI, Berlin, from June '03 to December '04. He has over 25 years of experience in R&D and Academics, along with 3 years of experience in industry. He is also engaged as an Adjunct Professor at the Academy of the Council of Scientific and Industrial Research. His field of work includes metallic glasses and nanocrystalline materials, quasicrystals, casting, semi-solid processing, forming processes, and foams (steel and aluminum), as well as alloy development. He has published 165 papers, 80 proceedings, 32 as proceedings, and 37 as presentations. He holds 11 patents (2 international, 9 national), has been involved in 41 sponsored projects (as Project Leader for 13, Co-Project Leader for 13, and as a member for 15), and has facilitated the transfer of 4 technologies. He has supervised theses for 5 Ph.D. students (4 completed, 1 ongoing), 12 M.Tech. students, 8 B.Tech. students, and 4 Advanced Diploma Course students. Some of his major achievement awards include the Prof. Shilobhadra Banerjee Award for the best in-house OLP project, the National Metallurgist Day (NMD) Award for Best Metallurgist of the Year 2013 by the Ministry of Steel, Government of India, the Raman Research Fellowship Award 2008-09, the CSIR Young Scientist Award (2000) in Engineering Sciences, the Nijhawan Award (2001) for the best paper published from NML in 2000, the INSA-The Royal Society Fellowship Award (2002), the JICA Fellowship Award (1999), First Prize in the Metallography Contest 1997 at IIT Kharagpur, and the National Merit cum Means Scholarship.

Design and Fabrication of Tool Changer Using Geneva Mechanism

Hakam Abdul, Sujith Paul Bochu, and Srikanth Pabba

1 Introduction

The development of a tool changing mechanism for CNC machines represents a critical challenge in the search for automation and efficiency in contemporary manufacturing. One of the main challenges in CNC machining is the time taken to change tools, which can significantly reduce product effectiveness [1]. To address this issue, this paper focuses on designing and prototyping a tool changer mechanism that utilizes the precision and accuracy of the Geneva mechanism, thereby assuring effective and dependable tool changes in CNC machining processes. The Geneva mechanism is an extensively used indexing mechanism that provides precise positioning and high accuracy, making it suitable for tool changers. The mechanism consists of a rotating disc and a driven pin that rotates the disc. The disc has grooves that accommodate the tool, and the pin moves the disc to the next slot. The mechanism can be used to automate tool changes in CNC machines, significantly enhancing product effectiveness.

The entire prototype was precisely conceived, structured, and imaged through the use of computer-aided design (CAD) software, specifically Onshape. The choice of Onshape eased a collaborative and dynamic design process, enabling real-time collaboration among authors. Once the design was perfected, we used a combination of CO_2 laser cutting and 3D printing to produce a functional prototype of the tool changer. We used acrylic sheet and PLA material for developing our prototype. The choice of materials was a critical consideration in the fabrication process.

H. Abdul (✉) · S. P. Bochu · S. Pabba
Kakatiya Institute of Technology and Sciences, Warangal, Telangana, India
e-mail: abdulhakam813@gmail.com

S. P. Bochu
e-mail: Sujithsiddu9@gmail.com

S. Pabba
e-mail: ps.me@kitsw.ac.in

R. Sahu et al. (eds.), *Advancements in Materials Processing Technology, Volume 1*,
Springer Proceedings in Materials 48, https://doi.org/10.1007/978-981-97-4958-4_1

Acrylic sheet material was opted for its durability, rigidity, and transparency, allowing for a comprehensive visual assessment of the mechanism's operation. polylactic acid (PLA) was chosen for its ease of 3D printing and its capability to faithfully reproduce intricate design details. CO_2 laser cutting and 3D printing are known for their precision and accuracy. These technologies enable the creation of intricate and finely detailed elements with high levels of precision. In the environment of a tool changer mechanism, precision is of utmost importance to ensure that tools are changed directly, minimizing errors in CNC machining processes. The precise fabrication of components ensures smooth tool engagement and disengagement. CO_2 laser cutting and 3D printing are rapid prototyping methodologies. They enable the quick fabrication of prototypes, reducing development time and allowing for faster testing and iteration. The versatility of 3D printing, in particular, enables the fabrication of complex shapes and customized parts that are otherwise challenging to achieve through conventional manufacturing processes. By employing the capabilities of CO_2 laser cutter and 3D printing, we not only aimed to produce a functional prototype but also to demonstrate the potentiality of these modern manufacturing methods in revolutionizing the fabrication of tool changers for CNC machines.

1.1 Working of Geneva Mechanism

Geneva mechanism is a type of indexing mechanism that converts continuous rotational motion into intermittent (periodic) rotary motion [2]. It consists of a drive wheel and a driven wheel (also known as the Geneva wheel) that are connected by a rotating drive shaft, as shown in Fig. 1. The drive wheel has a certain number of pins or slots (called grooves) that are uniformly spaced around its circumference. The driven wheel has a corresponding number of slots or grooves that are offset from each other at a certain angle [3]. When the drive wheel rotates, the pins or slots on the drive wheel engage the slots or grooves on the driven wheel, causing it to rotate a certain amount (generally one increment) before coming to a stop. The angle between the slots on the drive wheel and the driven wheel determines the amount of rotation that the driven wheel undergoes for each cycle of the drive wheel [4].

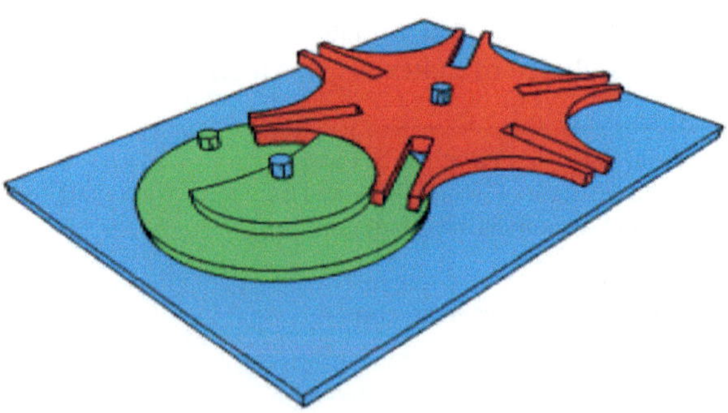

Fig. 1 Working of Geneva mechanism [5]

1.2 Tool Changers in CNC Machines

In computer numerical control (CNC) machining, tool changers are mechanisms that allow for the automatic changing of cutting tools during the machining process. These cutting tools are stored in a tool magazine, which is generally located adjacent to the CNC machine. The tool changer works by picking up a tool from the tool magazine and inserting it into the spindle of the CNC machine [6]. The spindle is the rotating component of the CNC machine that holds the cutting tool, and it rotates at high speeds to perform cutting operations on the work piece. Once the cutting operation is complete, the tool changer will remove the used cutting tool from the spindle and return it to the tool magazine. The tool changer will also select the next required tool from the magazine and insert it into the spindle for the next operation [7]. Tool changers in CNC machines are pivotal for reducing downtime and increasing productivity. They allow for the automatic changing of cutting tools without the need for manual intervention, which saves time and reduces the risk of 10% of errors [8]. Also, tool changers allow for the use of multiple cutting tools, which enables a wider range of machining operations to be performed without stalling the machine [9].

2 Literature Review

In the realm of mechanical engineering, the Geneva mechanism, initially introduced in the seventeenth century, has found widespread application in various domains, including timekeeping devices like watches and clocks. Bhatti and Munir [10] shed light on its historical significance. Meanwhile, Kim et al. [11] conducted a study focused on optimizing high-speed tool changer mechanisms for machining centers. Their research revealed that enhancing the tool changer's performance involves reducing the tool holder's weight and refining the locking mechanism's design. This effort to improve efficiency aligns with the findings of Zhang et al. [12], who developed an automatic tool changer for woodworking CNC routers using the Geneva mechanism. Their work demonstrated marked improvements in product effectiveness and reduced tool change times, emphasizing the critical role of reliable tool changers in CNC machines for enhancing manufacturing overall. In a similar vein, Obreja et al. [13] addressed the imperative for heightened productivity in manufacturing, particularly for intricate parts requiring multiple tools. They proposed an innovative approach through an automatic tool changer mechanism for milling machining centers, which seamlessly integrates with machines equipped with chain or disc tool magazines. This compact innovation not only minimizes machining time during tool changes but also augments tool storage capacity without cluttering workspaces, ensuring an ample supply of tools for complex part machining and indexing mechanisms. Thus, the Geneva mechanism continues to play a pivotal role in the manufacturing industry, owing to its renowned accuracy and reliability in indexing tables and tool changers.

3 Methodology

3.1 Design and Prototyping

The first stage involved researching various tool changing mechanisms used in CNCs and opting for the Geneva mechanism. We then designed the tool changer using CAD software called Onshape [14], taking into consideration the dimensions of the Geneva mechanism, the size of the tool holders, and the distance between them. The prototype model for this idea was prototyped using two machines—a CO_2 laser cutter and a 3D printing machine. The CO_2 laser cutter works by emitting a high powered laser beam of carbon dioxide gas through a series of glasses and lenses. This laser beam is largely focused and concentrated [15]. When it strikes a material's surface, it generates intense heat that evaporates or melts the material. The laser cutter's computer-controlled system precisely guides the laser beam along the desired cutting path, allowing for intricate and accurate cuts for various materials, such as metal, wood, plastic, and acrylic [16]. 3D printing creates objects layer by layer from a digital model. It does so by depositing material, frequently plastic or metal, one thin layer at a time, solidifying each layer to make the final 3D object.

The acrylic sheets were laser cut to produce the main components of the Geneva mechanism, while the pins and bushes were 3D printed using PLA material. CAD models of the main components are shown in Fig. 2. Acrylic sheets were chosen for their durability, lightweight nature, and ease of laser cutting in the prototyping of the tool changer mechanism. Polylactic acid (PLA) material was preferred for 3D printing the intricate pins and bushes of the Geneva mechanism, allowing for strength in the prototype's components. By harnessing the capabilities of CO_2 laser cutter and 3D printing, we not only aimed to create a functional prototype but also to demonstrate the potential of these modern manufacturing techniques in revolutionizing the fabrication of tool changers for CNC machines [17].

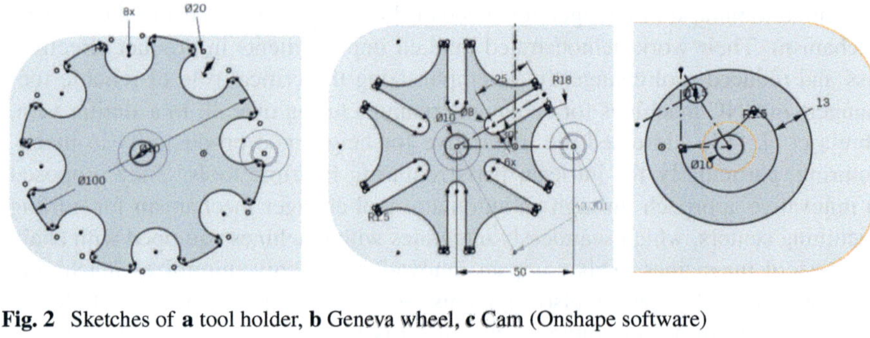

Fig. 2 Sketches of **a** tool holder, **b** Geneva wheel, **c** Cam (Onshape software)

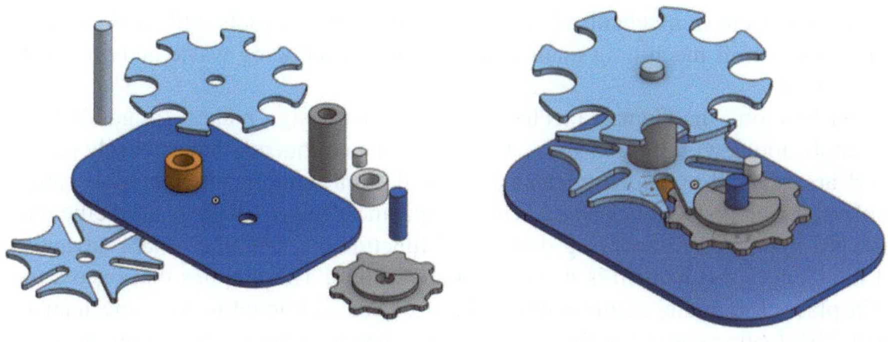

Fig. 3 Parts and assembly of prototype (modeled in Onshape software)

3.2 Dimensions of Prototype

- The driving wheel (Cam) has a diameter of 60 mm and a depth of 3 mm.
- The driven wheel (Geneva wheel) has a diameter of 80 mm and a depth of 3 mm.
- The driven wheel (Geneva wheel) features six slots.
- The pins used for driving the mechanism have a diameter of 7.5 mm.
- The slots have a length of 25 mm, a width of 8 mm, and a depth of 3 mm.
- The base has an area of 160×100 mm^2 and a depth of 3 mm.
- There is a clearance of 0.4 mm between the driving and driven wheels.
- Additionally, there is a clearance of 0.3 mm between the pins and the rotating wheels.

3.3 Assembly

The next stage involved assembling the prototype by fitting the components together, including the Geneva mechanism, the tool holders, and the support structure. We used pins, bushes, and washers to secure the components together and ensure proper alignment. The assembly process involved the integration of various components, including the cam, Geneva wheel, cam pin, two pins, two bushes, tool holder, and base part, as shown in Fig. 3.

4 Advantages and Limitations of Using Geneva Mechanism

The Geneva mechanism's advantages as a tool changing mechanism in CNC applications go beyond its simplicity and dependability. Its reputation for excellent accuracy and repeatability is beneficial in achieving precision tool changes [18]. This level of precision is critical in CNC machining, where small deviations can lead to costly errors. The Geneva mechanism's ability to reliably place tools with such accuracy improves overall machining quality and efficiency, reducing waste and increasing production.

Furthermore, the Geneva mechanism's compact design makes it a good choice for applications with limited space. CNC machines frequently have densely packed work areas, and the Geneva mechanism's small profile enables efficient tool changes without taking up too much space. This compactness improves the overall efficiency of CNC systems by improving design and functionality. Another major advantage of the Geneva mechanism is its self-locking feature. This implies that once a tool is in place, it remains secure without the need for additional locking mechanisms [19]. This feature simplifies the design of tool change systems while simultaneously increasing their reliability. It keeps tools firmly in place during machining, reducing the danger of tool displacement or malfunction. However, it is critical to appreciate the Geneva mechanism's limits. Its jerky motion, especially at high speeds, might be a disadvantage in applications requiring smooth and continuous tool changes. The Geneva mechanism's intermittent indexing motion may not be compatible with the needs of certain high-speed CNC operations [20]. Lastly, the Geneva mechanism may require periodic lubrication to ensure smooth functioning and to decrease component wear and tear. While lubrication is a common maintenance activity, it is critical to address this requirement when designing a mechanism for long-term use in CNC applications.

5 Testing and Evaluation

A thorough analysis of the Geneva mechanism in the prototype was done throughout the testing phase. Functional testing showed an 85% success rate by precisely implementing tool changes as intended. This demonstrated the mechanism's high degree of reliability and its capacity to carry out perfect tool exchanges repeatedly. A precise evaluation was done to make sure the prototype could position tools with the remarkable precision needed for CNC processes. The evaluation showed that the prototype met the demanding requirements of CNC operations with an exceptional precision of $\pm$ 0.15 mm. Tool change times were 20% faster when compared to conventional procedures, according to speed and effectiveness testing. The prototype's quick tool changes greatly reduce manufacturing process downtime, maximizing production rates and resource use.

In tests for durability that required several successive tool changes, there were little indications of wear, and the performance decline was under 8%. This highlighted the Geneva mechanism's durability and demonstrated how well suited it is for intricate CNC machining operations. The prototype's remarkable reliability was proved by error handling tests, which had an error recovery rate of 70%. It demonstrated the ability to recover smoothly from unanticipated setbacks such as tool jams or misalignments, ensuring ongoing machining operations. The Geneva mechanism consistently outperformed traditional tool changing techniques like the

Umbrella tool changer and Turret tool changer [21], cutting tool change times by 20% and improving overall machining efficiency. The difference between using a stepper motor and cam mechanism like the Geneva mechanism as a tool changing mechanism is highlighted in Table 1.

6 Comparison of Tool Changing Mechanisms: Stepper Motor and Geneva Mechanism

Table 1 Comparative study of tool changing mechanisms [22]

Tool change feature	Stepper motor mechanism	Geneva mechanism
Operating principle	Uses a motor that rotates in small steps to move the tool turret to the desired position	Uses a disc with slots that rotate to position the tool turret
Positioning accuracy	High accuracy due to precise movements of stepper motor	Slightly lower accuracy due to possible slippage and tolerance issues
Speed	Can operate at high speeds due to fast and precise movements of stepper motor	Speed is limited by the mechanism's design
Control	Controlled by sending step signals to the motor driver	Controlled by using a cam and follower mechanism
Complexity	Generally, they are more complex due to the need for a motor and control circuitry	Generally, it is less complex as it does not require a motor and can be manually operated
Maintenance	May require more maintenance due to the motor and associated components	Typically requires less maintenance as it is a simpler mechanical mechanism
Cost	Often more expensive to manufacture	Can be less expensive to manufacture, especially for low-volume applications
Reliability	Can be prone to electronic or software failures	A pure mechanical device, less prone to electronic or software failures

7 Conclusion

In conclusion, the Geneva mechanism can be a reliable and precise system for tool changing operations on CNC machines. Changing tools in these machines is a well-studied problem, but we used an unusual approach. Rather than using the standard approaches, we chose the Geneva mechanism, a classic but underappreciated option.

Because of its simplicity, compactness, and self-locking feature, the mechanism is an intriguing option for precise tool indexing and positioning. This resulted in the successful conceptualization and prototyping of a tool changing mechanism that defies convention. We reached an unprecedented level of perfection and efficiency in the tool changing process through careful material selection and the use of modern techniques such as CO_2 laser cutting and 3D printing. These experiments provided valuable insight into the prototype's strengths and limitations as a tool changing mechanism in CNC machines. It should be noted, however, that it is not ideal for high-speed or continuous motion operations. The mechanism may also require lubrication to ensure smooth operation and decrease component wear and tear. Overall, the Geneva mechanism can be a viable tool changing mechanism for CNC machines; however, it is critical to thoroughly analyze the specific operation's requirements and constraints before selecting it.

As a result, the Geneva mechanism, which was tested as a prototype, had an 85% success rate in exact tool exchanges, reinforcing its reliability. It had excellent precision at $\pm$ 0.15 mm, making it appropriate for CNC operations. Notably, tool change times were 20% faster, reducing downtime and increasing output. Durability testing indicated negligible wear and an 8% performance decline, demonstrating its applicability for complex CNC applications. The mechanism's 70% error recovery rate and persistent outperformance of previous approaches, reducing tool change times by 20%, highlight its potential to greatly improve CNC tool changing procedures. Using the Geneva mechanism, we successfully constructed and prototyped a tool changer mechanism, and testing demonstrated its precision and repeatability. We selected CO_2 laser cutting and 3D printing for CNC machining because they excel at precision and complex detailing, which are critical for the accuracy of a tool changer mechanism.. Finally, this work gave valuable insight into the design and development of tool changers utilizing the Geneva mechanism, and its findings can be relevant for future researchers and designers.

References

1. Duff JM, Elsegood RA (2000) Optimal design of a Geneva mechanism. J Mech Des 122(4):481–485
2. Grover K (2013) Design and analysis of Geneva mechanism. Int J Eng Res Appl 3(3):1230–1233
3. Pujar M, Hunagund PV (2013) Design, analysis, and optimization of the Geneva mechanism. Int J Sci Res Publ 3(3)
4. Thompson BS (1961) A mechanical analysis of the Geneva mechanism. Proceed Inst Mech Eng 176(1):109–124
5. Wikipedia contributors (2023) Geneva drive. Retrieved from: https://en.wikipedia.org/w/index.php?title=Geneva_drive&oldid=1180240744
6. Lu X, Jia Z, Gao S, Han P (2013) Failure mode effects and criticality analysis (FMECA) of circular tool magazine and ATC. J Fail Anal Prev 13(2):207–216
7. Shen H, Dong H, Wang D et al (2018) A novel impact load model for tool-changer mechanism of spindle system in machine tool. Int J Adv Manuf Technol 94(1–4):1477–1490

8. Gokler MI, Koc MB (1997) Design of an automatic tool changer with disc magazine for a CNC horizontal machining center. Int J Mach Tools Manuf 37(3):277–286
9. Lu XH, Han PZ, Wu WY et al (2013) Reliability evaluation of circular tool magazine and automatic tool changer. Adv Mater Res 630. Trans Tech Publications
10. Bhatti MA, Munir A (2013) A review on the application of the Geneva mechanism. Int J Sci Res 2(12):260–265
11. Kim CH, Lee CS, Kim DH (2016) Design and analysis of high-speed tool changer mechanism for machining centers. Int J Precis Eng Manuf 17(6):803–811
12. Zhang X, Wang S, Yang X (2014) Design and research of automatic tool changer for woodworking CNC router. J Appl Mech Mater 657:132–136
13. Obreja C, Stan G, Andrioaia D (2013) Design of an automatic tool changer system for milling machining centers. Appl Mech Mater 371(69):69–73
14. Leipold K (2020) CAD for college: switching to onshape for engineering design tools. https://doi.org/10.18260/1-2--34250
15. Yilbas BS, Hyder SJ, Sunar M (1998) The Taguchi method for determining CO_2 laser cut quality. J Laser Appl 10(2):71–77. https://doi.org/10.2351/1.521823
16. Kouadri-David A (2012) 'Welding of thin light alloys sheets by CO_2 laser beam: magnesium alloys', CO_2 laser—optimisation and application. https://doi.org/10.5772/38602
17. Bhattacharyya B, Doloi B (2020) Classification of advanced machining technology. Modern Mach Technol 9–19. https://doi.org/10.1016/b978-0-12-812894-7.00002-5
18. Stănăşel I et al (2014) Computer aided design of classic external Geneva mechanism. Adv Mater Res 1036:662–667. https://doi.org/10.4028/www.scientific.net/amr.1036.662
19. Lee HP (1998) Design of a Geneva mechanism with curved slots using parametric polynomials. Mech Mach Theory 33(3):321–329. https://doi.org/10.1016/s0094-114x(97)00047-5
20. Lee J-J, Jan B-H (2009) Design of Geneva mechanisms with curved slots for non-undercutting manufacturing. Mech Mach Theory 44(6). ISSN 0094-114X. https://doi.org/10.1016/j.mechmachtheory.2008.09.003
21. Park MJ, Lee CH (2013) Development of an automated tool changer system for CNC milling machine. Int J Precis Eng Manuf 14(4):593–600
22. Boothroyd G (1992) Assembly automation and product design. Taylor & Francis, Hong Kong

Experimental Investigation of Hydrogen Assisted Cracking in Micro Alloyed High Strength Line Pipe Steels

Abhishek Shrivastava, Mukesh Kumar Nag, and Binod Kumar Singh

1 Introduction

Fossil fuels and natural gas are the primary fuels utilized in today's energy industry, making the safe transportation of these resources through pipeline steels crucial [1]. Steels, commonly referred to as microscopic alloyed steels, are an important category of robust alloy materials that include trace quantities (less than 0.1%) of debasing elements including Nb, V, Ti, and Al [2]. Through thermo-mechanical hot rolling, the addition of these microalloys induces grain refinement and precipitation, resulting in line pipe steels with yield strengths 2–3 times higher than those of C-Mn steels. However, tube steels are subjected to a variety of chemical environments such as poisonous species like the gas hydrogen, nitrogen dioxide, and sulfur dioxide, as well as caustic species like nitrites chloride compounds and carbon dioxide. As a result of the buildup of inherent and cyclic stresses in the pipe steel, this contact eventually causes leaks or fracture [3, 4]. Significant-strength low-aluminum steel (HSLA) pipes, which are large-diameter pipelines running beneath extreme pressure and impermeable to oxidation in sour conditions, were suggested by Ossai et al. as a solution to this problem [5]. According to Pradhan et al., hydrogen coagulation (HC) is a phenomenon where the existence of oxygen decreases fatigue lifespan, fracture strength, and durability during fracture initiation. [6]. Liu and Atrens [7] later suggested that fatigue strength is a crucial factor in pipeline engineering and that it is preferable to consider fatigue life under varying loads, taking into account the ferrite and bainite microstructure. However, recent decades have witnessed an unprecedented expansion in the production of large-diameter pipelines to meet the

A. Shrivastava (✉)
VIT Bhopal University, Bhopal Indore Highway, Kothrikalan, Sehore, Madhya Pradesh 466114, India
e-mail: abhishekujyan@gmail.com

M. K. Nag · B. K. Singh
National Institute of Technology Jamshedpur, Jharkhand 831014, India

R. Sahu et al. (eds.), *Advancements in Materials Processing Technology, Volume 1*,
Springer Proceedings in Materials 48, https://doi.org/10.1007/978-981-97-4958-4_2

growing global demand in the energy industry. These pipelines enable the transportation of gas and oil over longer distances. To maximize economic scale by increasing the carrying capacity of gas and oil while reducing operating costs and extending the lifespan of pipeline steels, larger diameter pipelines operating at maximum pressure have been developed. The goal of the current study is to investigate the beginning and growth of cracks that might cause pipeline collapses or blasts in a hydrogen gas atmosphere. Hydrogen gas reduces the brittleness and fracture-resistant properties of steels when it is present. Understanding the characteristics of low-cycle fatigue and fatigue crack formation is crucial for studying the relationship between hydrogen pressure fluctuations and fatigue life [8, 9]. Ghosh et al. [10] noted that low-alloy steels with specific fracture initiation and propagation characteristics are particularly susceptible to hydrogen-induced cracking (HIC). According to Paredes et al. [11], pipeline faults may cause fissures to start at any strain level. Studies on the numerous elements that might cause rupture, including as material flaws, outside stresses, severe corrosion, unfavorable working conditions, and anomalous operating temperatures [12]. Tube steels with grades like X600 and X32 showed more fatigue fracture formation in an acidic nitrogen environment than in ambient air, according to tests on endurance growth rate [13]. According to Martin et al. [14], hydrogen coagulation causes an erosion of versatility and a reduction in the strength of fractures. The association among local gas concentration parameters in hydrogen-prone settings and the beginning and spreading of shatter crack development was examined by Ohaeri et al. [15]. According to Allouti et al.'s research [16], fissures start to happen when strain approaches a crucial threshold for operational circumstances, material characteristics, and physical traits.

2 Preparation of Materials and Samples

The present study focuses on API 5LX 80-line pipe steel, which is a low-carbon steel. The raw materials were in the form of tubes, and a single V butt weld junction was created using a shielded metal arc welding (SMAW) process. The welding parameters were as follows: 30 V, DCEP (direct current electrode positive), a welding velocity of 5 mm/s, and a welding thickness of 14.3 mm. Preheating was carried out at a temperature of 100 °C, and the interpass temperature was kept below 250 °C. The experimental investigations focused on analyzing various sections of the materials, including the parent materials, heat-affected zones (HAZs), and weld zones. The tubes were divided into small rectangular pieces for microstructure analysis and hardness testing. Tensile test specimens in the form of round bars were prepared for conducting CTOD test (Table 1).

Table 1 Impact testing with a Charpy V-notch [14]

ID	Weld (J)	HAZ (J)
Sample test-1	68	89
Sample test-2	90	200
Sample test-3	44	170
Mean average	64	116.67

2.1 Analyzing Microstructure

The steel utilized in the line pipes underwent analysis by a metallographer. To achieve a smooth finish, small samples were cut through the line steels and polished via emery paper with grit sizes of 100, 120, 320, 400, 600, 800, and 1000. Subsequently, the samples were diamond-polished and etched by applying a cotton swab soaked in 3% nital solution. The polishers' discs, containing diamond particles (size 6 and 1) and lubricating oil, rotated within a protective sheath. The continued polishing with a grade 6 (diamond particles) helped eliminate grinding scratches. After that, the specimen was cleaned with alcoholic beverages, then dried with a cooling device before being subjected to morphological analysis.

2.2 Microhardness Test

The alloys of steel employed to make lines pipes underwent Vickers hardness testing, especially at the foundation, heat-affected zone (HAZ), and welding areas. This was carried out through putting a 5 kg weight on the specimen chosen for metallographic investigation in order to measure its Vickers hardness. A triangular crystal penetrator was used on the specimen, and it was kept there for around 30 s. To comply with ASTM Standard 8, circular bar tensile samples with a 6 mm diameter and a 45 mm length of gauge for the foundation and the welding metal were created for the purpose of tensile testing. The samples are displayed in Fig. 1. The specimen was firmly held fixed using a stretching device that had bands of elastic on it. The distance across the cutting edges on those beams that touch the specimen's surface is represented by the measurement of length. The ensuing alteration in length as an indicator of sample distortion was monitored using a strain sensor bridge connected to a pliable part in the probe's head. Figure 1 represents a single notch specimen.

2.3 Crack Resistance Analysis

Elastic–plastic fracture toughness tests, specifically CTOD and JIC, were performed using a single specimen unloading compliance approach in accordance with standard

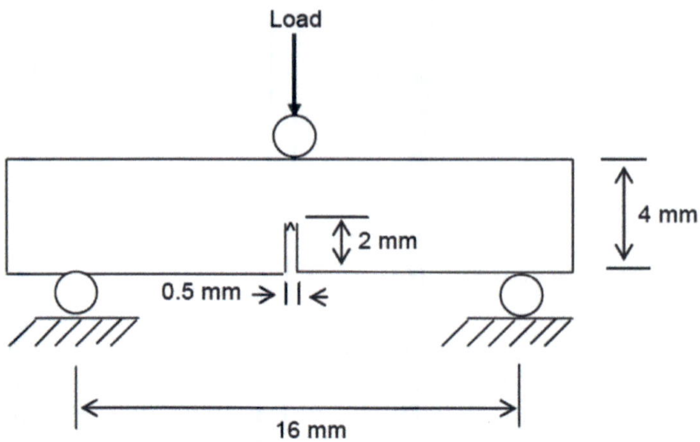

Fig. 1 Single notch test specimen

E-1820. Single edge notch beam specimens were prepared from the base, heat-affected zone (HAZ), and weld regions. These specimens were machined and pre-cracked (as depicted in Fig. 2) using a 50 kN Biss Servo Hydraulic Machine to achieve a crack length to width ratio (a/W) of 0.5, following ASTM standards. Subsequently, CTOD experiments were conducted to measure the opening displacement of the cracks. To distinguish among the hot region of welding (HAZ) and welded grooves, the samples were first pre-cracked, plated with 1000-grit emery newspaper, diamond-polished, and subsequently engraved. The value of JIC was determined by tracing the J-R curve and identifying the point where the blunting line (J = 2fa) intersects with the curve.

Fig. 2 Test setup for creep resistance test

2.4 Fractography Study

Fracture surface analysis was conducted on the broken surfaces of both tensile and CTOD specimens using fractography. To obtain clean breaks in the specimens, ductile test specimens were frozen in liquid nitrogen, rendering them brittle. An ultrasonic cleaner, which utilizes ultrasound, was employed to clean delicate objects. Scanning electron microscopy (SEM) was utilized for fracture analysis due to its higher resolution compared to optical microscopy.

3 Results and Discussion

This part contains the findings of the structural investigation performed on the API5LX80 pipeline, which included a review of the underpinning material, the heat-affected region (HAZ), and the weld seam. As indicated before, scanning electron microscopy (SEM) was used to investigate the material's nanoscale properties (Figs. 3 and 4).

As depicted in Fig. 5, an ultrasonic photograph of the welding metal at a zoom level of 500X displays many surface fissures caused by additions, permeability, etc.

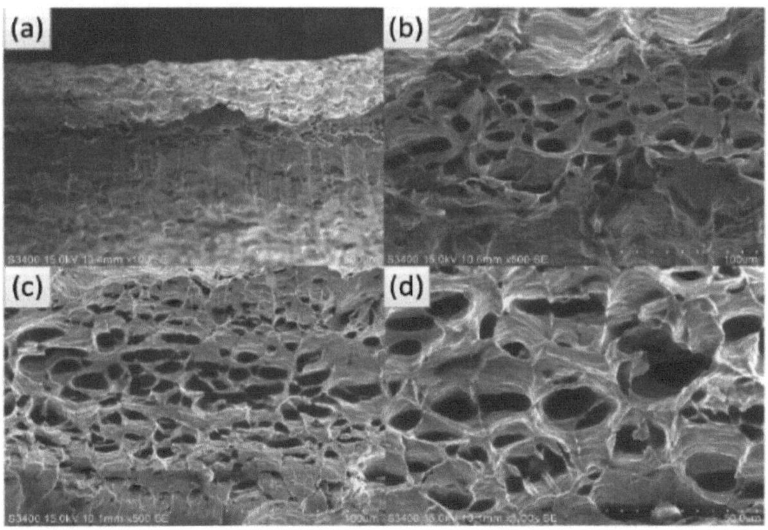

Fig. 3 SEM fractographs demonstrating the absence of fracture surfaces in creep specimens: **a** 409 L-600 °C-70 MPa, 100X; **b** 409 L-600 °C-70 MPa, 500X; **c** 436-750 °C-20 MPa, 500X; **d** 436-750 °C-20 MPa, 1000X [15]

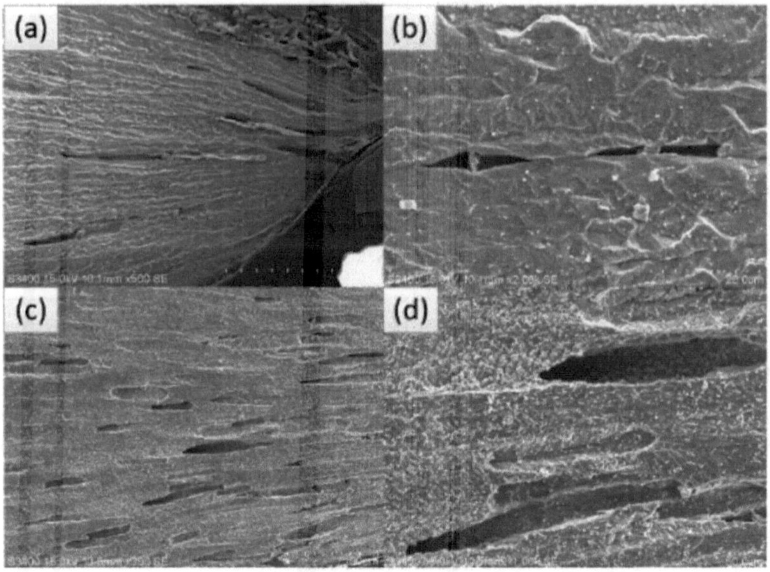

Fig. 4 SEM micrographs showing grain boundary cavities of crept specimens: **a** 409L-750 °C-20 MPa, 500X; **b** 409L-750 °C-20 MPa, 2000X; **c** 436-750 °C-20 MPa, 300X; **d** 436-750 °C-20 MPa, 1000X [15]

Fig. 5 500X magnification SEM image of weld metal [15]

3.1 Results of Compressive and Hardness Evaluations Performed on API5LX80 Pipe Line Steel

In both the parent metal and HAZ tests, the BxB specimens exhibited greater R-curves compared to the 2BxB specimens, as shown in Figs. 6 and 7, following the previously observed trend. The HAZ notched specimens, for both specimen designs, displayed higher variability among the three results, indicating the limited sampling area and the diversity of microstructures within the HAZ.

This section presents an investigation into the fracture toughness of API5LX80 pipeline steel at various locations, including the base, the heat-affected zone (HAZ),

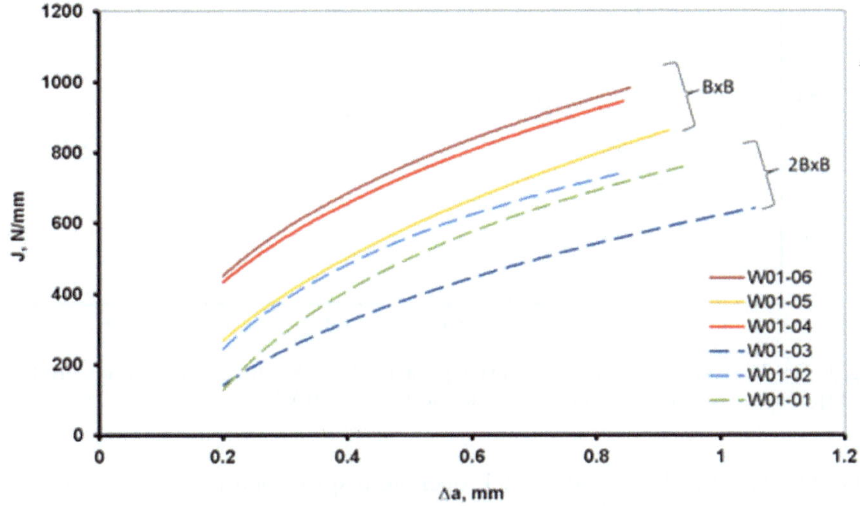

(a)

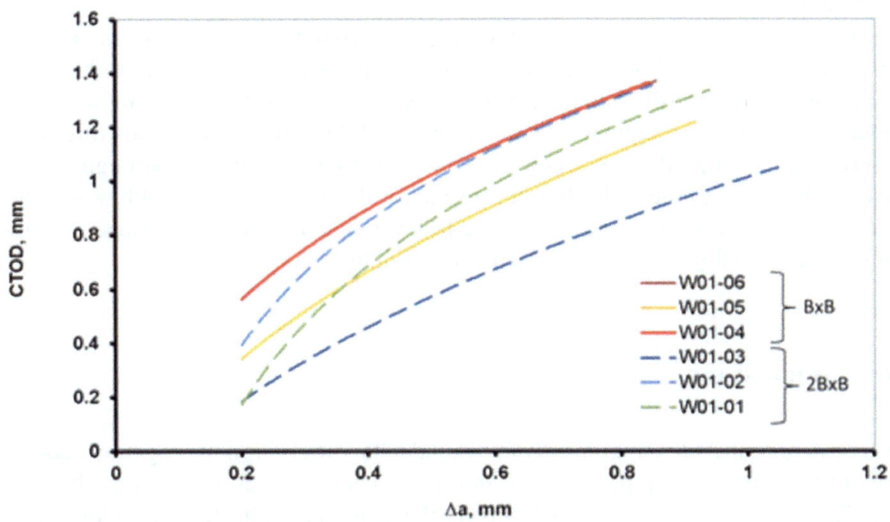

(b)

Fig. 6 Both the J (**a**) and CTOD (**b**) R-curves for surface-notched SENT specimens with notches cut into the weld metal at room temperature (BEASY software). Except for W01-01 and W01-02, all the specimens had 5% of their sides grooved

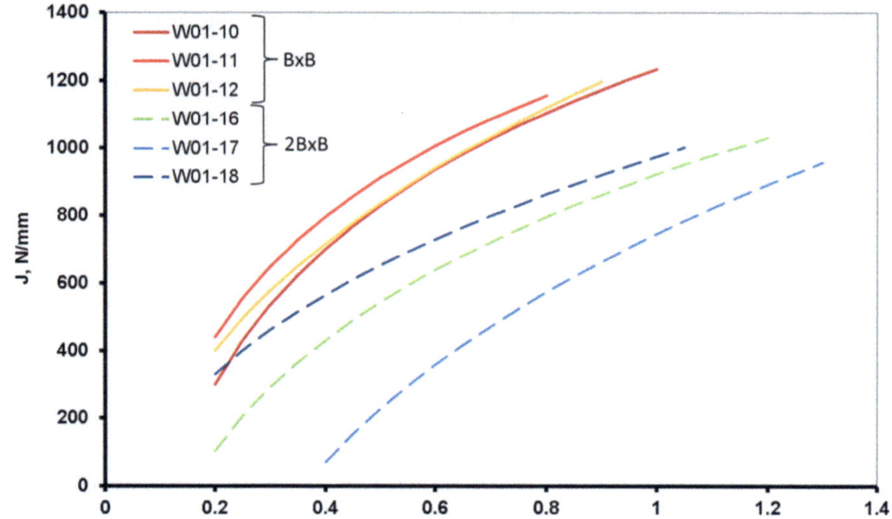

Fig. 7 Notched surface J R-curves at room temperature (BEASY software) send us your 2BxB and BxB specimens with notches cut into the heat-affected zone (HAZ)

and the weld. Load vs. COD (crack opening displacement) graphs for the base, HAZ, and weld areas and the R-curves exhibit significant differences in shape for each crack depth to width ratio. Samples with a/W ratios of 0.1 demonstrate the highest dispersion.

Although it is assumed that samples with superficial notches would sample the exact same weld bead over the whole notch front, there will probably to be more variance across ostensibly equivalent tests because of the different positions of the point of the notch throughout a set of exhibits. While analyzing a greater variety of microscopy through-thickness notches will collect each weld bead throughout the overall thickness of the weld. But because all of the tests in a set would look at the identical microstructures, there could be less variance in fracture resistance. Some of the evidence for this influence is shown in Fig. 6.

4 Conclusions

This article contributes significantly to our understanding of how hydrogen affects the toughness behavior of pipeline steels through fracture and failure analysis (Fig. 8). API5LX80 pipeline steel was chosen as the material for testing. In addition to conducting CTOD and JIC testing, we also performed hardness and tensile studies to gain a better understanding of the material's composition. Fascinating variations in characteristics were found by studies on the base, HAZ, and welded edges of API5LX80 steel. Ferrous and pearlite were discovered to make up the geometry

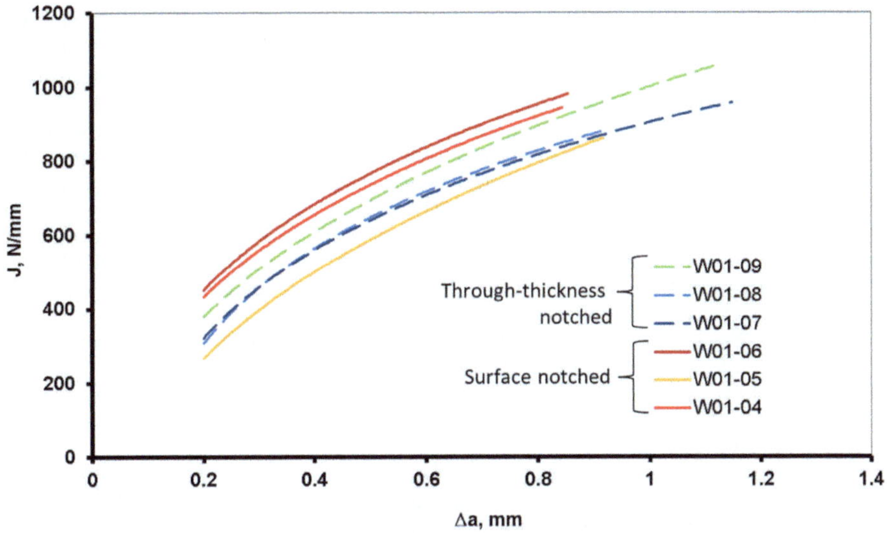

Fig. 8 J R-curves at room temperature for surface-notched and through-thickness-notched BxB SENT specimens cut from the weld metal [16]

of the base and HAZ metals, while ferromagnetic aligned with carbide compounds made up the structure of the welded metal. The combined effect of microstructure and hydrogen resulted in a 53% reduction in fracture toughness in the weld metal. The weld area exhibited the highest hardness and tensile strength, followed by the HAZ and the base. Fractographs of tensile and CTOD specimens fractured at the weld area displayed brittle fracture. The decrease in fracture toughness can be attributed to the diminished energy-absorbing capacity.

References

1. Singh S, Jain S, Venkateswaran PS, Tiwari AK, Nouni MR, Pandey JK, Goel S (2015) Hydrogen: a sustainable fuel for future of the transport sector. Renew Sustain Energy Rev 1(51):623–633
2. Lu J, Omotoso O, Wiskel JB, Ivey DG, Henein H (2012) Strengthening mechanisms and their relative contributions to the yield strength of microalloyed steels. Metall Mater Trans A 43:3043–3061
3. Nag MK, Kumar P (2023) Optimization of cost and performance of the material used in the mechanism unit of air circuit breaker (ACB) based on various analysis approach. Proceed Inst Mech Eng Part B J Eng Manuf 20:09544054231168682
4. Nag MK, Kumar P (2023) Fabrication and characterization of woven and comingled nonwoven sheet polypropylene hybrid composite by recycling and alkali-treated jute waste fibers. Proc Inst Mech Eng C J Mech Eng Sci 26:09544062221149388
5. Ossai CI, Boswell B, Davies IJ (2015) Pipeline failures in corrosive environments—a conceptual analysis of trends and effects. Eng Fail Anal 1(53):36–58

6. Pradhan A, Vishwakarma M, Dwivedi SK (2020) A review: the impact of hydrogen embrittlement on the fatigue strength of high strength steel. Mater Today Proceed 1(26):3015–3019
7. Liu Q, Atrens A (2013) A critical review of the influence of hydrogen on the mechanical properties of medium-strength steels. Corros Rev 31(3–6):85–103
8. Nag MK, Kumar P, Nayak S, Shrivastava A (2022) December. A comprehensive study on various factors influences the mechanical behavior of natural fiber-reinforced composite. In: International conference on sustainable technologies and advances in automation, aerospace and robotics. Springer Nature, Singapore, pp 471–481
9. Shrivastava A (2023) Trajectory tracking control with steady error minimization of multi-axes space robot based on amnesia feedback controller. J Braz Soc Mech Sci Eng 45(6):292
10. Ghosh G, Rostron P, Garg R, Panday A (2018) Hydrogen induced cracking of pipeline and pressure vessel steels: a review. Eng Fract Mech 1(199):609–618
11. Paredes M, Wierzbicki T, Zelenak P (2016) Prediction of crack initiation and propagation in X70 pipeline steels. Eng Fract Mech 1(168):92–111
12. Nag MK, Shrivastava A (2024) Dodecyl methacrylate (DM) dispersion-assisted surface modification approach for increasing crystallinity of coir fibers. Proc IMechE Part E: J Process Mech Eng 09544089241226762
13. Nag MK (2024) Lifetime estimation and surface degradation of MCCB contact tips due to multiple electrical interruptions. Proc IMechE Part C: J Mech Eng Sci 238(5):1466–82
14. Martin ML, Dadfarnia M, Nagao A, Wang S, Sofronis P (2019) Enumeration of the hydrogen-enhanced localized plasticity mechanism for hydrogen embrittlement in structural materials. Acta Mater 15(165):734–750
15. Ohaeri E, Eduok U, Szpunar J (2018) Hydrogen related degradation in pipeline steel: a review. Int J Hydrogen Energy 43(31):14584–14617
16. Nayak S, Nag MK, Shrivastava A, Paswan M (2022) A comprehensive study on performance enhancement analysis and environmental impact of flat-plate solar water heater integrated with nanofluids. In: international conference on sustainable technologies and advances in automation, aerospace and robotics. Singapore: Springer Nature Singapore, pp 385–394

An Investigation on the Factors Responsible for High Rate of Product Rejection in a Assembling Plant at Cluster Level and Critical Technique to Minimize Rejection Count

Ajay Kushwah and Abhishek Kumar Jain

1 Introduction

The rate of product rejection (RTRG) in any production and manufacturing process (PRMNPS) is one of the critical and significant factor as it impacts product quality (PQ), production costs efficiency (PCE), customer satisfaction, reputation, and regulatory compliance (CMPLS) [1]. The key motivation behind any manufacturing installation or production companies or organization is to maximal the profit margins on their products; therefore, the production process or assembling procedure of organization holds the origin to the conquest of these organizations at global scenario. The overall cost of the final product incorporates numbers of parameter and variables. Few of them can be managed by the implications of optimal scientific procedure. Rate of rejection is one such factor which can be optimized by the implementation of advance manufacturing application such as Six Sigma or lean manufacturing applications [2].

By striving for a low rejection rate (RTRG), companies can improve their overall performance (OEF), competitiveness, and long-term success. Maintaining a low rate of product (LOWC) or item rejection in the production process is crucial for number of reasons such as cost efficiency (CEF), time savings, customer satisfaction, product reputation, product brand image, and compliance and regulatory requirements [3]. Rejected products (RTRG) represent wasted resources, including raw materials, labor, and energy. By reducing the rejection rate, a company can minimize these costs and improve its overall efficiency. Rejected products often require rework or repair, which adds to the production time. A low rejection rate allows for smoother production flow, reducing delays and ensuring timely delivery to customers [4]. High

A. Kushwah (✉) · A. K. Jain
MATS University, Raipur, Chhattisgarh, India
e-mail: ajay221064@gmail.com; ajay.kushwahadurg@gmail.com

R. Sahu et al. (eds.), *Advancements in Materials Processing Technology, Volume 1*,
Springer Proceedings in Materials 48, https://doi.org/10.1007/978-981-97-4958-4_3

rejection rates (RTRG) can result in poor product quality, leading to dissatisfaction of customers. By minimizing rejects (RTRG), a company can provide consistent and reliable products that meet or exceed customer expectations. Satisfied customers are more likely to become repeat customers and recommend the company to others. Consistently producing high-quality products with a low rejection rate contributes to a positive reputation and a strong brand image. A reputation for quality and reliability enhances customer trust and can differentiate a company from its competitors [5]. Monitoring and reducing rejection rates (RTRG) can drive a culture of continuous improvement within the organization. By identifying the root causes of rejections and implementing corrective actions, companies can optimize their production processes, enhance quality control measures, and improve overall efficiency [6].

2 Literature Survey

Pikul Vejjanugraha, Kanda Tiwatthanont and Napaphat Vichaidis published an article in IEEE [7] in year 2022 concerning to rate of rejection in an automobile plant. This article suggests techniques to minimize the rate of rejection in an assembling line production process of a manufacturing installation. The articles introduce an automatic perspective for investigating and sensing the overall equipment effectiveness that is (OEE) in a manufacturing plant by implementing a live wire feedback system to monitor process continuously. Power BI virtual software environment was used and successfully implemented on mathematical model of the objective application.

A. Paramasivam, P. Vidhushini, and S. P. Srinivasan in their article [8] elaborated about the rate of rejection in assembling line in the manufacturing process of a gear reduction starter motor [9]. The article covers the basic reason for greater rate of rejection which is outcome at assembling line and the critical root accountable is eliminated by advancing the present machine with advance automatic sophisticated machineries which generates batter quality product with exact dimensional precision. The dimensional correctness is maintained on each every the item produced in the line by making cross second level cross check the same with dimension measurement system assembled with the machine, the operation is performed automatically and odd item are picked out of the line [10].

J. P. Costa, I. S. Lopes and J. P. Brito published an article in science direct entitled "Six Sigma application for quality improvement of the pin insertion process" [11], explains about the Six Sigma technique in the context of a specific industry. The article identifies and discovers the possible application of Six Sigma technique in different assembling process present in a manufacturing plant. As per the article conclusion, Six Sigma is a potential technique to detect the factors, responsible for high rate of rejection, and at the same time suggests counter measures which can be taken to address the possible losses due to the same.

3 Technique to Address High Rate of Rejection

Addressing a high rate of rejection in production is crucial for improving product quality, reducing costs, and increasing customer satisfaction. Here are some techniques that can help to address the issue. Figure 1 depicts the techniques which can be used to address high rate of rejection [12].

Following are the key areas needed to be focus to minimize high rate of rejection;

1. Root Cause Analysis: Perform a thorough analysis to identify the root causes of rejection. Look beyond the symptoms and examine underlying factors that contribute to the problem.
2. Process Optimization: Review production processes and identify areas where improvements can be made. This may involve streamlining workflows, eliminating bottlenecks, or implementing automation to reduce human error.
3. Training and Skill Development: Ensure that workforce is adequately trained and possesses the necessary skills to perform their tasks effectively.
4. Quality Control and Testing: Implement robust quality control measures throughout the production process. Regularly inspect and test products at various stages to catch potential issues early on.
5. Statistical Process Control (SPC): Use SPC techniques to monitor production processes in real time. By collecting and analyzing data, you can identify trends, patterns, and deviations from normal operations. This allows you to take corrective actions promptly.
6. Continuous Enhancement: Following a practice of continuous improvement within objective organization will positively address high rejection rate and high rework rate which will ultimately adversely affect the CTC of product to the company and hence overall price at market. Encourage employees to contribute ideas for process enhancements, and implement a system for capturing and evaluating these suggestions. Regularly review and refine your production processes to reduce rejection rates.

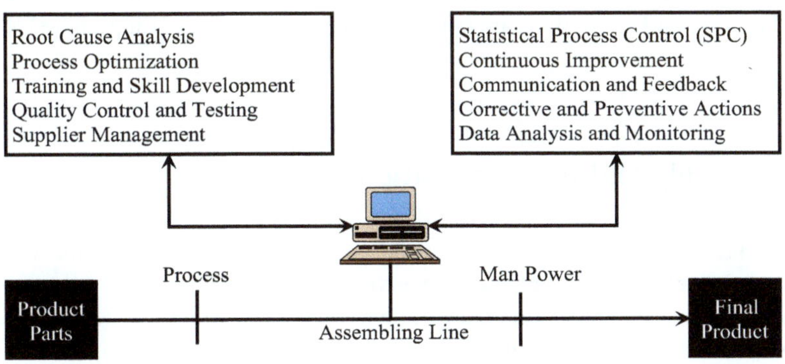

Fig. 1 Technique to address high rate of rejection in assembling line of a production process

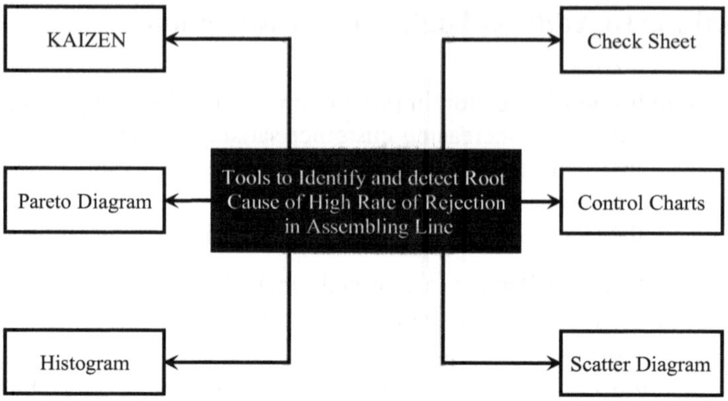

Fig. 2 Tools and methods to identify high rate of rejection in assembling line of a production process

7. Corrective and Preventive Actions: Whenever a rejection occurs, investigate the cause thoroughly, and implement appropriate corrective actions. Additionally, identify preventive measures to avoid similar rejections in the future.
8. Data Analysis and Monitoring: Collect and analyze data on rejection rates, quality metrics, and production parameters. Monitor key performance indicators (KPIs) to identify trends and patterns. This data-driven approach will help you make informed decisions and prioritize areas for improvement.

4 Tools to Investigate Root Cause of High Rejection Rate

Figure 2 depicts the commonly used technique to address high rate of rejection in assembling line.

The tools and methods which are commonly utilized for investigating root cause of high rejection rates are (1) Flowchart (2) Pareto diagram (3) Histogram (4) Check sheet (5) Cause and effect diagram (6) Control charts (7) Scatter diagram (8) Kaizen [13].

5 Case Study and Implementation

The presented research article has considered a case of assembling line which is manufacturing electronic glucometer which gives blood sugar level with 5% (±) accuracy. The instrument is composite of 15 mini devices which needed to assemble in a compact cabinet. The instrument cabinet is needed to pass through 20 different processes, and at each process halt, a device is added to the cabinet. Figure 3 depicts the assembling process of glucometer [14].

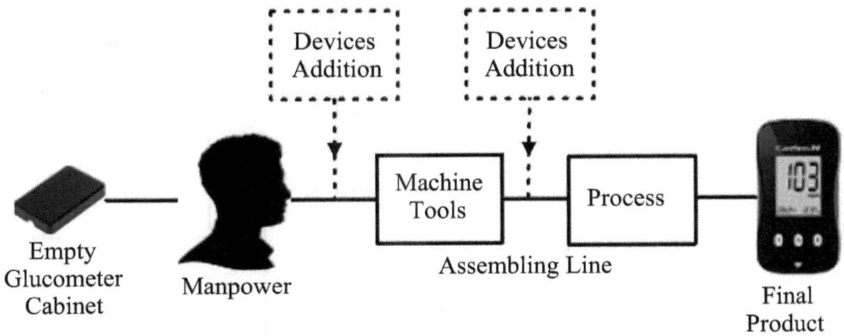

Fig. 3 Glucometer assembling line—a case study under the presented research article

5.1 Problem Identification

The glucometer assembling line under consideration is a segment or a part of complete glucometer manufacturing procedure which is taken into account at cluster level. Value addition process at different halt during movement of product cabinet encounters human error, machine error, and wastes at different sub-procedure halt. This will result in high rate of product rejection, poor product quality, low production quantity, and high cost to the company for the product under process.

5.2 Objectives

The objective of the case research article is to detect minor and major human errors, machine errors at micro-level, monitoring errors, poor quality accessories errors, and root causes of rejection of product. To model the procedure in mathematical formulation so as to implement Six Sigma and Kaizen to minimize rate of rejection is first objective of the research article. Control the rework procedure and scrape management also included in the objectives. A standard procedure formation with most advance and optimal solution is final objective of the article [15].

5.3 Research Methodology Adopted

The presented research article adopted a proven and tested systematic methodology test the hypothesis suggested with proper parameter and boundary conditions. Figure 4 depicts the summary of the research methodology adopted in the article.

The first step which is taken is to mathematically calculate the rate of product rejection in the present assembling line. All parameter, boundary conditions, constrains, and condition must be taken into account. After estimating RTRJ of the process, next

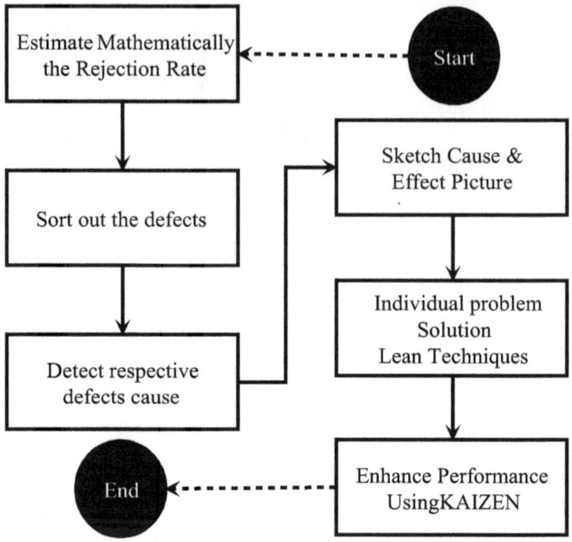

Fig. 4 Research methodology adopted in the present research article

step is to detect causes of the defect one by one at each process halt that at each value addition hot spot. Finds solution for each cause individually and implement the same in the process using Six Sigma and Kaizen technique in the present line. In the next step, a flowchart is suggested for hypothesis testing in assembling line [16].

5.4 Procedure Flowchart

Scientifically optimal flowchart has been adopted for implementing the suggested hypothesis in the object assembling line. Figure 5 below depicts the flowchart for the same.

6 Result and Analysis

A test system has been considered which is basically an assembling line and assembling glucometer. Glucometer is a device which is used to measure current sugar level in the blood sample using strip. This is assembled using tiny electromechanical devices [17]. Test system taken 15 electromechanical devices under consideration. Total 20 subprocesses were considered in assembly line. Following is the detail of test system:

Name of item: Glucometer

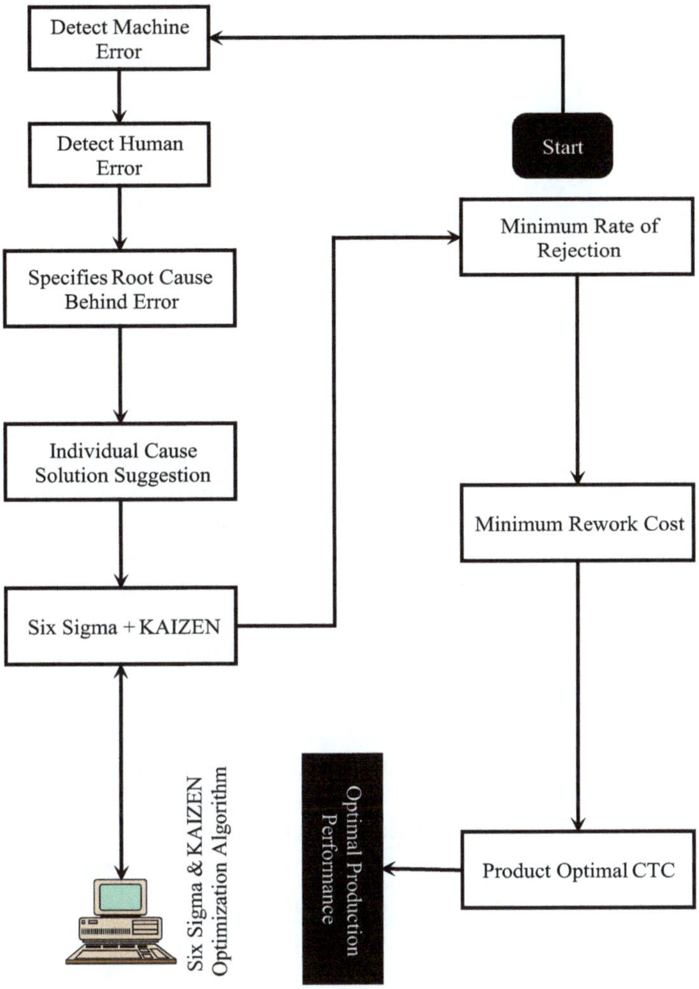

Fig. 5 Flowchart for implementing suggested hypothesis in the object assembling line procedure of manufacturing application

Number of parts likely to be assembled on empty cabinet of glucometer = 15

Number of subprocess participated in line = 20

Sample size count = 100 Item

Total time taken to pass through complete assembling line = 45 min

$$\text{Rate of rejection} = \frac{\text{Waste Material Cost}}{\text{Total Cost of Production}} \times 100 \quad (1)$$

$$\text{Rate of rejection} = \frac{\sum_{\text{Item Count}=1}^{\text{Item Count}=20} \text{CTC of Rejected Item}}{\sum_{\text{Item Count}=1}^{\text{Item Count}=20} \text{CTC of Total Input Item}} \times 100 \tag{2}$$

$$\text{Rework rate} = \frac{\text{rework hours}}{\text{productive labor hours}} \times 100 \tag{3}$$

$$\text{Rework rate} = \frac{\sum_{\text{Item Count}=1}^{\text{Item Count}=20} \text{CTC of Rework for reprocessing}}{\sum_{\text{Item Count}=1}^{\text{Item Count}=20} \text{CTC of Total Productive Time}} \times 100 \tag{4}$$

Table 1 presents the rate of rejection in two different cases. First one column demonstrated the rate of rejection without application of any scientific technique. The second column shows the improvement in the product quality as the rate of rejection is decreased. The rejected item count decreases as soon the Six Sigma technique is applied on the assembly line of the industrial installation. The enhancement in the overall performance is significant and accountable. As the number count of the assembled item increases, the rate of rejecting is simultaneously affected by the item count in the assembling line [18]. Next table will display the improvement in the outcome after implementation of the Kaizen on assembling line.

Table 2 presents the change in the rate of rejection and rejected item count after implementation Kaizen technique in assembling line. The table clearly indicates that the applications of Kaizen will surely decrease the count of the rejected item, and rate of rejection will be decreased [19, 20].

Table 1 Effect on rate of rejection before and after implementation of six sigma technique

No	Item count	Rate of rejection No technique applied	Rate of rejection Six sigma applied	Improvement (%)
1	20	3 (15.0%)	2 (10.0%)	5.00
2	40	5 (12.5%)	4 (10.0%)	2.50
3	60	8 (13.4%)	7 (11.66%)	1.74
4	80	14 (16.25%)	12 (15.0%)	1.25
5	100	17 (17.0%)	16 (16.0%)	1.00

Table 2 Effect on rate of rejection before and after implementation of Kaizen on assembling line under consideration

No	Item count	Rate of rejection No technique applied	Rate of rejection Kaizen Applied	Improvement (%)
1	20	3 (15.0%)	02 (10.00%)	5.00
2	40	4 (10.0%)	03 (07.50%)	3.00
3	60	8 (13.4%)	06 (10.00%)	3.40
4	80	14 (16.25%)	11 (13.75%)	2.50
5	100	17 (17.0%)	15 (15.00%)	2.00

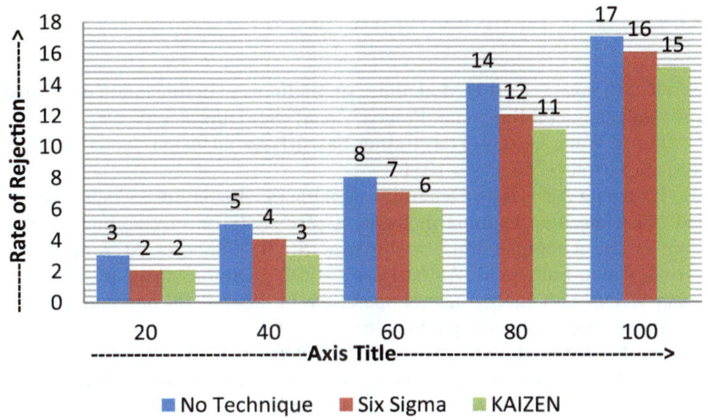

Fig. 6 Bar plot for rejection item count in all the three different cases, without any technique, with six sigma technique and with Kaizen technique

Figure 6 represents the data obtained by the implementation of different technique with same input parameter, same boundary condition, similar limits, and same constraints.

7 Conclusion

The article formulates a mathematical model of assembling line to investigate the root causes high rate of rejection at cluster level using Kaizen process. Prime reason of high rate of rejection is primarily human error, which is nearly 40% of the total causes responsible for high rejection rate. Machine error and poor quality part devices are second factor which is responsible for high rate of rejection. These two factors together share nearly 30% of the accountability. Both of these factors were addressed using Six Sigma and Kaizen technique on mathematical model, and the result was found to be impressive.

8 Future Scope

The suggested technique can be implemented in different section of the industries. A hybrid technique which can be combination of Six Sigma and Kaizen can be created which will surely inherit the advantages from both the individual techniques. The proposed work can be implemented in multiple domains of different manufacturing and assembling installations. The technique can be more comprehensive in case these will be integrated AI-based complex hybrid algorithm.

References

1. Nyemba WR, Mbohwa C (2017) Modelling, simulation and optimization of the materials flow of a multi-product assembling plant. Proced Manuf (Sci Direct) 8:59–66
2. Zhang C, Zhao Y, Wang J (2022) Transformer-based dynamic fusion clustering network. Knowl Based Syst (Sci Direct) 258:1012–1031
3. Valdés JJ, Tchagang AB (2020) Deterministic numeric simulation and surrogate models with white and black machine learning methods: a case study on inverse mappings. 2020 IEEE symposium series on computational intelligence (SSCI). Canberra, ACT, pp 2495–2503
4. Matamoros J, Calvo-Fullana M, Antón-Haro C (2015) On the impact of correlated sampling processes in WSNs with energy-neutral operation. In: 2015 IEEE international conference on communications (ICC), London, pp 258–263
5. Calvo-Fullana M, Antón-Haro C, Matamoros J, Ribeiro A (2017) Random access policies for wireless networked control systems with energy harvesting sensors. 2017 American Control Conference (ACC). Seattle, WA, USA, pp 3042–3047
6. Shan HY, Li LN, Yuan Y, Wang C (2018) Simulation and optimization of production line in Em-plant based assembly workshop. In: 2018 IEEE international conference on industrial engineering and engineering management (IEEM), Bangkok, Thailand, pp 1381–1385
7. Jia Z, Zhang L, Arinez J, Xiao G (2016) Finite production run-based serial lines with Bernoulli machines: performance analysis, bottleneck, and case study. IEEE Trans Autom Sci Eng 13(1):134–148
8. Kersten M (2017) The end of the manufacturing-line analogy. IEEE Softw 34(6):89–93
9. Gericke GA, Kuriakose RB, Vermaak HJ, Mardsen O (2019) Design of digital twins for optimization of a water bottling plant. In: Annual conference of the IEEE industrial electronics society, Portugal, pp 5204–5210
10. Paramasivam A, Vidhushini P, Srinivasan SP, Arunprakash N, Kakulapati V (2022)Manufacturing process rejection analysis of heavy duty gear reduction starter motor. Mater Today Proceed 59:1295–1300
11. Prateek G, Pathania A, Bhatti H, Chand Rojhe K (2021)Leveraging lean six sigma: reducing defects and rejections in filter manufacturing industry. Mater Today Proceed 46:5–8
12. Costa JP, Lopes IS, Brito JP (2019) Six sigma application for quality improvement of the pin insertion process. Proced Manuf Proced Manuf 38:1592–1599
13. Vejjanugraha P, Tiwatthanont K, Vichaidis N, Yatsungnoen T (2022) An automated data analytics and overall equipment effectiveness visualization technique for assembly line on continuous manufacturing system using power BI. In: 7th international conference, IEEE
14. Aldurgam MM, Alghadeer MY, Abdel-Aal MAM, Selim SZ (2019) Productivity improvement through multi-objective simulation optimization—a case study. IEEE Access 7:40230–40239
15. Ma H, Lee HK, Shi Z, Li J (2020) Workforce allocation in motorcycle transmission assembly lines: a case study on modeling, analysis, and improvement. IEEE Robot Autom Lett 5(3):4164–4171
16. Zhang L, Wang Y, Wu X (2021) Cluster-based information fusion for probabilistic risk analysis in complex projects under uncertainty. Appl Soft Comput Sci Direct 104:1320–1335
17. Shafronenko A, Bodyanskiy Y, Pliss I, Irina K (2021) Online credibilistic fuzzy clustering method based on cauchy density distribution function. In: 11th international conference on advanced computer information technologies (ACIT), Deggendorf, Germany, pp 704–707
18. Heitkamper K, Reissig L, Bravin E, Glück S, Mann S (2023) Digital technology adoption for plant protection: assembling the environmental, labour, economic and social pieces of the puzzle. Smart Agric Technol 4:2122–2134
19. Lager T, Don Simms C (2023)From customer understanding to design for processability: reconceptualizing the formal product innovation work process for non-assembled products. Technovation (Sci Direct) 125:1217–1231
20. Zhang Z, Dershan R, Enayati AMS, Yaghoubi M, Richert D, Najjaran H (2022) A high-fidelity simulation platform for industrial manufacturing by incorporating robotic dynamics into an industrial simulation tool. IEEE Robot Autom Lett 7(4):9123–9128

Simulation of Gasification Parameters Using Phenomenological Study for MSW Feedstock Based on Plasma Technique: A Review

Aman Minhas and R. S. Banshtu

1 Introduction

Waste generation is a basic and inevitable activity that belongs to every individual. Urbanization, modernization, and population growth are some of the stimulating factors which are increasing waste on a daily basis. The world bank report estimated that the waste generation will increase from 2.01 billion tons in 2016 to 3.40 billion tons in 2050. About 33% of waste goes untreated in the environment as open dumping and open burning. From 267.8 million metric tons of waste generated in 2017, only 12.7% utilized for energy recovery, and 52% goes to landfill [1, 2]. Landfill disposal causes nuisance in environment like odor, flies, harmful gases, and leachate problems. Economy influences waste management practices to great extent as discovered from Fig. 1. High- income group countries produced nearly double the waste produced by low-income group countries as reported in Figs. 2 and 3. Gasification of MSW often viewed as alternative of incineration process. However, it releases pollutants like dioxins and furans in the environment. Plasma-embedded gasification for MSW is an emerging technology with promising results. In this technique, highly flammable plasma torches are used to gasify MSW with suitable oxidants like ionized feed, steam, oxy-rich air, etc., to produce syngas.

The paper presents a detailed review of simulation techniques used to model a gasification process efficiently by evaluating the effects of different governing parameters under a wide range of environmental conditions.

A. Minhas (✉) · R. S. Banshtu
Department of Civil Engineering, National Institute of Technology, Hamirpur, Himachal Pradesh 177005, India
e-mail: amanminhas416@gmail.com

R. S. Banshtu
e-mail: banshtu@nith.ac.in

R. Sahu et al. (eds.), *Advancements in Materials Processing Technology, Volume 1*, Springer Proceedings in Materials 48, https://doi.org/10.1007/978-981-97-4958-4_4

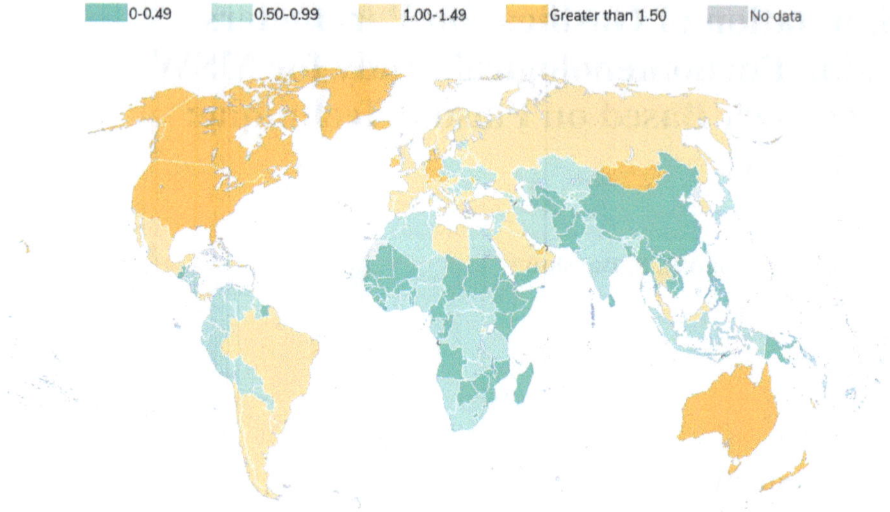

Fig. 1 Annual municipal solids waste generated (kg/capita/day) [1]

Fig. 2 Waste generation by different income groups [28]

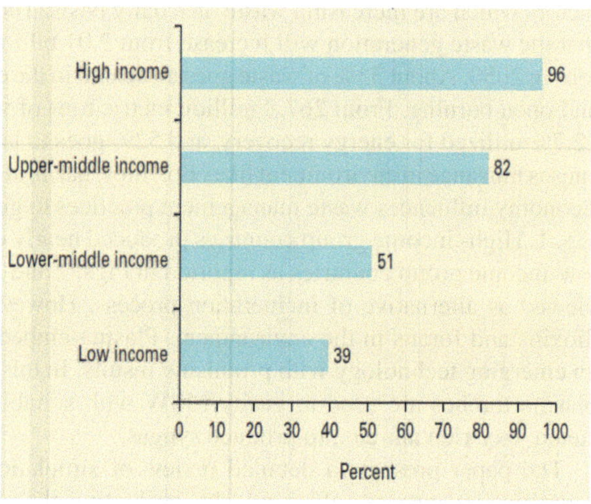

2 Trends of Waste Management in World Economies

The world generates nearly 2.01 billion tons of waste every year which is expected to grow to 3.40 billion tons by 2050 nearly twice the average population growth rate over the same duration [3]. High-income country contributes nearly 34% to total waste generated annually although they account for 16% in global population which depicted a positive correlation between income and waste generation. Daily per capita waste is expected to rise by 19% in high-income countries by 2050.

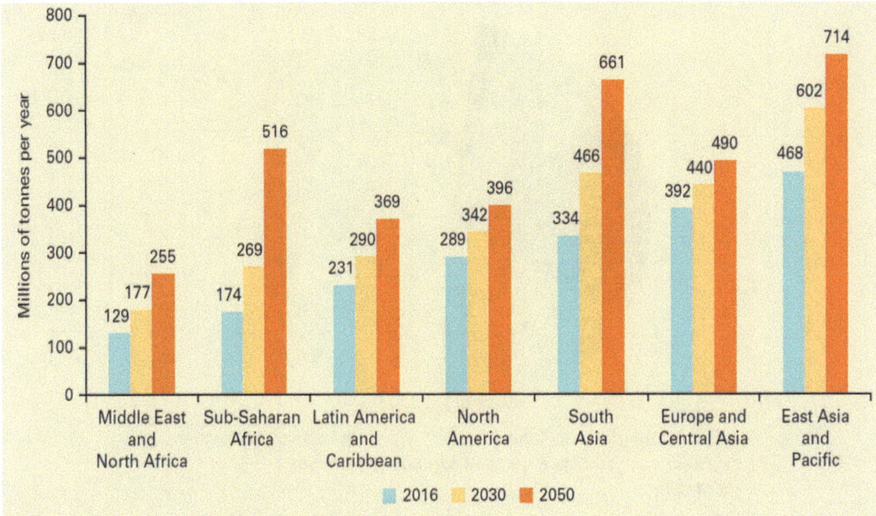

Fig. 3 Growth trends of waste generated in subsequent years [28]

Waste differs in terms of income level of countries. High-income countries account for less than 32% for food and green waste and nearly 51% for recyclable and dry waste like paper, metals, cardboard, glass, etc. Low-income countries on other hand account for 53–57% green and food waste and less than 20% for recycle cum dry waste as depicted in Fig. 4. Low- and middle-income countries rely mostly on open dumping. Incineration is adopted mostly in high capacity, high income, and land constrained countries Fig. 5. Based upon total waste generated annually, it is estimated nearly 1.6 billion tons of total CO_2 in the atmosphere equivalent to net greenhouse gas emission. Most of it contributed through open dumping and landfills without proper gas collection systems. As per ongoing trends, anticipated value is expected to rise to 2.38 billion by 2050 if no improvement made in current scenario [4].

3 Plasma Gasification: An Emerging Technology in Field of Gasification

3.1 Introduction to Plasma Gasification

Plasma is usually termed as the fourth state of matter first progressed by NASA in 1970s during testing of heat shield materials for Apollo and Mercury spacecrafts. Later, after exhaustive research and testing, plasma technique evolved as heating source. Since then, it is believed to have widespread environmental application. In

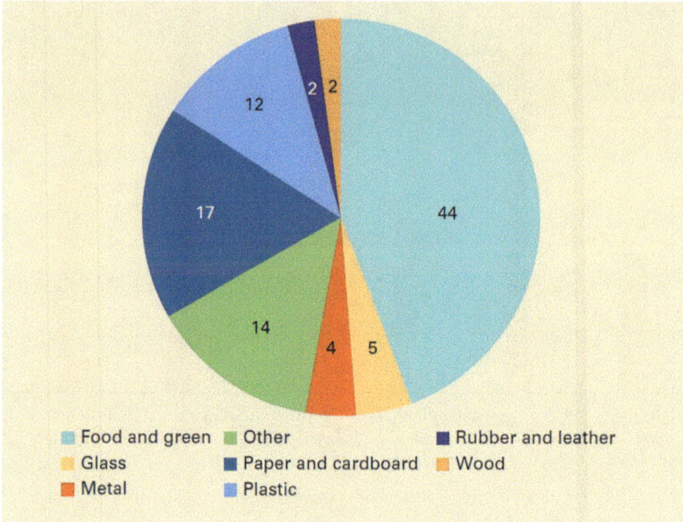

Fig. 4 Global waste composition [28]

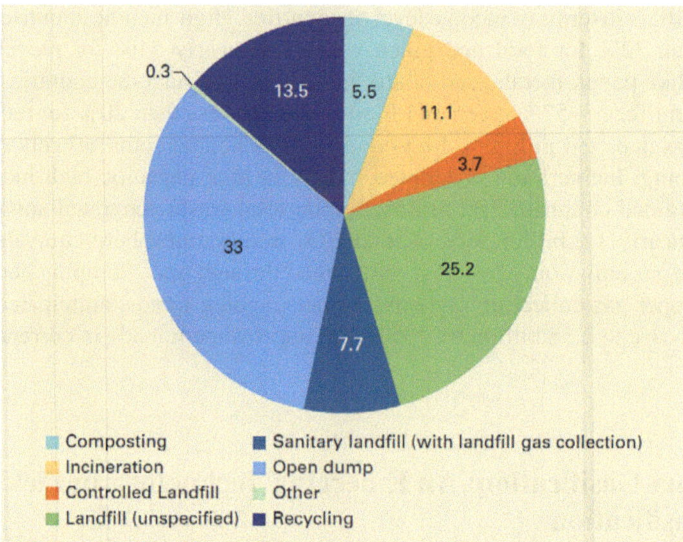

Fig. 5 Global waste treatment and disposal [28]

this technique, a strong electric current passes through inert gas at reduced pressure to ionize them into a highly flammable plasma arc. A schematic of plasma gasification is shown in Fig. 6.

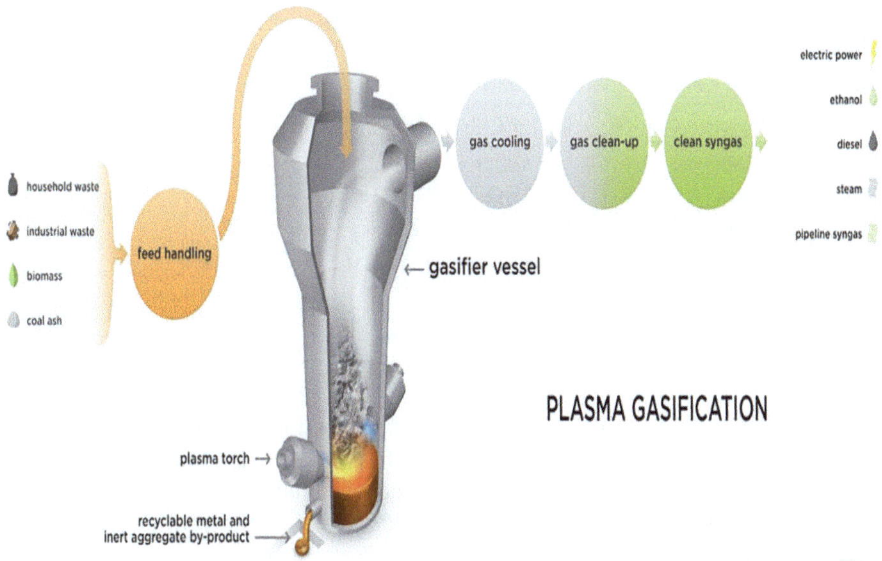

Fig. 6 Schematic of plasma gasification (*ADA ce.eco chemicals Switzerland*)

During this process, atoms and molecules of constituting gases get converted into ionic state at high temperature (2000–14,000 °C), hence named plasma [5]. A number of unit process are involved during MSW gasification ranging from waste handling, plasma enrichment to syngas cleaning, and cooling operation [6]. After waste handling, RDF feedstock is exposed to highly ionized plasma torch to produce syngas (typical constituents CO_2, H_2, CH_4, C_2H_6, and HC) and inert gas in the presence of suitable oxidants [7, 8]. Enrichment of feed with oxidants during gasification is done in various mixing proportion, viz. 75% O_2 + 25% steam or 50% O_2 + 25% steam + 25% air, etc. Proportioning of oxidants helps in efficient gasification by establishing direct relationship with stoichiometry of reaction required to produce syngas of high calorific value.

3.2 Chemical Reaction Involved in Plasma Gasification

Gasification process is accompanied by series of redox reactions taking place at different zones of gasifier reactor. Zones in gasifier are distinguished by nature of reaction and temperature profile of different reactions taking place inside reactor [9]. Syngas evolved from RDF feed synced with plasma torch to ionize it at high temperature. Atoms or molecular bonds break down at higher temperature ranging from 2500 °C to 4500 °C to form new ionic products and by-products like molten slag. These by-products may find their application in construction and ceramic industry [10]. One of major extract in the process is tar which consumes substantial amounts

Table 1 Showing reactions, type of reaction involved, and heat of formation (ΔH_f) of various reactions involved in biomass gasification reactor [27]

S.no.	Reaction name	Reaction equation	The heat of formation (ΔH_f)
1	Incomplete oxidation	$CO \rightarrow C + 0.5\,O_2$	– 283
2	Oxidation	$C + O_2 \rightarrow CO_2$	– 394
3	Water gas	$C + H_2O \rightarrow CO + H_2$	+ 131
4	Boudouard	$C + CO_2 \rightarrow 2\,CO$	+ 172
5	Shift	$CO + H_2O \rightarrow CO_2 + H_2$	– 41.2
6	Hydrogasification	$C + 2\,H_2 \rightarrow CH_4$	– 74.8
7	Ammonia formation	$N_2 + 3\,H_2 \rightarrow 2\,NH_3$	– 46.1

of heat present in fuel. Removal of liquid tar is often done during gas cooling and cleaning stage at primary and secondary collection points [11]. The combination of plasma and gasification is a novel technique and sustainable way of MSW disposal and harnessing energy (syngas) for societal needs [9, 12]. Table 1 shows gasification reaction with their heat of formation ΔH_f [9].

3.3 Computational Modeling Approach for Plasma Gasification

Many researchers have made computational modeling of gasification systems as foundation of their research, over past few decades. Co-gasification of MSW with plasma using Aspen Plus was studied by Mazzoni et al. [13]. However, lower efficiency from the process highlighted a number of constraints involved in building effective gasification systems. Effects of gasification parameter like residence time, equivalence, steam/biomass ratio, and temperature were studied for syngas yield using different computational techniques by several authors [14, 15]. Galeno et al. studied performance of plasma-integrated fuel cell system on RDF waste using Aspen Plus to determine environmental feasibility of plasma technology [16]. Improved system efficiency using computational models has paved the way for new researches to come in this field with significant improvement in process design and efficiency. Figure 7 depicts one of the computational-based approaches for modeling biomass gasification systems.

3.3.1 Principle of Gasification Modeling

The basic modeling approach followed by most authors is Gibbs free energy minimization principle. During the modeling of heat processes, free energy of the reactor is kept minimum which is achieved when the reaction approaches the state of dynamic equilibrium. The free energy of the reaction is minimum during the state of dynamic

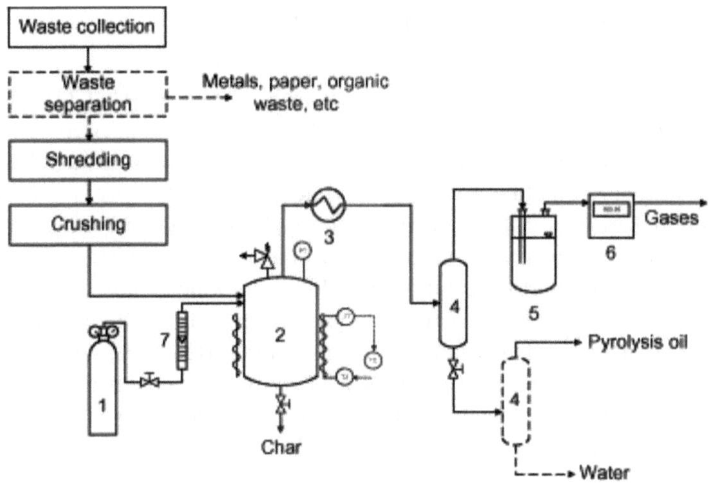

Layout of the pyrolysis process

(1. Gas bottle, 2. Reactor, 3. Cooling, 4. Separation, 5. Water trap, 6. Gas flow meter, 7. Rotameter)

Fig. 7 Computational modeling approach for MSW-based gasification systems using Aspen Plus [17]

equilibrium explained by the equation [18].

$$G^t = \sum_{i=1}^{n} n_i u_i \tag{1}$$

where n_i stands for the number of moles, and μ_i stands for the chemical potential of ith species of fuel gas. In the above equation, μ_i is estimated using the ideal gas equation at low pressure to determine the equilibrium constant using fugacity coefficients.

$$\mu_i = G_i^0 + RT \ln\left(\frac{f_i}{f_i^0}\right) \tag{2}$$

The Lagrange multiplier method may be used to determine n_i when the total free energy of the system is minimized [19]

$$\frac{\mathrm{d}L}{\mathrm{d}n_i} = \Delta G^0(f, i) + n_i RT \ln\left(\frac{n_i}{n_{\text{tot}}}\right) + \sum_{j=1}^{k} \lambda_i a_{ij} = 0 \tag{3}$$

Here, L is Lagrange function, λ_j is Lagrange multiplier, and a_{ij} is the jth atom of ith species.

3.4 Effects of Catalyst in Plasma Gasification

The primary objective of gasification is a high yield of syngas (H_2/CO). Methane and thermodynamically stable tar products can be effectively converted by various organic and inorganic mineral catalysts. Dolomite is an attractive mineral catalyst with CO as a primary constituent which can yield high-quality syngas at relatively lower cost and high electric potential [20]. Similarly, Olivine is another mineral catalyst with lower cost having structural formula $Mg_{16}Fe_{0.4}SiO_4$ and $Mg_{1.24}Fe_{0.76}SiO_4$ [21]. It is similar to supporting catalyst like dolomite which finds its application in fluidized and circulating bed reactor with promising results. Other than mineral catalysts, various metals and alkali metals are also being used in the co-incineration and gasification of refuse-derived fuel (RDF), MSW waste, and dewatered sludges. Figure 8 shows noble plasma catalysis approach for biomass gasification applied in various configurations.

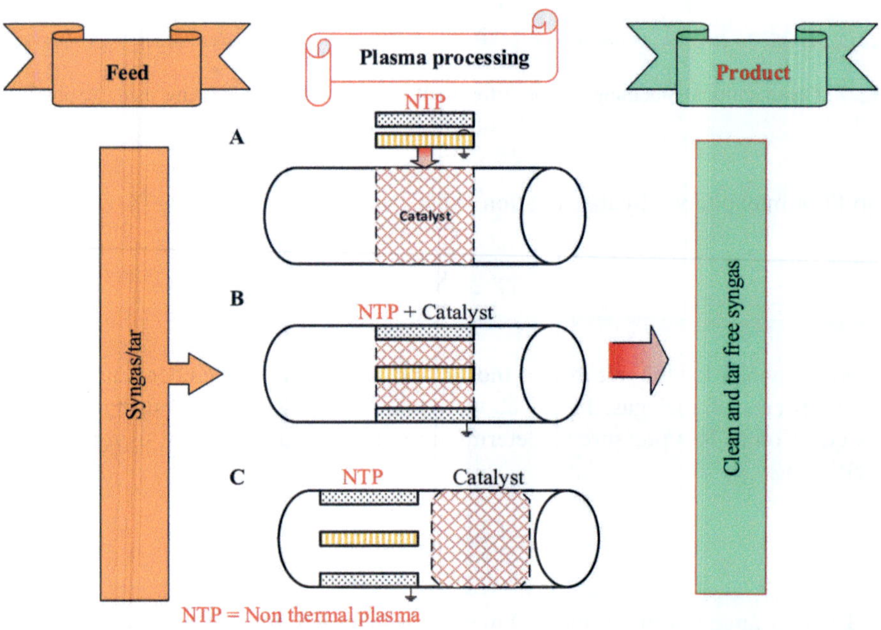

Fig. 8 Plasma-catalytic-based hybrid process schematic (**a** Pre-treatment stage, **b** during plasma, **c** post-plasma stage) [2]

4 Major Challenges Associated with Plasma Gasification

Despite key findings and recent advancements in technological processes, there are several techno-economical constraints involved in plasma gasification. Plasma technology is relatively expensive when compared with pyrolysis and gasification. Various challenges involved in these system are depicted in Fig. 9. The economy of a plasma plant depends upon several parameters like local environment, characteristics of MSW, and process feasibility. Plasma gasification being highly-energy-intensive technology requires high power inputs to produce relatively cleaner syngas and to vaporize molten slag waste for molecular dissociation to break molecular bonds into ions and atoms at elevated temperatures in the gaseous phase [22]. The plasma technique requires highly-energy-intensive DC power inputs to produce plasma flame which can put extra load on the existing power grid system. Moreover, operational and maintenance cost associated with plasma gasification is quite high which may be another cause for the lesser commercialization of this technology especially in Europe [23]. Energy-intensive processing industries (EPIs) that work on plasma building technologies are another major contributor of greenhouse gas (approx. 30%) [24] which has to cut down GHG's emissions to net zero by 2070 (Paris Agreement).

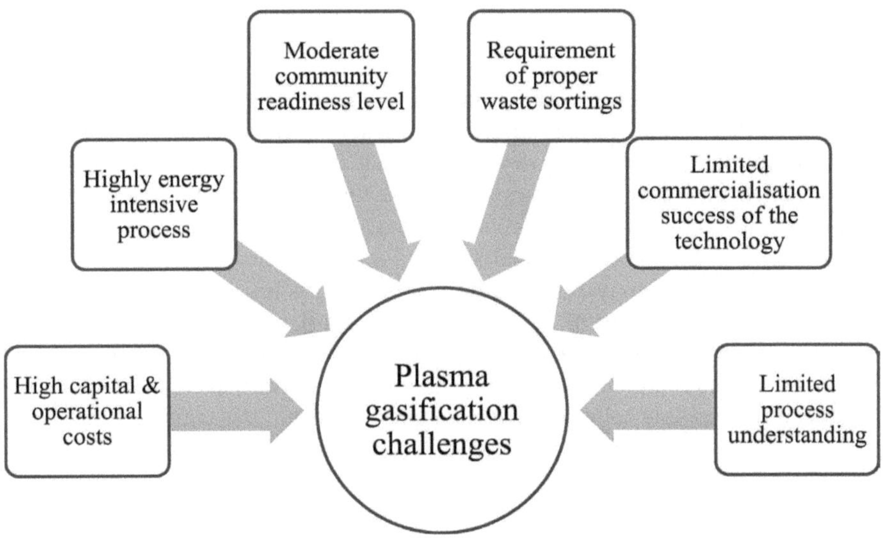

Fig. 9 Schematic of major challenges faced in plasma gasification technology [12]

5 Environmental Effects of Plasma Gasification

Plasma-embedded gasification during pyrolysis forms volatile products like Cl, S, and N. At a temperature range of 300–600 °C, large amount of SO_2 was detected, and NH_3 was the major product at 260 °C and goes to a much higher temperature range. Various Cl and S species like HCl and SO_2 (or H_2S) are volatile products of pyrolysis that leads to corrosion of reactor and liquid/gas contamination of syngas. PVC is a non-avoidable constituent of MSW which releases HCl during melting and can corrode the walls of the reactor [25]. The char obtained from co-pyrolysis has a lower fuel value than the waste liquid fraction (tar and slag). Moreover, the char produced may be toxic to some extent. Some researchers reported that char produced from MSW (plastic, pine, biomass, and rubber) contained inorganic contaminants (Pb, Zn, Hg, and Ar) and organic contaminants (alkyl phenols, aromatics, and HC) which are harmful ecotoxic waste [26]. Therefore, MSW feed to be used for gasification should not be commingled with industrial waste. To economize commercial gasification technology, environmental impacts of both char and liquid contaminants should be studied in detail.

6 Conclusion

Recent reviews and studies on waste-to-energy conversion technology have stipulated that most of the combustion, co-gasification, and pyrolysis processes need to be interconnected with gas cleaning devices like heat exchangers, wet scrubbers, ESP, etc., for which cost may sometimes rise to the total investment cost incurred for the parent systems. Mostly rotary kilns, fluidized bed, updraft, and circulating bed gasifiers are used for the commercial scale-up of gasification technology. Plasma-embedded gasification may prove to be a promising solution in waste-to-energy conversion systems despite of its high initial investment and energy costs. New techniques in process design and better quality control may be the future of this technology. Various simulation software and effective performance improving tools are some of the possible solutions that can make effective contribution in improving science associated with this field of study.

References

1. Chen D, Yin L, Wang H, He P (2014) Pyrolysis technologies for municipal solid waste: a review. Waste Manage 34(12):2466–2486. https://doi.org/10.1016/j.wasman.2014.08.004
2. Sajid M, Raheem A, Ullah N, Asim M, Ur Rehman MS, Ali N (2022) Gasification of municipal solid waste: progress, challenges, and prospects. Renew Sustain Energy Rev 168:112815. https://doi.org/10.1016/j.rser.2022.112815

3. Shah AV, Srivastava VK, Mohanty SS, Varjani S (2021) Municipal solid waste as a sustainable resource for energy production: state-of-the-art review. J Environ Chem Eng 9(4):105717. https://doi.org/10.1016/j.jece.2021.105717
4. Perkins G (2020) Chapter 1—production of electricity and chemicals using gasification of municipal solid wastes. In: Bhaskar T, Pandey A, Rene ER, Tsang DCW (eds) Waste biorefinery. Elsevier, pp 3–39. https://doi.org/10.1016/B978-0-12-818228-4.00001-0. https://www.sciencedirect.com/science/article/pii/B9780128182284000010
5. Ibrahimoglu B, Yilmazoglu MZ (2020) Numerical modeling of a downdraft plasma coal gasifier with plasma reactions. Int J Hydrogen Energy 45(5):3532–3548. https://doi.org/10.1016/j.ijhydene.2018.12.198
6. Zhang Q, Dor L, Fenigshtein D, Yang W, Blasiak W (2012) Gasification of municipal solid waste in the plasma gasification melting process. Appl Energy 90(1):106–112. https://doi.org/10.1016/j.apenergy.2011.01.041
7. Sanlisoy A, Carpinlioglu MO (2017) A review on plasma gasification for solid waste disposal. Int J Hydrogen Energy 42(2):1361–1365. https://doi.org/10.1016/j.ijhydene.2016.06.008
8. Cudjoe D, Wang H (2022) Plasma gasification versus incineration of plastic waste: energy, economic and environmental analysis. Fuel Process Technol 237:107470. https://doi.org/10.1016/j.fuproc.2022.107470
9. Tavares R, Ramos A, Rouboa A (2019) A theoretical study on municipal solid waste plasma gasification. Waste Manage 90:37–45
10. Tang B, Lin J, Qian S, Wang J, Zhang S (2014) Preparation of glass–ceramic foams from the municipal solid waste slag produced by plasma gasification process. Mater Lett 128:68–70. https://doi.org/10.1016/j.matlet.2014.04.097
11. Nakamura S, Kitano S, Yoshikawa K (2016) Biomass gasification process with the tar removal technologies utilizing bio-oil scrubber and char bed. Appl Energy 170:186–192. https://doi.org/10.1016/j.apenergy.2016.02.113
12. Munir M, Mardon I, Al-Zuhair S, Shawabkeh A, Saqib N (2019) Plasma gasification of municipal solid waste for waste-to-value processing. Renew Sustain Energy Rev 116:109461
13. Mazzoni L, Janajreh I (2017) Plasma gasification of municipal solid waste with variable content of plastic solid waste for enhanced energy recovery. Int J Hydrogen Energy 42(30):19446–19457. https://doi.org/10.1016/j.ijhydene.2017.06.069
14. Marias F, Demarthon R, Bloas A, Robert-arnouil JP (2016) Modeling of tar thermal cracking in a plasma reactor. Fuel Process Technol 149:139–152. https://doi.org/10.1016/j.fuproc.2016.04.001
15. Guan Y, Luo S, Liu S, Xiao B, Cai L (2009) Steam catalytic gasification of municipal solid waste for producing tar-free fuel gas. Int J Hydrogen Energy 34(23):9341–9346. https://doi.org/10.1016/j.ijhydene.2009.09.050
16. Galeno G, Minutillo M, Perna A (2011) From waste to electricity through integrated plasma gasification/fuel cell (ipgfc) system. Int J Hydro Energy 36(2):1692–1701. https://doi.org/10.1016/j.ijhydene.2010.11.008
17. Ate¸s F, Miskolczi N, Borsodi N (2013) Comparision of real waste (msw and mpw) pyrolysis in batch reactor over different catalysts. Part i: Product yields, gas and pyrolysis oil properties. Bioresour Technol 133:443–454. https://doi.org/10.1016/j.biortech.2013.01.112
18. Favas J, Monteiro E, Rouboa A (2017) Hydrogen production using plasma gasification with steam injection. Int J Hydrogen Energy 42(16):10997–11005. https://doi.org/10.1016/j.ijhydene.2017.03.109
19. Koukkari P, Pajarre R (2006) Introducing mechanistic kinetics to the Lagrangian Gibbs energy calculation. Comput Chem Eng 30(6):1189–1196. https://doi.org/10.1016/j.compchemeng.2006.03.001
20. Prabhansu, Karmakar MK, Chandra P, Chatterjee PK (2015) A review on the fuel gas cleaning technologies in gasification process. J Environ Chem Eng 3(2):689–702. https://doi.org/10.1016/j.jece.2015.02.011
21. Zhu HL, Zhang YS, Materazzi M, Aranda G, Brett DJL, Shearing PR, Manos G (2019) Co-gasification of beech-wood and polyethylene in a fluidized-bed reactor. Fuel Process Technol 190:29–37. https://doi.org/10.1016/j.fuproc.2019.03.010

22. Breeze P (2018) Chapter 7—advanced waste-to-energy technologies: gasification, pyrolysis, and plasma gasification. In: Breeze P (ed) Energy from waste. Academic Press, pp 65–75. https://doi.org/10.1016/B978-0-08-101042-6.000078. https://www.sciencedirect.com/science/article/pii/B9780081010426000078
23. Fischedick M, Roy J, Acquaye A, Allwood J, Ceron J-P, Geng Y, Kheshgi H, Lanza A, Perczyk D, Price L et al (2014) Industry in: climate change 2014: mitigation of climate change. Contribution of working group iii to the fifth assessment report of the intergovernmental panel on climate change. Technical report
24. UNFCCC, V (2015) Adoption of the Paris agreement. Proposal by the President 282:2
25. Yuan G, Chen D, Yin L, Wang Z, Zhao L, Wang JY (2014) High efficiency chlorine removal from polyvinyl chloride (pvc) pyrolysis with a gas–liquid fluidized bed reactor. Waste Manage 34(6):1045–1050. https://doi.org/10.1016/j.wasman.2013.08.021
26. Bernardo M, Lapa N, Gonçalves M, Barbosa R, Mendes B, Pinto F, Gulyurtlu I (2010) Toxicity of char residues produced in the co-pyrolysis of different wastes. Waste Manag 30(4):628–635. https://doi.org/10.1016/j.wasman.2009.10.015
27. Bassyouni M et al (2014) Date palm waste gasification in downdraft gasifier and simulation using ASPEN HYSYS. Energy Convers Manag 88:693–699. https://doi.org/10.1016/j.enconman.2014.08.061
28. Amankwah, Akuffo et al (2024) World Bank Open Data, 19 Jan 2024. www.data.worldbank.org/

Finite Element Analysis of Sandwich Composite Plate with Viscoelastic Frequency-Dependent Damped Core

Baibhav Kumar, Vishesh Ranjan Kar, Mrityunjay Kumar, and M. L. Chandravanshi

1 Introduction

In engineering applications, those structures are considered as effective which possesses high load-carrying capacity, are light in weight, have high stiffness, and damage tolerant. With conventional simple materials or alloys, the application was limited and costly [1]. But with the advancement of technology, hybrid type of materials like composites, functionally graded materials, nanocomposites, etc., came into picture. Advantages of these materials are that they are lightweight, have high flexural strength, and alterable properties according to the need. Among these novel materials, our area of interest is sandwich composite structures. Talking about the design of the plate, it consists of face layers made up of tough materials and a soft core sandwiched between the tough layers. Due to its lightweight, high strength, and damping properties, the sandwich composites find its use in various applications like automobiles, aeronautics, ships, civil engineering, etc. A viscoelastic polymer has been used as a soft core by the author in the paper. Viscoelastic materials are those which have both elastic and viscous properties.

Generally viscoelastic materials are used in areas where damping is required for example, in exhaust system, engine mount system, etc., of a vehicle. An interesting feature of viscoelastic materials is that by increasing its thickness up to a certain limit, it does not affect the weight of the whole composite. An increase in demand for these composites motivated researchers to develop different types of structures,

B. Kumar (✉) · V. R. Kar
National Institute of Technology, Jamshedpur, India
e-mail: baibhav.shriv@gmail.com

V. R. Kar
e-mail: vishesh.me@nitjsr.ac.in

M. Kumar · M. L. Chandravanshi
Indian Institute of Technology (Indian School of Mines), Dhanbad, India
e-mail: madan@iitism.ac.in

R. Sahu et al. (eds.), *Advancements in Materials Processing Technology, Volume 1*,
Springer Proceedings in Materials 48, https://doi.org/10.1007/978-981-97-4958-4_5

for instance, plates, shells, beams, etc., depending upon the application. To evaluate the different properties of these composite materials and for optimization of structures, various mathematical models and numerous finite element methods (FEM) have been adopted. To determine the static and free vibrational response of a simply supported sandwich plate, Moita et al. [2] adopted Reissener–Mindlin theory and classical plate theory for modeling a simply supported sandwich plate and used a simple triangular plate finite element model (FEM) for analysis. The response of a given sandwich composite also depends upon its geometry. Mohammadi and Sedaghati [3] have compared the transient vibration response of a composite shell having a thick and thin viscoelastic core for both linear and nonlinear displacements. To measure the steady-state response of a constant and frequency-dependent viscoelastic sandwich beam model for linear and nonlinear vibration, Bilasse et al. [4] coupled one mode Galerkin's procedure with the harmonic balance technique. To analyze the effect of moderate thickness on the damping characteristics of a viscoelastic core in a sandwich plate, Rena et al. [5] adopted a FEM formulation based upon Kirchhoff's plate theory and a general high order deformation theory. To check the effect of stability of a sandwich plate with a viscoelastic core, a bending, free vibration, and buckling analysis was conducted by Arani et al. [6] using a new first-order shear deformation theory (FSDT) and Kelvin-Voigt model. Investigation of the effect of different parameters like geometry, core stiffness, boundary condition, nonlocal parameter, and the damping effect of structures on the bending, free vibration, and buckling of viscoelastic sandwich nanoplates was done by Kolahchi [7] using various nonlocal theories, novel numerical methods, and Kelvin-Voigt model. An analytical study was done by Mahmoudkhani et al. [8] to estimate nonlinear free and forced vibration response due to bending vibration of sandwich plates with the viscoelastic core using the single integral viscoelastic model and fifth-order multiple scales method (MMS). A comparison between different viscoelastic theories has been done by Permoon and Farsadi [9] based on natural frequency and loss factors of a sandwich plate with a viscoelastic core. A Biot model was combined with FEM model was used to determine the natural frequency and loss factor of a viscoelastic cored sandwiched plate by Huang et al. [10], and a comparison was made between the obtained values with experimental results. The relationship between loss factor, natural frequency, thickness, and displacement relationship between layers of a PCLD plate having viscoelastic core has been investigated by Huang et al. [11] by using FSDT, Hamilton's principle, and eigenvalue problems. A modified Fourier-Ritz method was adopted to find the effects of boundary conditions, dimensional ratios, material loss factors on frequency-dependent properties of porous functionally graded sandwich plate with viscoelastic core is done by Zhang et al. [12]. For evaluation process, FSDT combined with modified Fourier method was adopted.

The objective of this work is to do a comparison between viscoelastic and elastic material models for same material by performing a free vibration analysis. Also, free vibration analysis for a sandwich composite having a viscoelastic core and metal core is done considering viscoelastic behavior for both is done.

2 Mathematical Formulation

A sandwich composite plate with a viscoelastic core has been taken from Huang et al. [10], as shown in Fig. 1. For vibrational analysis of the plate, linear distribution of displacement is considered. All assumptions for the analysis are taken in accordance with FSDT. The displacement is assumed to be linear and strain displacement related as:

$$\begin{aligned} X_i(\alpha, \beta, k, t) &= x_i(\alpha, \beta, t) + z^i \lambda_i(\alpha, \beta, t) \\ Y_i(\alpha, \beta, k, t) &= y_i(\alpha, \beta, t) + z^i \xi_i(\alpha, \beta, t) \\ W_i(\alpha, \beta, k, t) &= w_i(\alpha, \beta, t) \end{aligned} \tag{1}$$

where X, Y, W are time-dependent displacement in α, β, k directions, respectively.

$$\{\varepsilon\} = \begin{bmatrix} {}^{\delta}/_{\delta\alpha} & 0 & 0 \\ 0 & {}^{\delta}/_{\delta\beta} & 0 \\ {}^{\delta}/_{\delta\beta} & {}^{\delta}/_{\delta\alpha} & 0 \\ 0 & {}^{\delta}/_{\delta k} & {}^{\delta}/_{\delta\beta} \\ {}^{\delta}/_{\delta k} & 0 & {}^{\delta}/_{\delta\alpha} \end{bmatrix} \times \begin{Bmatrix} X_i \\ Y_i \\ W_i \end{Bmatrix} \tag{2}$$

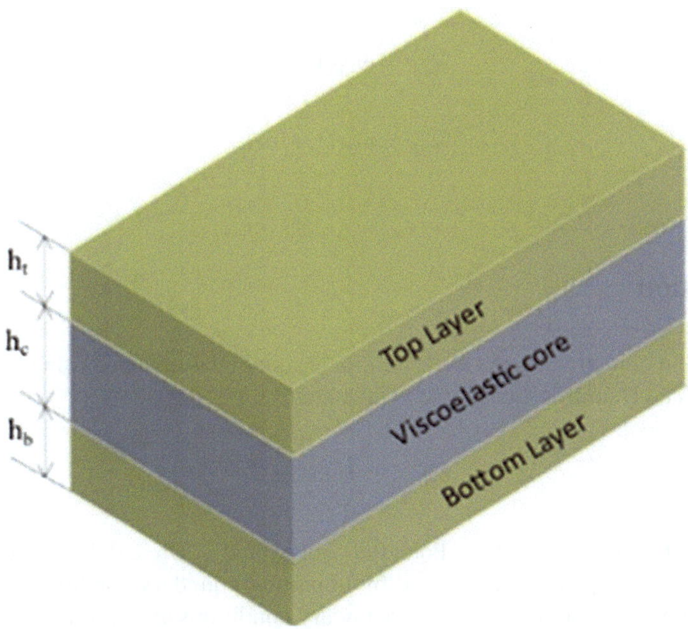

Fig. 1 Sandwich composite having viscoelastic core. h_c = core thickness, h_t = top face thickness, and h_b = bottom face thickness

According to Hooke's law, stress–strain relation for an isotropic material is given as follows:

$$\{\sigma\}^{(i)} = \begin{bmatrix} Q_{11}^i\ Q_{12}^i\ Q_{13}^i & 0\ 0 \\ Q_{12}^i\ Q_{22}^i\ Q_{23}^i & 0\ 0 \\ Q_{13}^i\ Q_{23}^i\ Q_{33}^i & 0\ 0 \\ 0\ 0\ 0 & Q_{44}^i\ Q_{45}^i \\ 0\ 0\ 0 & Q_{45}^i\ Q_{55}^i \end{bmatrix} \times \begin{Bmatrix} \varepsilon_x^{(i)} \\ \varepsilon_y^{(i)} \\ \gamma_{xy}^{(i)} \\ \gamma_{yz}^{(i)} \\ \gamma_{xz}^{(i)} \end{Bmatrix} \tag{3}$$

$$\{\sigma^-\} = [T]^{-1}[Q][R][T][R]^{-1}\{\varepsilon\} \tag{4}$$

$\{\sigma\}^{(i)}$ = Stress developed, $\varepsilon_x^{(i)}$ = Strain of the ith layer, $Q(i)$ = Stiffness matrix of ith layer, $[T]$ = Transformation matrix, $[R]$ = Reuter matrix, and σ^- = Stress in another direction. As the core shows a viscoelastic behavior, the material will have complex stiffness coefficient (Q_{ij}). The complex modulus is represented as follows:

$$\Omega_1^i = \omega_1^o(1 + i\eta) \tag{5}$$

$$\Omega_2^i = \omega_2^o(1 + i\eta) \tag{6}$$

$$\Omega_{12}^i = \omega_{12}^o(1 + i\eta) \tag{7}$$

$$\Omega_{13}^i = \omega_{13}^o(1 + i\eta) \tag{8}$$

$$\Omega_{23}^i = \omega_{23}^o(1 + i\eta) \tag{9}$$

3 Solution

The equation of motion for the free vibrational situation is obtained using Hamilton's principle. The Hamilton's principle is given as

$$\delta \int_{t_1}^{t_o} (T - U)\mathrm{d}t = 0 \tag{10}$$

where T = Kinetic energy, U = Potential energy, and = Dirac delta function.

The obtained equations are solved by imparting boundary conditions. Since the plate is simply supported; hence, the boundary conditions are as follows:

$$([K] + l[C])X = (\omega^o)^2[M]X \tag{11}$$

where $[K]$ = Stiffness matrix, $[M]$ = Mass matrix, $[C]$ = Complex stiffness matrix, and = Complex circular frequency.

$$\omega_n = \sqrt{\mathrm{Re}(\omega^o)^2},\ \eta = \sqrt{\mathrm{Im}(\omega^o)^2}/\sqrt{\mathrm{Re}(\omega^o)^2} \quad (12)$$

Equation 12 gives us the natural frequency and damping coefficient.

4 Result and Discussion

The analysis was performed for two different materials using Ansys. Before analysis, a comparative study between elastic and viscoelastic analysis was done for a viscoelastic material, and the result is shown in Fig. 2a. The variation of natural frequency is also plotted with thickness ratio as shown in Fig. 2b. The analysis of the loss factor of the epoxy core shows that as the loss factor converges to a constant value as the sandwich becomes thinner. Also, it is observed that the variation in the loss factor is not that much significant. Hence, to determine the damping behavior, a constant value of the loss factor for the epoxy was taken as 0.034. Further, the free vibration analysis of the sandwich composite with *a*/*h* ratio 50 was performed, and modal damping ratio of the sandwich composite at three modes was evaluated shown in Table 1. The modal damping ratio is increasing with frequency.

The mesh convergence for the sandwich composite was found at a mesh size of 30 × 30. In Fig. 3, comparison is presented with presented model, and fixed equivalent single layer (ESL) model presented by Rao et al. [13]. The result is compared with respect to six vibrational modes. The ESL model has achieved natural frequency (NF) of 23.25, 44.61, 70.77, 79.97, 91.78, and 126.59, respectively. Whereas, the presented model has achieved NF of 23.129, 44.428, 70.17, 80.055, 90.805, and

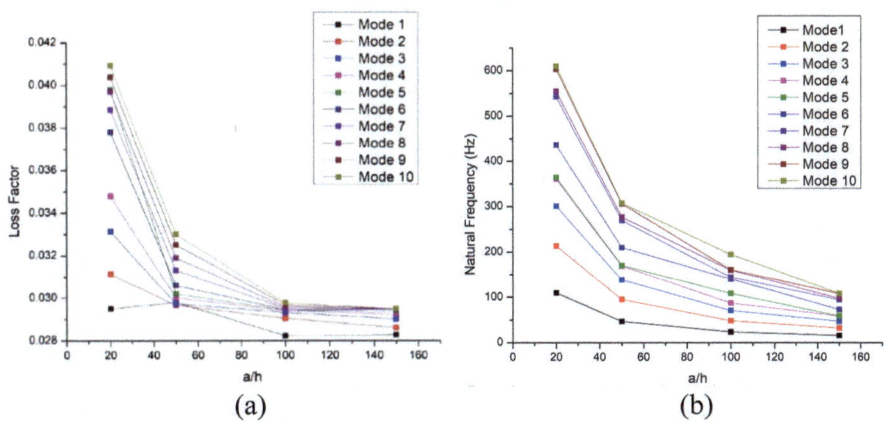

Fig. 2 **a** Loss factor variation, **b** natural frequency variation at different thickness ratios

Table 1 Modal damping ratio and damped natural frequency at three modes for the sandwich composite

Modes	Natural frequency (Hz)	Modal damping ratio $\times 10^{-6}$
1	72.95328	4.64
2	85.549447	5.95
3	108.9579	8.45

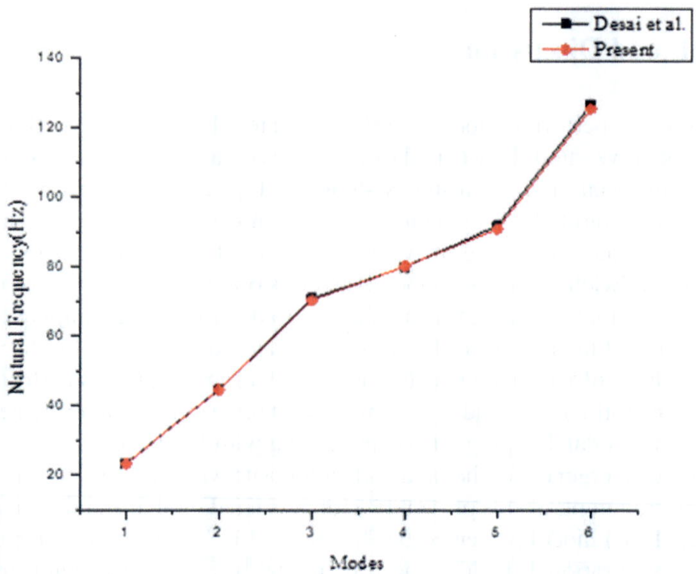

Fig. 3 Validation of present result with Rao et al. [13]

125.36, respectively. Therefore, from result, it can be observed that the presented model outperforms better as compared to Rao et al. [13].

5 Conclusion

In this paper, the damping properties of a sandwich plate with viscoelastic core are obtained based upon Hamilton's principle using FSDT. The value of NF at different modes is obtained for sandwich composite plate was obtained, and the loss factor values of the core are obtained. It is found that the variation in the loss factor of the core at different thickness ratios is not very significant, and a constant value can be taken for analysis. Further with a constant value of the loss factor for the core of the sandwich composite, free vibration of the composite was performed on ANSYS and modal damping ratio was evaluated for three modes and it was found that as the frequency increases the damping behavior of the sandwich composite increases.

References

1. Uzal A, Sonmez FO, Oz FE et al (2018) A composite sandwich plate with a novel core design. Compos Struct 193:198–211. https://doi.org/10.1016/J.COMPSTRUCT.2018.03.047
2. Moita JS, Araújo AL, Martins P et al (2011) A finite element model for the analysis of viscoelastic sandwich structures. Comput Struct 89:1874–1881. https://doi.org/10.1016/J.COMPSTRUC.2011.05.008
3. Mohammadi F, Sedaghati R (2012) Linear and nonlinear vibration analysis of sandwich cylindrical shell with constrained viscoelastic core layer. Int J Mech Sci 54:156–171. https://doi.org/10.1016/J.IJMECSCI.2011.10.006
4. Bilasse M, Daya EM, Azrar L (2010) Linear and nonlinear vibrations analysis of viscoelastic sandwich beams. J Sound Vib 329:4950–4969. https://doi.org/10.1016/J.JSV.2010.06.012
5. Ren S, Zhao G, Zhang S (2019) A layerwise finite element formulation for vibration and damping analysis of sandwich plate with moderately thick viscoelastic core. https://doi.org/10.1080/15376494.2018.1504360
6. Ghorbanpour Arani A, Emdadi M, Ashrafi H et al (2019) Analysis of viscoelastic functionally graded sandwich plates with CNT reinforced composite face sheets on viscoelastic foundation. J Solid Mech 11:690–706. https://doi.org/10.22034/JSM.2019.668608
7. Kolahchi R (2017) A comparative study on the bending, vibration and buckling of viscoelastic sandwich nano-plates based on different nonlocal theories using DC, HDQ and DQ methods. Aerosp Sci Technol 66:235–248. https://doi.org/10.1016/J.AST.2017.03.016
8. Mahmoudkhani S, Haddadpour H, Navazi HM (2014) The effects of nonlinearities on the vibration of viscoelastic sandwich plates. Int J Non Linear Mech 62:41–57. https://doi.org/10.1016/J.IJNONLINMEC.2014.01.002
9. Permoon MR, Farsadi T (2021) Free vibration of three-layer sandwich plate with viscoelastic core modelled with fractional theory. Mech Res Commun 116:103766. https://doi.org/10.1016/J.MECHRESCOM.2021.103766
10. Huang Z, Wang X, Wu N et al (2020) The finite element modeling and experimental study of sandwich plates with frequency-dependent viscoelastic material model. Mater 2020 13:2296. https://doi.org/10.3390/MA13102296
11. Huang Z, Qin Z, Chu F (2014) Vibration and damping characteristics of sandwich plates with viscoelastic core. https://doi.org/10.1177/1077546314545527
12. Zhang Y, Jin G, Chen M et al (2020) Free vibration and damping analysis of porous functionally graded sandwich plates with a viscoelastic core. Compos Struct 244:112298. https://doi.org/10.1016/J.COMPSTRUCT.2020.112298
13. Rao MK, Desai YM (2004) Analytical solutions for vibrations of laminated and sandwich plates using mixed theory. Compos Struct 63:361–373. https://doi.org/10.1016/S0263-8223(03)00185-5

Thermoelectric Power Generation from Low-Cost Carbon-Based Materials

Chandra Shekhar Choudhary, Avithi Desappan Dhass, Ranjit Prasad, and Ram Krishna

1 Introduction

Carbon materials including CNTs, graphene, and carbon-based composites are extensively explored for thermoelectric (TE) energy generation. These materials possess unique properties, including exceptionally high electrical conductivity and tunable thermal characteristics, which make them highly suitable for thermoelectric applications. This study delves into an in-depth exploration of these carbon-based materials, meticulously elucidating their properties and highlighting their potential in thermoelectric generation. Commercial TE materials are PbTe, Bi_2Te_3, and related alloys due to their low cost and high thermoelectric figure of merit (ZT) of 0.5–1 and power factor (PF) of 1500–4500 W/mK^2 at ambient temperature. Polymers and other organic compounds are interesting TE materials since they lack Te and Pb but are less efficient.

Low thermal conductivity (1.5 W/mK) and low PF (78–500 W/mK^2) make semiconducting polymers desirable, with room-temperature ZT values of 0.1–0.42. Comparing carbon-based materials with different thermoelectric materials, we discovered that they have potential for use in thermoelectric power generation as an alternative material. CNTs are one of the most intriguing TE materials in carbon-based materials. Depending on their composition, CNTs have varying band gaps. Their natural quantum confinement effect improves TE efficiency. These materials have mechanical characteristics that allow them to be used in lightweight and flexible TE devices. TE power factors for p- and n-type CNTs without doped semiconducting species can be in the range of 700–1000 W/mK^2. CNTs are suitable for applications

C. S. Choudhary · R. Prasad · R. Krishna (✉)
Department of Metallurgical and Materials Engineering, National Institute of Technology Jamshedpur, Jamshedpur, Jharkhand, India
e-mail: krishnamme@gmail.com

A. D. Dhass
Department of Mechanical Engineering, Indus University, Ahmedabad, India

R. Sahu et al. (eds.), *Advancements in Materials Processing Technology, Volume 1*, Springer Proceedings in Materials 48, https://doi.org/10.1007/978-981-97-4958-4_6

with infinite heat sources, such as waste or sunlight, despite their low ZT (0.05 at 300 K) [1]. A high percentage of lignin in carbon nanofibers makes them more prone to fusing together and enhances the Seebeck coefficient and electrical conductivity of the product. When lignin-based carbon nanofibers were treated with hydrazine, their Seebeck coefficient changed from p-type to n-type.

Hydrazine contains amino groups that provide electrons. A lignin-based carbon nanofiber carbonized at 900 °C has an n-type power factor of 10.2 $Wcm^{-1}\ K^{-2}$, 11 times greater than a nanofiber carbonized without lignin. There is preliminary evidence that low-cost carbon-based thermoelectric materials enhance lignin conversion efficiency. There is a lack of research on low-cost carbon-based thermoelectric materials, resulting in an unknown mechanism [2].

A traditional TE material is less flexible, fracture-resistant, robust, and temperature-stable than a carbon-based one. Due to its ability to hybridize in *sp*, sp^2, and sp^3 bonds, carbon is one of the most adaptable elements. Fullerenes, carbon nanotubes, graphene, and diamonds all form from carbon. Carbon is one of the most abundant elements on Earth and is nontoxic in most forms. Since diamond and graphene have an ultra-high cost and varying thermal conductivity (Fig. 1), they are not suitable for use as TE materials, but nanoengineering can significantly reduce the cost of certain carbon allotropes.

Carbon-based nanoparticles are also promising candidates for TE development. Low-cost, structurally flexible, printable, and scalable TE materials are being explored for their potential as TE materials. In addition to band structure modification and electron energy filtering, materials can have higher Seebeck coefficients (S). This approach is unable to improve material ZT due to the strong link between S and ZT. It is usually necessary to give up in the process of reduction. Based on Fig. 2, it can be seen that carbon nanotube network TE properties are qualitatively interdependent.

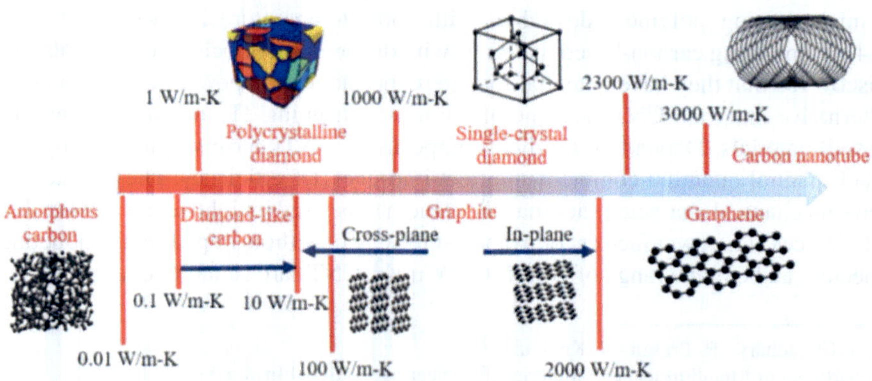

Fig. 1 Diagram showing carbon-based materials' wide heat conductivity range [3]

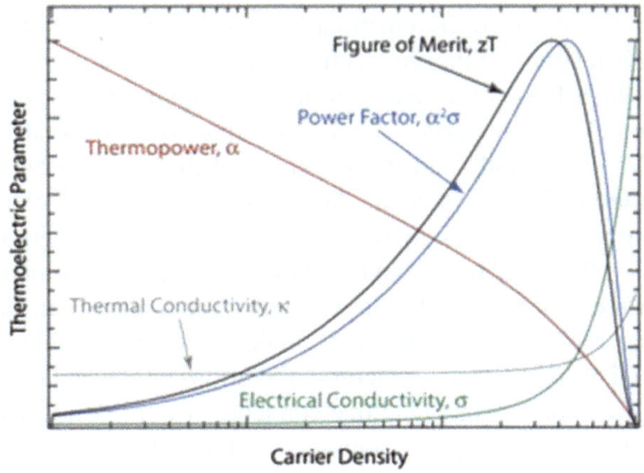

Fig. 2 TE characteristics are coupled as a function of carrier density in this design. Data on single-walled carbon nanotube networks was used to create each curve [3]

2 Carbon Materials and Methodology

It is essential to fabricate and characterize carbon-based materials for TE generation applications, especially those belonging to the graphite family, in order to understand their structure, textures, and properties. This understanding is crucial not only for improving existing TE properties but also for creating entirely novel ones. There are five key steps involved in fabrication and characterization that include bonding nature, structure, nanotexture, microtexture, and properties/functions.

Carbon materials have been fabricated and characterized using a variety of techniques, ranging from classic to modern, for a variety of reasons, including the development of new materials, the understanding of fabrication processes, and the enhancement of material properties. They can be studied with the help of spectroscopy techniques, imaging techniques, X-ray techniques, mechanical testing, thermal and chemical analysis, and electrical and thermal conductivity measurement. These diverse techniques give us a comprehensive understanding on the carbon materials, paving the way for innovation, improvement, and the creation of new applications and functions.

3 Advantages of Carbon Materials

Utilizing carbon materials in thermoelectric generation offers numerous advantages. Firstly, carbon materials are abundant and readily available, making them a sustainable choice for energy generation. Additionally, these materials are lightweight and highly compatible with flexible and lightweight applications, allowing for their integration into various devices and systems. Furthermore, carbon-based thermoelectric devices have the potential for cost-effective production methods, ensuring their economic viability.

Large-scale 2D chemical synthesis for bulk thermos-electrics is now possible. Nanostructure thermoelectric materials decrease grain boundary phonon scattering, which lowers thermal conductivity, making them good for high-performance thermoelectric materials. High ZT values of 2D metal chalcogenides like group IV-VI compounds and transition metal dichalcogenide monolayers (TMDCs) attract thermoelectric interest. Anisotropic crystal and electronic topologies limit heat conductivity, making these materials promising thermos-electrics [4].

High in-plane thermal conductivities make diamond, graphene, and graphite ideal for electrical and photonic system temperature control. Their high heat conductivity decreases in TE applications, making them unsuitable. Some allotropes keep promise TE properties after structural alterations or clever TE device design. Carbon atoms in graphene form a 2D lattice connected by sp^2 bonds and can exist alone. Graphene's 2D structure improves charge transport with Dirac cone-like linear electronic bands. Graphene hung at 2×10^{11} cm^2 electron density has mobility (cm^2/V-S) more, indicating substantial mobility. Despite being mobile, intrinsic graphene's low Seebeck coefficient and high prevent it from being a viable TE material [3].

4 Challenges and Limitations

While carbon materials offer immense potential in thermoelectric generation, there are also challenges and limitations associated with their use. Compared to traditional materials like bismuth telluride, carbon materials currently exhibit lower thermoelectric efficiency. However, the review confidently asserts that these obstacles can be effectively surmounted through ongoing efforts to enhance material performance. Continuous advancements in doping strategies and nanostructuring techniques are pushing the boundaries of carbon-based thermoelectric materials.

There are few organic and hybrid organic–inorganic high-performance thermoelectric materials. Doping, thermal, and electrical transport mechanisms in complex systems with multiple components and ordered and disordered phases must be understood to create next-generation thermoelectric materials. Organic semiconductors are difficult to manage and understand since they depend on pure phases or composites like CNTs' molecular structure and form. Polymers' molecular weight–thermoelectric performance relationship can guide future materials. Understanding the structure–property interactions between the semiconducting polymer and composite components will aid processing, which involves several input parameters. Composite thermoelectrics may benefit from bespoke semiconducting polymers with unique CNT interactions due to the development of organic semiconductors for thermoelectric applications [5].

5 Recent Developments and Research Trends

The review expounds upon recent developments and emerging research trends in the realm of carbon-based thermoelectric materials. Breakthroughs in doping strategies and nanostructuring techniques have improved the performance of these materials, making them more suitable for practical applications. Moreover, researchers are actively exploring novel carbon materials that offer unique properties and capabilities, expanding the possibilities of thermoelectric generation.

Graphene equivalent g-C_3N_4 could replace carbon in many applications. g-C_3N_4, made feasible by N and H atoms, is a promising carbon-based material alternative. Predicted g-C_3N_4 thermoelectric, electronic, and transport properties. Large surface areas increase phonon scattering surfaces, lowering thermal conductivity, while charge carrier mobility promotes electrical transport. The maximum figure-of-merit (ZT) value of 0.51 is achieved by the bismuth antimony telluride ($Bi_{0.4}Sb_{1.6}Te_3$) alloy at 375 K, and it exhibits excellent thermoelectric conversion efficiencies of up to 7.5%.

Lattice and material thermal conductivity decreases with g-C3N4 weight percent. In 'nano-bulk' materials, photonon scattering at various length scales can lower kl and k by introducing nanostructures into the bulk p-type $Bi_xSb_{2-x}Te_3$ matrix.

Phonon phase boundary scattering greatly lowers thermal conductivity in Bi $_{0.4}Sb_{1.6}Te_3$/g-C_3N_4 nanocomposites. Adding g-C_3N_4 to nanostructured Bi $_{0.4}Sb_{1.6}Te_3$ scatters phonons and lowers lattice and total thermal conductivity by creating interfaces, defects, and grain boundaries. Because of its extremely high specific surface area and thermal conductivity, graphene nanomaterial proved perfect for accelerating the freezing process of water.

The thermal conductivity of BSb1CN with 1 wt% g-C_3N_4 sample at 420 K is 0.33 W/m K, substantially lower than pure Bi $_{0.4}Sb_{1.6}$ Te_3's 0.721 W/m K and comparable to graphene and multi-walled carbon nanotube.

For $Bi_{0.4}Sb_{1.6}Te_3$ with x wt.% g-C_3N_4 (× 14 0, 0.25, 0.5, and 1.0), compute the figure of merit (ZT) as a function of temperature using electronic and phonon transport coefficient (Fig. 3 To boost thermoelectric performance, add g-C_3N_4 to $Bi_{0.4}Sb_{1.6}Te_3$. The peak ZT of $BSb_{0.50}CN$ with 0.5 wt% g-C_3N_4 is 1.09, four times that of pure $Bi_{0.4}Sb_{1.6}Te_3$. All samples' maximum merit is 420 K. Increased power factor and thermal conductivity damping raise ZT [6]. Isolated semiconducting CNT Li-$MNSiO_4$(lms) had 340 mW cm^{-1} K^{-2}. It has been reported that thermoelectric CNTs can be used to create films, fibers, and even shape-deformable dough. CNTs can generate multiple bundle junctions due to their one-dimensional structure and pliability. Well-debundled CNT networks have high Seebeck coefficient (S). Because nanotube morphology, including diameter and rolling direction, cannot be regulated, single-walled CNTs are a combination of semiconductors and metals.

CNPs filter energy because they can establish intimate contact with the CNT surface, match carriers' mean free paths, and efficiently restrict low-energy carrier transit. To create heterostructures, a petrol lighter burned CNT-polymer composite sheets. Interleaved CNTs result from polymer chain burning. S and PF values

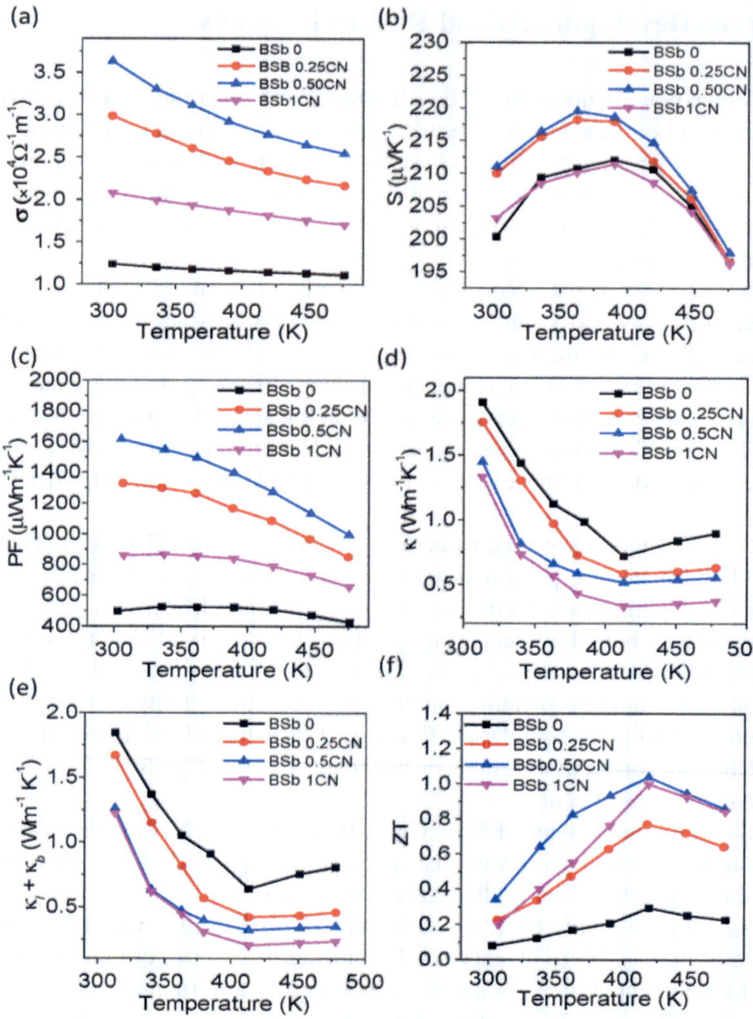

Fig. 3 $Bi_{0.4}Sb_{1.6}Te_3$'s thermoelectric characteristics change from 0% (BSb_0) to 0.25% ($BSb_{0.25}CN$), 0.5% ($BSb_{05}CN$), and 1% (BSb1CN) g-C_3N_4. There are five measures of thermal conductivity: **a** electrical, **b** Seebeck, **c** power factor, **d** total thermal, **e** lattice, and **f** figure of merit [6]

of combusted all-carbon heterostructure lms were 54.0, 1.4 mV K^{-1} and 503, 49 mW m^{-1} K^{-2}, 2.8 and 5.5 times higher than as-prepared CNT lms. A flexible thermoelectric generator with 3D stacked elements efficiently harvests energy from vertical temperature variations ($\triangle$T) using all-carbon heterostructures [7].

Latest TE and TEG technology uses organic thermoelectrics. OTEGs can be deposited on flexible glass, PET, PEN, etc., via vacuum or simple and scalable solution deposition. Organic semiconductors are used to build personal, portable, and flexible thermal modules due to their low cost, lightweight, mechanical flexibility, and low-temperature solution processability over wide areas. TE materials include conducting and nonconducting polymers, nanocarbons (carbon nanotubes, graphene, fullerenes, etc.), and nanocomposite systems, with many prototype TEGs constructed to demonstrate their power generation potential.

Organic TE materials are limited by low temperatures and longevity. Since TE and PV devices are being combined to boost electricity generation, organic TE materials and OTEGs have a bright future. Figure 4 displays a flexible Kapton polymeric film substrate with SWCNT f-TEG. Kapton engineering thermoplastic made of polyimide resins can withstand 500 °C. At 110 K, the f-TEG generates 41.1 mV of thermoelectric voltage.

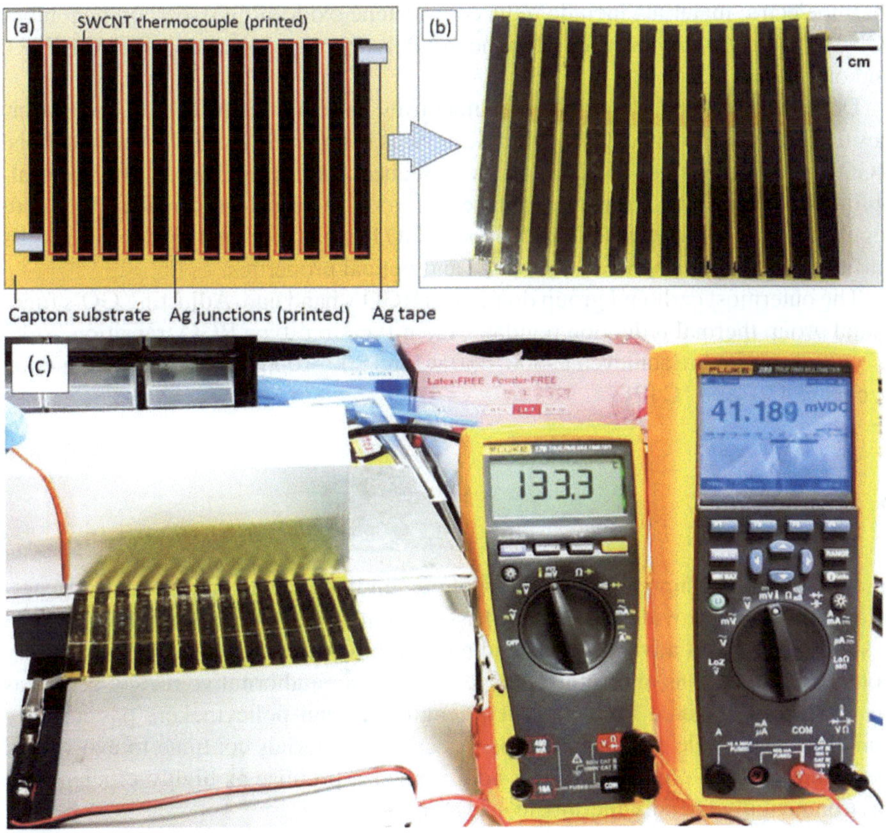

Fig. 4 **a** 14 p-type serially connected SWCNT thermocouples with Ag junctions on a fully printed flexible OTEG, **b** digital picture of the device, and **c** an operational f-TEG producing 41.1 mV at 110 K [8]

6 Applications and Future Prospects

Applications for carbon-based thermoelectric materials are numerous. From wearable devices that can capture body heat to automotive waste heat recovery systems and remote power generation, these materials offer versatile solutions to energy harvesting challenges. Additionally, the scalability and commercial viability of carbon-based thermoelectric devices provide promising prospects for widespread adoption in various industries.

New semiconducting materials and composites combined with modern production methods should make thermoelectricity generation more appealing and useful. For inexpensive, lightweight, nontoxic, flexible, scalable, and eco-friendly organic thermoelectric materials, researchers are developing new materials. Popular organic thermoelectric materials include reduced graphene oxide (RGO), carbon nanotubes (CNTs), poly(3,4-ethylenedioxythiophene): Poly(styrenensulfonate) (PEDOT:PSS), and others.

Doping and hybrid composite manufacturing can improve n- and p-type properties for high-performance thermoelectric devices in organic thermoelectric materials. RGO's strong electrical conductivity, low Seebeck coefficient, processability, scalability, and device manufacturing make it a good thermoelectric material. Due to regulated graphene oxide (GO) reduction, RGO has oxygen functional groups on its surface that alter mechanical, electrical, and optical properties.

The outermost carbonyl group dominates RGO's band gap. Adjusting GO's functional group thermal reduction regulates the n-type to p-type RGO transition. RGO n-type properties are achieved at 300–450 °C and 800–1000 °C, but p-type reduction occurs at 450–800 °C [9].

7 Conclusion

In conclusion, a comprehensive review of carbon materials in thermoelectric generation confidently provides an unbiased assessment of their strengths and limitations. It showcases recent advancements and emerging research trends while envisioning potential applications and future prospects. Such an authoritative review serves as an invaluable resource for researchers, engineers, and policymakers passionately pursuing sustainable energy solutions. As carbon materials continue to evolve and overcome existing challenges, they hold immense promise as highly efficient and environmentally friendly thermoelectric materials.

References

1. Hung N, Nugraha A, Saito R (2019) Thermoelectric properties of carbon nanotubes. Energies 12(23):4561

2. Yao M, Bi X, Wang Z, Yu P, Dufresne A, Jiang C (2022) Recent advances in lignin-based carbon materials and their applications: a review. Int J Biol Macromolecules
3. Chakraborty P, Ma T, Zahiri AH, Cao L, Wang Y (2018) Carbon-based materials for thermoelectrics. Adv Condens Matter Phys
4. Li D, Gong Y, Chen Y, Lin J, Khan Q, Zhang Y, Xie H (2020) Recent progress of two-dimensional thermoelectric materials. Nano-Micro Lett 12:1–40
5. Finn PA, Asker C, Wan K, Bilotti E, Fenwick O, Nielsen CB (2021) Thermoelectric materials: current status and future challenges. Front Electron Mater 1:677845
6. Shyni P, Pradyumnan PP, Rajasekar P, Narayanan AM, Umarji AM (2021) Graphitic carbon nitride-bismuth antimony telluride nanocomposites: a potential material for thermoelectric applications. J Alloy Compd 853:156872
7. Hwang H, Jang KS (2021) Thermoelectric all-carbon heterostructures for a flexible thermoelectric generator. Sustain Energy Fuels 5(1):267–273
8. Tzounis L (2019) Organic thermoelectrics and thermoelectric generators (TEGs). Advanced thermoelectric materials for energy harvesting applications 7
9. Mehmood T, Kim JH, Lee DJ, Dizhur S, Hirst ES, Osgood RM, Xu J (2020) A flexible, printable, thin-film thermoelectric generator based on reduced graphene oxide–carbon nanotubes composites. J Mater Sci 55:10572–10581

Experimental Investigation on Fresh and Mechanical Properties of Waste Plastic Aggregate-Based Concrete

Debasis Sau, Amit Shiuly, and Tumpa Hazra

1 Introduction

The world is constantly changing and building structures are increasingly defining each nation's identity. The construction industry is having a difficult time fulfilling consumer demand. Numerous new materials and technologies have been developed in this area but the environmental problems still exist as a result of the earlier construction methods still holding a bigger market proportion. As a result, efforts have been undertaken to change classic building materials like concrete in order to gain an important position and alleviate the negative environmental effects of the construction sector. In order to reduce the amount of carbon dioxide that concrete contributes to the world's atmosphere, as a benefit researchers have added a variety of waste elements into the concrete matrix. To determine whether they could replace one of the constituents in concrete partially or entirely, the wastes from various sectors have been assessed. The main challenge was to choose recycled waste for concrete without sacrificing the material's essential mechanical and durability features. In addition, different recycled/waste materials were used in place of the coarse aggregate, which makes up the majority of the concrete mix design volume, to increase the amount of waste that is recycled. One or more of the recycled materials utilized in place of coarse aggregates in concrete include coconut shells [1], used tire rubber [2], recycled coarse aggregates obtained from crushed concrete [3], and plastic waste [4]. For concrete, plastic was primarily employed in two separate ways: first, as broken-down bits from sources like polyethylene terephthalate bottles, and second, as a raw material for

D. Sau (✉) · A. Shiuly · T. Hazra
Civil Engineering Department, Jadavpur University, Kolkata, India
e-mail: debasiss.civil.rs@jadavpuruniversity.in

A. Shiuly
e-mail: amit.shiuly@jadavpuruniversity.in

T. Hazra
e-mail: tumpa.hazra@jadavpuruniversity.in

R. Sahu et al. (eds.), *Advancements in Materials Processing Technology, Volume 1*,
Springer Proceedings in Materials 48, https://doi.org/10.1007/978-981-97-4958-4_7

creating coarse aggregate that is based on plastic. An experimental investigation was conducted to determine, PET aggregates are used to replace 0, 10, 20, 30, and 50% of the sand in self-compacting mortar. The findings showed that adding more plastic waste aggregate improved the consistency of the mortar [4, 5], noted that when the quantity of plastic waste in concrete mixtures increased, both fresh and dry densities declined. This is indicated by the low plastic aggregate density. Further, according to the findings, adding PET aggregates to concrete mixtures resulted in a minor drop in compressive strength; this reduction was more pronounced at lower cement content and higher w/c ratios [6]. Furthermore, the results showed that concrete using flexible particles waste had a very minor improvement in flexural strength. Moreover, according to [7], this is because adding plastic particles to the mixture increased the tensile strength of concrete because their tensile strength was higher than that of other concrete components.

The literature review conclusively very less shows that recycled aggregate manufactured from waste polyethylene terephthalate (WPET) can be incorporated up to 40% of the concrete mix. The bulk of past research only used plastic aggregate (PA) to substitute coarse aggregate in concrete of the M20 grade. Few articles on concrete of the M25 or M30 grades documented experiments on the substitution of plastic aggregate for natural coarse aggregate (NCA). To show how lightweight concrete built with WPET reacts differently from concrete made with natural particles, extensive testing is required. Additionally, additional research needs to be done on optimizing various concrete mixes for waste plastic-based concrete using diverse software and multi-objective analytical approaches to obtain desired features at the lowest possible cost. As a result, in the current investigation, the performance of M30 grade concretes containing waste plastic-based aggregates was first evaluated by looking at the fresh and hardened properties such as workability or slump, fresh density, dry density, ultrasonic pulse velocity, compressive strength (CS), flexural strength (FS), and split tensile strength (STS) as partial replacements for NCA. The best mixture proportions of lightweight waste plastic-based aggregate sustainable concrete were then compared in terms of cost.

2 Experimental Plan and Materials

2.1 Material

The following materials are acceptable for use in concrete according to the IS 10262-2019 [8] code for this study. According to IS-456:2000 [9], water was utilized in concrete. Portland Pozzolana Cement (PPC) was used in accordance with IS 1489 (Table 1) [10]. In the current investigation, natural fine aggregate (NFA) conforming to IS 383-1970 [11] Zone II was utilized (Table 2). In accordance with IS 383-1970 [11], this study shows how to employ NCA of size 20 mm down (Table 2). The WPET plastics (Fig. 1) obtained from regional cutting units for segregation ranged in size

Table 1 Properties of PPC cement

Properties	Average	Limitation [10]
Normal consistency (%)	34	–
Initial setting time (min)	120	Min 30
Final setting time (min)	300	Max 600
Soundness (mm)	2	Max 10
Cement fineness (cm^2/g)	350	Min 300, Max 450
Compressive strength 7 days (MPa) 28 days (MPa)	25 42.85	Min 22 Min 33

Table 2 Properties of NCA and NFA

Properties	Fine aggregate	Coarse aggregate
Specific gravity	2.66	2.76
Bulk density (kg/m^3)	1600	1650
Water absorption (%)	1.10	0.70
Fineness modulus	2.84	6.6

less than 20 mm (Table 3). WPET was used in concrete as a partial replacement for NCA. As a chemical admixture, CONPLAST SP 430 DIS was utilized in accordance with IS 9103-1999 [12].

Fig. 1 Photocopy of WPET

Table 3 Properties of WPET

Properties	WPET
Specific gravity	0.85
Absorption %	0.45
Start melting point (°C)	92
Impact strength (J/m)	110–200
Modulus of elasticity (MPa)	1900
Size (mm)	< 20
Color	Mixed

2.2 *Experimental Plan*

2.2.1 Mix Distribution of WPET-Based Concrete

In accordance with IS-456-2000 (IS-456:2000), the conventional concrete of M30 grade in accordance with IS 10262:2019 (IS 10262 (REVISED) 2019) mix containing cement, NFA, and NCA in the ratio of 1:1.68:2.85 with 0.37 W/C ratio was created. One percent of the total cementitious material of super plasticizer was used. Table 4 of the mixing proportions for concrete volumetric replacement using WPET as an alternative to NCA. The WPET was heated to boost binding strength, and as it began to melt, a particular amount of sand (adjusted every mix proportion) was scattered around the container. Iron Mason Brick Trowel was then used to shake the mixture inside the container until the majority of the WPET were attached with NFA. Then this container was then allowed to cool in the open air before being utilized to make concrete.

Table 4 Mix proportions of M30 grade of concrete with various replacement by WPET

Percentage of replacement	Cement (kg/m^3)	NFA (kg/m^3)	NCA (kg/m^3)	Chemical admixture (kg/m^3)	WPET (kg/m^3)	Water (kg/m^3)
0	410	690	1170	4.08	0	165
5	410	690	1143.417	4.08	21.8043	165
10	410	690	1083.238	4.08	43.6086	165
15	410	690	1023.058	4.08	65.4129	165
20	410	690	962.8779	4.08	87.2172	165
25	410	690	902.698	4.08	109.0215	165
30	410	690	842.5182	4.08	130.8258	165
35	410	690	782.3383	4.08	152.63	165
40	410	690	722.1584	4.08	174.4344	165

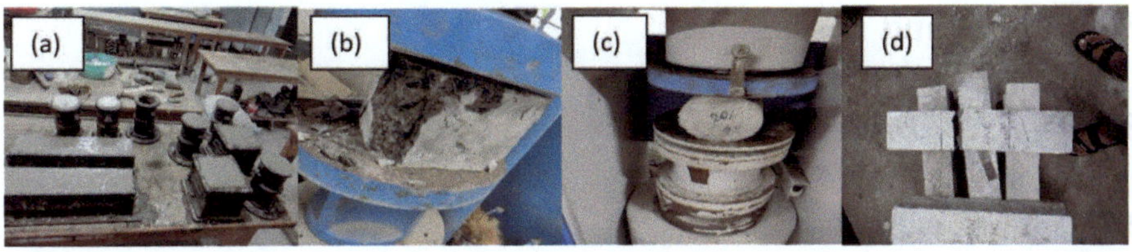

Fig. 2 **a** Sample preparation and failure pattern and experimental view of **b** CS, **c** STS, and **d** FS of WPET-based concrete

2.2.2 Preparation of Cube, Cylinder, and Beam and Curing

For testing specimens, the properties of concrete made with waste plastic aggregates were assessed using cast cubes of 100 mm × 100 mm × 100 mm [13], cylinders measuring 100 mm in diameter and 200 mm in height [14], and beams measuring 100 mm × 100 mm × 500 mm [15]. For a 28-day curing time, fresh qualities like slump, fresh density, and dry density as well as hardened properties like CS, FS, and STS were evaluated (Fig. 2). In order to assess the nature of the concrete, a non-destructive test using ultrasonic pulse velocity was conducted. Finally, cubes, beams, and cylinders with the same dimensions and curing durations were cast using an optimized mix proportion, and the necessary tests were then performed and compared using control concrete of experimental data.

2.2.3 Optimization Model and Cost Comparison

Once the technological feasibility of using WPET-based aggregates in lightweight concrete has been investigated, the economic viability must be considered [16]. An optimization study is therefore required to provide the finest concrete mix. The primary aim of optimization is to find the best mix ratio for a planned optimization method. By using Minitab 18 software's multi-factorial response optimizer method, it was designed to maximize 28-day CS, STS, and FS while minimizing slump and cost. In Minitab, optimize stat toolbox was used to create the optimization model, with the results shown in results and discussion section. Further, the cost comparison between control and the optimized mix proportion is done.

3 Results and Discussion

3.1 Fresh Property

3.1.1 Workability Test

In this investigation, the slump cone test was used to evaluate the flowability of the concrete mix. The slump cone relation is shown for various combinations in Fig. 3a. The test findings show that, in comparison with standard concrete, WPET boost workability. When NCA is partially replaced by WPET by 10–40%, respectively, compared to ordinary concrete, the increases in slump values are 0, 2, 5, 9, 12, 13, 16, and 16%. Other studies [5, 17] showed similar outcomes. This finding showed that plastic particles had uniform morphologies, a smooth surface and chemical admixture which led to enhanced fluidity.

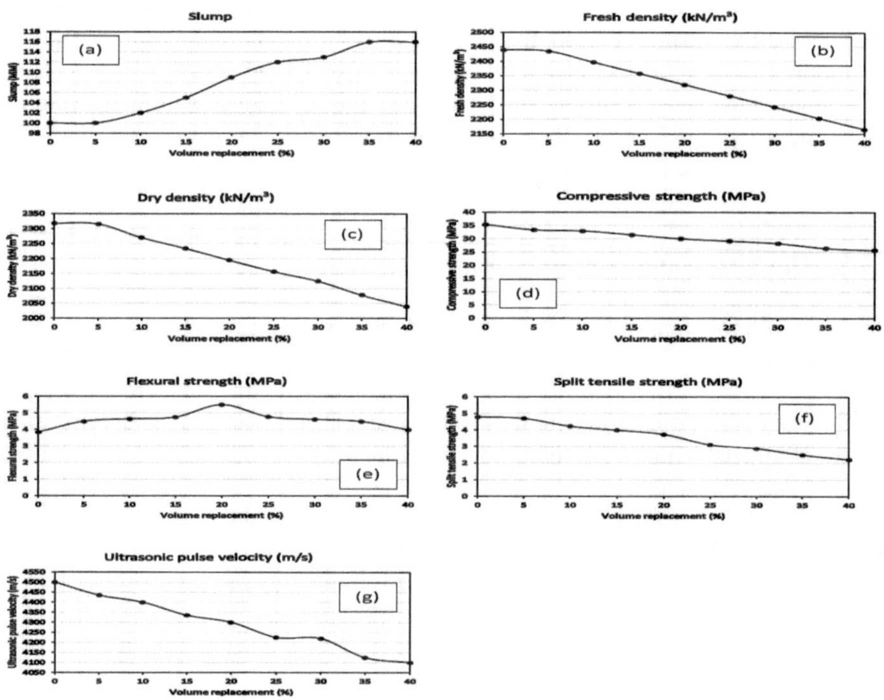

Fig. 3 Experimental results of various properties of plastic waste concrete by volumetric replacement of WPET: **a** slump, **b** fresh density, **c** dry density, **d** compressive strength, **e** flexural strength, **f** split tensile strength, and **g** ultrasonic pulse velocity

3.1.2 Fresh Density

Since the WPET aggregates in Table 3 have low specific gravities, they can be regarded as typical examples of low-weight aggregates. The new concrete's densities decreased as more waste PA was added to the control mix, as illustrated in Fig. 3b. The findings showed that, for partial replacement of NCA by WPET at a percentage of 10–40% compared to ordinary concrete, the dry density of concrete decreased by 0.19, 1.7, 3.34, 4.91, 6.48, 8.06, 9.63, and 11.2%. Others investigation [7, 17] presented similar outcomes.

3.2 Hardened Property

3.2.1 Dry Density

The dry densities of concretes for different replacement values are shown in Fig. 3c. After 28 days of curing, the findings showed that, for 10–40% partial replacement of NCA by WPET, the dry density of concrete decreased by 0.09, 2.04, 3.59, 5.28, 6.93, 8.30, 10.30, and 12% in comparison with conventional concrete. The dry density decreases as the proportion of PA replacement increases. The same actions were displayed in [17, 18]. The results reported in Fig. 3c indicate that the designed concrete can be referred to as lightweight aggregate concrete because dry densities were found to be less than 2200 kg/m^3 for replacement mix proportions of 20–40% [19, 20]. The prospective use of PA in concrete has the ability to considerably reduce the overall structure's dead load, an important design consideration.

3.2.2 Compressive Strength (CS)

The grade M30 waste plastic aggregate concrete mixtures' 28-day cured, and the results demonstrate decreasing trends in CS by increasing the amounts of WPET (Fig. 3d). For a partial replacement of NCA by WPET equivalent to regular concrete by 10–40%, the CS of the concretes reduced by 5.6, 6.88, 10.82, 15.01, 17.42, 19.97, 25.04, and 27.25%. When compared to the control mix proportion, the compressive strength (CS) of the PA concrete at a 20% mix percentage represents acceptability (compressive strength of concrete for M30 features). The drop in strength is principally caused by a decrease in the bulk density of the composites. Additionally, the weakening of the bond between the surfaces of the PA and cement paste may help to explain this pattern [17]. The hydrophobic characteristics of PA may also hinder the hydration of cement. The outcomes follow a similar pattern to those of [17, 21].

3.2.3 Split Tensile Strength (STS)

The STS of waste plastic concrete (Fig. 3e) was obtained in accordance with IS-516 [22]. When 10–40% of NCA was partially replaced by WPET instead of traditional concrete, the STS of the concrete fell by 2.08, 11.87, 16.66, 21.87, 35, 39.79, 47.91, and 53.54%. The decrease in tensile strength of concrete may be caused by the aggregate's altered form and stiffness. The angular, flat, elastic WPET granules have a structure similar to that of a range. Other researchers [17, 23] reported similar outcomes.

3.2.4 Flexural Strength (FS)

According to IS-516 [22], the FS (Fig. 3f) of plastic waste aggregate concrete was obtained. In comparison to ordinary concrete, the FS of concrete rose by 17.37, 21.28, 23.83, 43.45, 24.41, 20.50, 17.37, and 4.32% for partial substitution of NCA by WPET ranging from 10 to 40%. When compared to control, the tendency is stronger for mix proportions of up to 40%. The angular structure of the flaking particles encourages better interlock, indicating that the size and shape of the WPET aggregate directly affect the flexibility of the concrete. Additionally, PA's tensile strength behavior may result in an increase in bending strength under flexural stress [24]. Other researchers also reported similar results [17, 23, 25].

3.2.5 Ultrasonic Pulse Velocity Test (UPV)

The UPV measurements (Fig. 3g) provide important information on the pulse transfer rate in concrete specimens. The ultrasonic pulse velocity of concrete mixtures of grade M30 after 28 days of curing is shown in Fig. 3g. Wave reflection time increases with an increase in WPET, UPV decreases as a result. The UPV decreased by 8.88% at a 40% replacement ratio for WPET. The lowest UPV value across all proportions was 4100 m/s, which indicates high concrete quality, according to Almeshal et al. [26]. Given that the porosity of concrete has an inverse relationship to UPV, it may be inferred from UPV readings that the move from natural to PA had a detrimental effect on porosity. The pores produced by the WPET granules hindered the propagation of ultrasonic waves due to acoustic resistance. By partially transmitting the wave, WPET and pores all help to reduce the pulse velocity of the incident wave. Additionally, when NCA is substituted with PA, the UPV naturally decreases since it depends on the compressive strength and unit weight of the constituents [24]. Similar findings were documented in the literature [23, 25].

3.3 Optimization Model

The optimal mix ratio discovered by the optimization model (Fig. 4) [17, 23] is given in Table 5 with composite desirability 0.9784. Additionally, samples were made utilizing the aforementioned ideal mix proportion, and the results of the CS, FS, STS, and slump tests are given in Table 6. It should be noted that the optimize result has very low RMSE values near or equal to 1. Results indicated that employing a sizable amount of waste plastic as aggregate in concrete structures can reduce environmental pollution in a morally and sustainably acceptable manner.

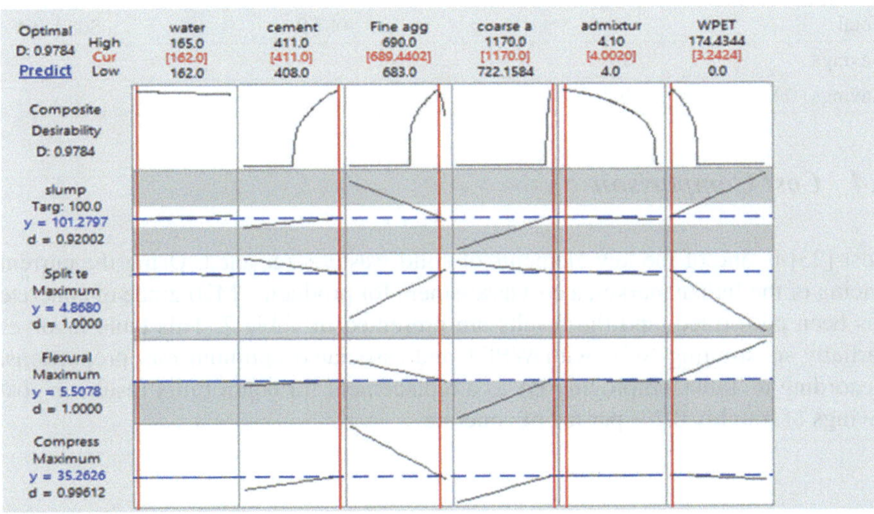

Fig. 4 Results of optimization model for optimum mix proportion

Table 5 Optimum mix proportion obtained from optimization model

Water (kg/m^3)	Cement (kg/m^3)	NFA (kg/m^3)	NCA (kg/m^3)	Chemical admixture (kg/m^3)	WPET (kg/m^3)
162	411	689.4402	1170	4.0020	3.2424

Table 6 Comparison between experimental results and optimization results obtained from optimization based-model

Property	CS (MPa)	FS (MPa)	STS (MPa)	Slump (mm)
Optimization results	35.26	5.51	4.9	101.3
Experimental results	34.26	5	4.6	101
RMSE	1	0.72	0.55	0.55

Table 7 Cost assessment for 1 m^3 WPET-based concrete of M30 grade

Ingredients	Cost (Rs)	Mix proportion of ordinary concrete		Optimum mix proportion of waste plastic concrete	
		Amount (kg)	Cost (Rs)	Amount (kg)	Cost (Rs)
Cement (bag)	350	410	2870	411	2877
NFA (tones)	600	690	414	689.44	413.664
NCA (tones)	900	1170	1053	1170	1053
WPET (tones)	300 (local)	0	0	3.2424	1
Water (lit)	0.1	165	16.5	162	16.2
Chemical admixture (lit)	180	4.08	734.4	4	720
Total			5087.9		5080.864
Savings	7				
Savings (%)	0.2				

3.4 *Cost Comparison*

Cost [23] is one of the key elements of solid waste is recycled. Using the current pricing of the Indian market, a cost assessment for producing M30 grade of concrete has been performed, and the results are presented in Table 7. This table involves partially substituting NCA with WPET and associated optimum mix proportions. According to Table, employing PA as a replacement for control mix results in cost savings of roughly 0.2% per m^3 of concrete.

4 Conclusion

In the current experiment, natural coarse aggregates are replaced with WPET-based aggregate to completely analyze the fresh and hardened features of M30 grade concrete. The best mix proportion was then obtained in order to maximize the fresh and hardened qualities of the concrete from optimization model. It is crucial to keep in mind that, for a specific water-cement ratio, the slump value increased as both WPET percentages increased up to 40% replacement level. This might be as a result of using plastic aggregate with less-rough surfaces and homogeneous size and shape. The increasing percentage replacement of WPET at constant water-cement ratios reduces the concrete density. It can therefore be light weight concrete. Additionally, the CS decreases as PA rises. Nevertheless, the CS is appropriate for structural uses up to a volumetric substitution of WPET aggregates of 20% (30 MPa) over natural aggregate. Additionally, WPET plastic aggregate concrete showed that after % replacement increase, the STS value decrease. For the FS of the waste plastic aggregate concrete increases in trend up to 20% (5.5 MPa) replacement. The UPV value declined steadily in the UPV experiment as well, reaching the same 40% replacement level as CS. Additionally, optimization was done to find the optimal mix proportion for maximizing CS, FS, STS, and minimize slump. Once more, samples were made using the optimized mix percentage & tested and compared against the CS, FS, STS and slump of control sample. The test result shows that the optimization solution produced very low RMSE values. Finally, cost comparison research confirms that 0.2% in cost savings can be achieved by using plastic aggregates in concrete in place of natural coarse aggregates.

Acknowledgements The authors gratefully acknowledge the financial support provided by the Department of Science and Technology and Biotechnology, West Bengal, India [1694(Sanc.)/ST/P/S&T/6G-9/2019 dated 06.02.2020] for financial support of the project work.

References

1. Gunasekaran K, Kumar PS, Lakshmipathy M (2011) Mechanical and bond properties of coconut shell concrete. Constr Build Mater 25(1):92–98. https://doi.org/10.1016/j.conbuildmat.2010.06.053
2. Miguel, Brito (2012) Concrete with used tire rubber aggregates: mechanical performance. ACI Mater J 109(3). https://doi.org/10.14359/51683818
3. Etxeberria M, Vázquez E, Marí A, Barra M (2007) Influence of amount of recycled coarse aggregates and production process on properties of recycled aggregate concrete. Cem Concr Res 37(5):735–742. https://doi.org/10.1016/j.cemconres.2007.02.002
4. Mohammed AA, Mohammed II, Mohammed SA (2019) Some properties of concrete with plastic aggregate derived from shredded PVC sheets. Constr Build Mater 201:232–245. https://doi.org/10.1016/j.conbuildmat.2018.12.145
5. Safi B, Saidi M, Aboutaleb D, Maallem M (2013) The use of plastic waste as fine aggregate in the self-compacting mortars: effect on physical and mechanical properties. Constr Build Mater 43:436–442. https://doi.org/10.1016/j.conbuildmat.2013.02.049
6. Frigione M (2010) Recycling of PET bottles as fine aggregate in concrete. Waste Manag 30(6):1101–1106. https://doi.org/10.1016/j.wasman.2010.01.030
7. Azhdarpour AM, Nikoudel MR, Taheri M (2016) The effect of using polyethylene terephthalate particles on physical and strength-related properties of concrete; a laboratory evaluation. Constr Build Mater 109:55–62. https://doi.org/10.1016/j.conbuildmat.2016.01.056
8. IS 10262(REVISED) (2019) Concrete mix proportioning-guidelines, New Delhi, India
9. IS-456:2000, Plain and reinforced concrete code of practice. BIS, New Delhi
10. IS-part-I, "ppcCem. Prop", 1489
11. IS 383 (1970) "IS- 383," Specif. coarse fine aggregates from Nat. Sources concr. New Delhi, India
12. IS 9103 (1999) Specification for concrete admixtures. Bur. Indian Stand. Delhi, pp 1–22
13. BS EN 12390-7 (2019) Testing hardened concrete. Density of Hardened Concrete
14. ASTM C496 (2004) Standard test method for splitting tensile strength of cylindrical concrete specimens. ASTM Int
15. ASTM C78/C78M-18 (2018) Standard test method for flexural strength of concrete using simple beam with third-point loading. ASTM Int
16. Leuca T, Novac M, Stanciu B, Burcă A, Codrean M (2014) Using minitab-box Benken software to optimize the induction heating process. J Electr Electron Eng 7(1):73–76
17. Shiuly A, Hazra T, Sau D, Maji D (2022) Performance and optimisation study of waste plastic aggregate based sustainable concrete—A machine learning approach. Clean Waste Syst 2:100014. https://doi.org/10.1016/j.clwas.2022.100014

18. Ismail ZZ, AL-Hashmi EA (2008) Use of waste plastic in concrete mixture as aggregate replacement. Waste Manag 28(11):2041–2047. https://doi.org/10.1016/j.wasman.2007.08.023
19. Hameed AM, Fatah Ahmed BA (2019) Employment the plastic waste to produce the light weight concrete. Energy Procedia 157:30–38. https://doi.org/10.1016/j.egypro.2018.11.160
20. Rumšys D, Bačinskas D, Spudulis E, Meškėnas A (2017) Comparison of material properties of lightweight concrete with recycled polyethylene and expanded clay aggregates. Procedia Eng 172:937–944. https://doi.org/10.1016/j.proeng.2017.02.105
21. Sau D, Shiuly A, Hazra T (2023) Utilization of plastic waste as replacement of natural aggregates in sustainable concrete: effects on mechanical and durability properties. Int J Environ Sci Technol. https://doi.org/10.1007/s13762-023-04946-1
22. Beurau of Indian Standard (1956) IS 516 (Indian Standard Methods of Tests for Strength of Concrete)
23. Sau D, Shiuly A, Hazra T (2023) Study on green concrete replacing natural fine and coarse aggregate by plastic waste—an experimental and machine learning approach. Mater Today Proc. https://doi.org/10.1016/j.matpr.2023.04.207
24. Tayeh BA, Almeshal I, Magbool HM, Alabduljabbar H, Alyousef R (2021) Performance of sustainable concrete containing different types of recycled plastic. J Clean Prod 328. https://doi.org/10.1016/j.jclepro.2021.129517
25. Sau D, Hazra T, Shiuly A (2023) Assessment of sustainable green concrete properties using recycled plastic waste as replacement for fine aggregate using machine learning technique. Compos Mech Comput Appl Int J 14(2):1–12. https://doi.org/10.1615/CompMechComputApplIntJ.2022044775
26. Almeshal I, Tayeh BA, Alyousef R, Alabduljabbar H, Mohamed AM (2020) Eco-friendly concrete containing recycled plastic as partial replacement for sand. J Mater Res Technol 9(3):4631–4643. https://doi.org/10.1016/j.jmrt.2020.02.090

Investigating the Influence of Chemical Treatment on the Mechanical Behavior of Kenaf/PP Composite

Deepak Kaushik, Sandeep Gairola, and Inderdeep Singh

1 Introduction

Natural fiber-reinforced composites (NFRCs) are the class of materials gaining increasing attention due to their potential to replace traditional synthetic fiber composites [1, 2]. These composites consist of a matrix material, typically a thermosetting or thermoplastic polymer, that is reinforced with fibers derived from natural sources such as flax, hemp, kenaf, sisal, bamboo, and jute. Incorporating natural fibers in composite materials offers several advantages, including lower cost, reduced environmental impact, and the ability to be easily sourced and processed locally [3–6]. As opposed to jute, hemp, or flax, kenaf allows for the utilization of up to 40% of the harvested stalk in the production of fiber [7]. Due to this quality, kenaf fiber is also more affordable than other natural fibers. The kenaf plant may reach heights of 3.6–4.3 m (12–14 ft) from seeds in about five to six months. Applications for natural fiber composites include the automotive, building, packaging, and consumer products industries [8]. They have good thermal and acoustic insulation characteristics, low density, excellent damping properties, and high strength-to-weight ratios. Natural fibers that are lignocellulosic are composed of cellulose, lignin, hemicellulose, waxes, pectin, and water-soluble materials. However, there are several challenges also related to natural fibers that must be addressed, including moisture absorption, poor adhesion to the polymer matrix, limited durability in severe environmental conditions, and difficulties in maintaining the consistent quality owing to natural fibers variability. To overcome these challenges, numerous types of treatments have been devised to strengthen the interfacial characteristics at the fiber-matrix interface along with increase in resistance to moisture and degradation.

Alkaline, typically NaOH (Sodium hydroxide) treatment of kenaf fiber, is an easiest method used to improve the interfacial adhesion between kenaf fibers and

D. Kaushik (✉) · S. Gairola · I. Singh
Indian Institute of Technology Roorkee, Roorkee, Uttarakhand, India
e-mail: deepak@me.iitr.ac.in

R. Sahu et al. (eds.), *Advancements in Materials Processing Technology, Volume 1*,
Springer Proceedings in Materials 48, https://doi.org/10.1007/978-981-97-4958-4_8

polymer matrices in NFRCs. Hemicellulose and lignin are broken down by the alkaline solution. It removes the oil and wax on the outer side of the fiber cell wall, exposing the cellulose microfibrils and creating a rough surface that enhances the fiber-polymer bonding. $NaHCO_3$(Sodium bicarbonate) treatment, on the other hand, is a simple, economical, and eco-friendly treatment process that alters the surface chemistry of the fibers and improves the tensile and flexural properties of the fibers and their composites. Besides, the aqueous solution of $NaHCO_3$ is marginally alkaline due to the presence of hydroxide ions (OH^-) and carbonic acid (H_2CO_3) [9]. The effectiveness of these treatments is determined through mechanical characterization, such as tensile, flexural, and impact testing. The treated fibers are then used in composite fabrication to develop composites with improved mechanical characteristics.

Samaei et al. [10] reported the influence of NaOH treatment on the sound absorption and tensile behavior of kenaf fibers. Fibers were treated with varying weight percentage (4–10) of NaOH for 4 h. Significantly enhanced tensile strength and sound absorption coefficient was noted for 6% NaOH-treated kenaf fibers. The diameter of treated fibers also reduced as compared to untreated fibers. Drastic decline in tensile strength was witnessed for higher concentrations (8% and 10%) of NaOH. Ismail et al. [11] investigated the effect of NaOH treatment on the characteristics of randomly oriented kenaf fiber mat and kenaf epoxy composites. Kenaf fiber mat was treated with 6 weight percentage NaOH solution for 24 h. The authors found that reinforcement of treated fibers reduced the mechanical performance of composite compared to untreated ones. FTIR analysis showed the disappearance of peaks because of breakdown of lignin present at the outside surface of the fiber. Fiore et al. [12] also reported a decrease in tensile strength of treated kenaf fibers which were treated using 6% NaOH for a long duration of 144 h. However, a duration of 44 h for the same concentration of NaOH yielded an improvement in mechanical behavior of both fibers and composites.

The alkali treatment of kenaf fiber has been explored by several other researchers [13–15]. However, depending on the type of kenaf bast and form of fiber used, optimum concentration, and duration of treatment may vary and, subsequently, the final effect on mechanical properties. However, when used at a large scale, most of the chemicals used for alkali treatment are hazardous to the environment. Therefore, researchers have made an attempt to employee a sustainable and eco-friendly treatment such as $NaHCO_3$ and borax to enhance the performance of natural fibers [16–18]. The authors reported an improvement in the interfacial bonding of fiber and matrix and, subsequently, in the mechanical characteristics of composites developed with the help of treated fibers.

The literature identified that a thorough analysis of eco-friendly fiber treatment and their consequences on the properties of kenaf fibers and their composites is still needed. Therefore, the primary goal of the current research investigation is to evaluate the effects of NaOH and $NaHCO_3$ treated fiber on the mechanical properties of kenaf/polypropylene (PP) composites.

2 Material and Methods

2.1 *Fiber Treatment*

Woven kenaf fiber (bidirectional) used for this experimentation were purchased from Go Green Products, Chennai (India). The fibers were subjected to clean with water followed by drying in a hot air oven at 50 °C for 24 h and designated as untreated fibers. For alkali treatment, the fibers were treated using 5% (w/v) NaOH solution for 2, 4, and 6 h. For eco-friendly treatment, fiber treatment was done with 10% (w/v) $NAHCO_3$ for 48, 96, and 144 h. Treated fibers were then neutralized using acetic acid, and pH strips were used to confirm the neutral behavior of fibers.

2.2 *Fabrication of Composite*

In order to fabricate the composites, PP was used as a matrix material and treated and untreated fiber mats were utilized as reinforcement. PP granules (Grade: 1110 MG) were procured from Indian Oil Corporation Limited, Panipat. PP sheets of 1 mm thickness were fabricated using a compression molding machine. After that 4 layers of fibers and 5 layers of PP sheet were arranged alternatively, to fabricate the composite plate using film stacking technique. A temperature of 160 °C and pressure of 20 bar was maintained for fabricating the composites using a compression molding machine. Samples for mechanical characterization were prepared from developed composites using high speed circular saw. Table 1 depicts the nomenclature of developed composites as per treatment type and duration of treatment.

Table 1 Nomenclature of developed composites

Treatment time (h)	Treatment type	Kenaf fiber-reinforced PP composite
0	–	UNKPPC
02	NAOH	KPPSH02
04		KPPSH04
06		KPPSH06
48	$NAHCO_3$	KPPSB48
96		KPPSB96
144		KPPSB144

2.3 Mechanical Characterization

Mechanical performance of developed composites was evaluated with the help of a universal testing machine (Make: Instron, Model:5982). Tensile and flexural properties of developed composites were determined in accordance with ASTM D3039 and ASTM D790, respectively, at a crosshead speed of 2 mm/min.

2.4 Morphological Analysis

The morphology of treated fibers, untreated fibers, and developed composites was studied via scanning electron microscope (Make: Carl Zeiss, Model; Carl Zeiss Ultra Plus).

3 Results and Discussions

3.1 Mechanical Characteristics

Figure 1a, b depicts the comparative graph of tensile and flexural properties of developed kenaf fiber-reinforced PP composites compared to PP. Reinforcing the PP with kenaf fibers resulted in significant improvement in tensile and flexural properties of UNKPPC, in comparison with neat PP. However, the improvement in mechanical properties was not that much significant, and therefore, chemical treatment of fibers has been employed that have shown further improvement in the tensile and flexural properties of KPPSH02, KPPSH04, KPPSB48, and KPPSB96 in comparison with UNKPPC. The modulus values were also improved with treatment duration, exhibiting a maximum tensile (1024.81 MPa) and flexural modulus (2350.36 MPa) for KPPSH04. The improvement in mechanical performance of developed composites after fiber treatment can be attributed to increased fiber strength, better interfacial bonding, and mechanical interlocking between the kenaf fiber and PP matrix. The chemical treatment of fiber removes the impurities from fiber surface, therefore, improves the surface adhesion characteristics of fiber by producing a rough surface topography. Moreover, treatment of fiber also results in partial elimination of hemicellulose content, and formation of hydrogen bond, thereby increasing the fiber strength. However, on further increase in treatment time duration, the tensile and flexural properties of KPPSH06 and KPPSB144 composites showed decrement in both tensile and flexural properties. The decrease in tensile and flexural properties of composites after 04 h of NaOH and 96 h of $NaHCO_3$ treatment can be ascribed to excessive fiber exposure to NaOH and $NaHCO_3$ solution. Excessive removal of the hemicellulose material from the kenaf fiber surface weakens the fibers. Aside from weakening the fibers, increasing the treatment duration beyond the optimal

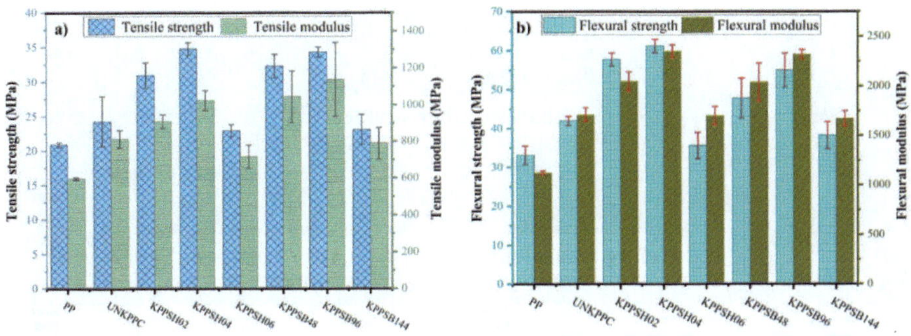

Fig. 1 **a** Tensile properties of developed composites compared to neat PP, **b** flexural properties of developed composites compared to neat PP

time produces severe fiber fibrillation. As a result, the mechanical interlocking sites between fibers and matrix are reduced. These findings were further substantiated by the fractography of fractured tensile samples shown in Fig. 2. Although both treatments (NaOH and $NaHCO_3$) enhanced the strength of developed composites up to optimal time duration, however, $NaHCO_3$ treatment exhibited a better tensile stiffness and approximately same tensile strength but at a higher time duration as compared to NaOH. The flexural properties of KPPSB96 were also found comparable to KPPSH04.

3.2 Failure Mechanism

The failure mechanism was further substantiated by the fractography analysis as depicted in Fig. 2. Figure 2a, b depicts that after chemical treatment, the fiber surface has shown removal of impurities and made the fiber surface rougher that can further contribute to the better mechanical interlocking between fiber-matrix interfaces. This revealed a gradual improvement in fiber-matrix adhesion and, therefore, have resulted in enhanced mechanical properties of the composites. From the SEM fractography, the improvement in fiber-matrix bonding can be ascertained from the presence of fiber breakages that could only be due to the enhanced interfacial characteristics between treated fiber and PP matrix. On the contrary, poor interfacial characteristics show the low stress transfer efficiency, and therefore, fiber pull-outs were observed that led to the premature failure of the specimens.

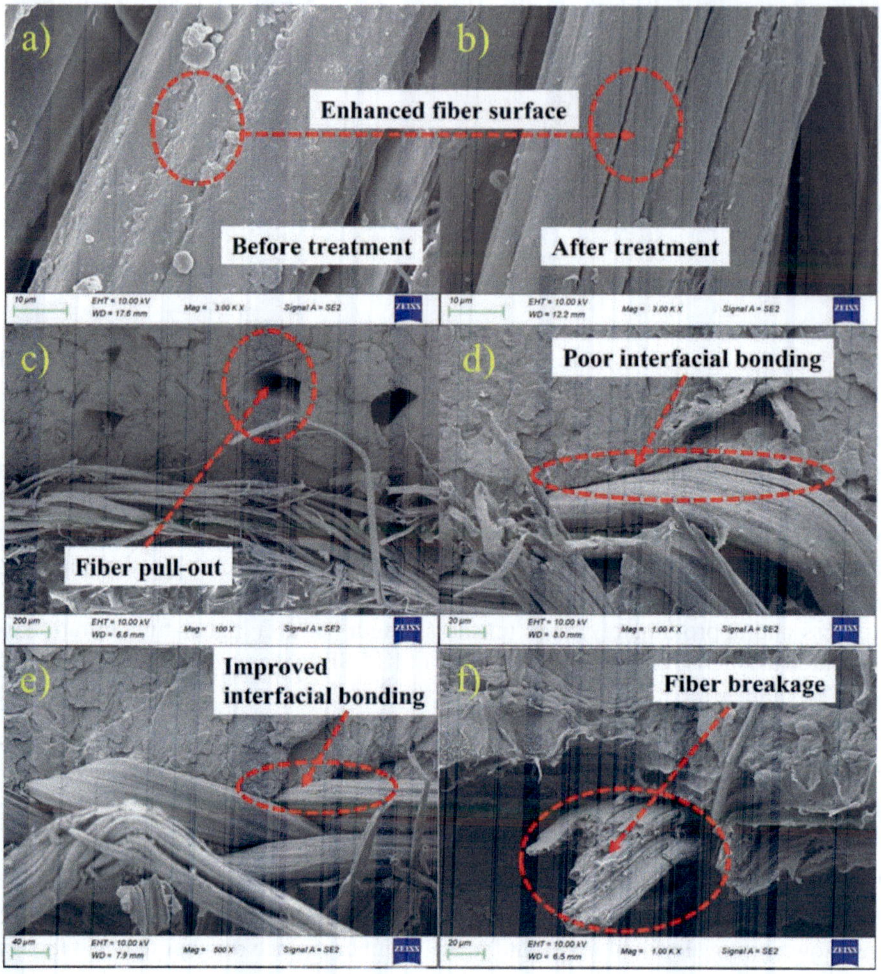

Fig. 2 **a, b** Surface morphology of fiber, **c–f** morphology of fractured specimens of developed composites

4 Conclusion

The experimental endeavor analyzed the influence of kenaf fiber treatment (NaOH and $NaHCO_3$) on the mechanical characteristics of kenaf/PP composites. It was found that both treatments modify the fiber surface, leading to enhanced fiber-matrix interaction and increased mechanical properties of developed composites. In comparison with UNKPPC, tensile strength, tensile modulus, and flexural modulus of KPPSH04 improved by 43.21%, 45.88%, 25.63%, and 37.38%, respectively, whereas corresponding increases for KPPSB96 were 41.28, 30.91, 39.61, and 35.51%. Therefore, it can be concluded that $NaHCO_3$, which is an environmentally acceptable treatment, may be used to treat kenaf fiber to develop composites with tensile characteristics that are similar to those of NaOH-treated kenaf fiber-based composites. However, longer duration of treatment is needed for the $NaHCO_3$ treatment compared to NaOH. Additionally, it was found that excessive exposure of fiber beyond the optimal time duration might lead to weakening of the fibers and as a result reduce the mechanical properties of developed composites. Hence, optimal concentration and duration of chemical treatment of natural fiber is a prerequisite to enhance the mechanical properties of natural fiber-reinforced composites.

References

1. Shah DU, Schubel PJ, Clifford MJ (2013) Can flax replace E-glass in structural composites? A small wind turbine blade case study. Compos Part B Eng 52:172–181. https://doi.org/10.1016/J.COMPOSITESB.2013.04.027
2. Wambua P, Ivens J, Verpoest I (2003) Natural fibres: can they replace glass in fibre reinforced plastics? Compos Sci Technol 63(9):1259–1264. https://doi.org/10.1016/S0266-3538(03)00096-4
3. Ku H, Wang H, Pattarachaiyakoop N, Trada M (2011) A review on the tensile properties of natural fiber reinforced polymer composites. Compos Part B Eng 42(4):856–873. https://doi.org/10.1016/j.compositesb.2011.01.010
4. John MJ, Thomas S (2008) Biofibres and biocomposites. Carbohydr Polym 71(3):343–364. https://doi.org/10.1016/j.carbpol.2007.05.040
5. Kumar S, Prasad L, Patel VK, Kumar V, Kumar A, Yadav A, Winczek J (2021) Physical and mechanical properties of natural leaf fiber-reinforced epoxy polyester composites. Polymers (Basel) 13(9):1369. https://doi.org/10.3390/polym13091369
6. Lotfi A, Li H, Dao DV, Prusty G (2021) Natural fiber-reinforced composites: a review on material, manufacturing, and machinability. J Thermoplast Compos Mater 34(2):238–284. https://doi.org/10.1177/0892705719844546
7. Hamidon MH, Sultan MTH, Ariffin AH, Shah AUM (2019) Effects of fibre treatment on mechanical properties of kenaf fibre reinforced composites: a review. J Mater Res Technol 8(3):3327–3337. https://doi.org/10.1016/j.jmrt.2019.04.012
8. Peças P, Carvalho H, Salman H, Leite M (2018) Natural fibre composites and their applications: a review. J Compos Sci 2(4):66. https://doi.org/10.3390/jcs2040066
9. Fiore V, Scalici T, Valenza A (2018) Effect of sodium bicarbonate treatment on mechanical properties of flax-reinforced epoxy composite materials. J Compos Mater 52(8):1061–1072. https://doi.org/10.1177/0021998317720009
10. Samaei SE, Mahabadi HA, Mousavi SM, Khavanin A, Faridan M, Taban E (2022) The influence of alkaline treatment on acoustical, morphological, tensile and thermal properties of kenaf natural fibers. J Ind Text 51(5_suppl):8601S–8625S. https://doi.org/10.1177/1528083720944240
11. Ismail NF, Mohd Radzuan NA, Sulong AB, Muhamad N, Che Haron CH (2021) The effect of alkali treatment on physical, mechanical and thermal properties of kenaf fiber and polymer epoxy composites. Polymers (Basel) 13(12). https://doi.org/10.3390/polym13122005
12. Fiore V, Di Bella G, Valenza A (2015) The effect of alkaline treatment on mechanical properties of kenaf fibers and their epoxy composites. Compos Part B Eng 68:14–21. https://doi.org/10.1016/j.compositesb.2014.08.025
13. Islam MR, Beg MDH, Gupta A (2014) Characterization of alkali-treated kenaf fibre-reinforced recycled polypropylene composites. J Thermoplast Compos Mater 27(7):909–932. https://doi.org/10.1177/0892705712461511

14. Ren Z, Wang C, Zuo Q, Yousfani SHS, Anuar NIS, Zakaria S, Liu X (2019) Effect of alkali treatment on interfacial and mechanical properties of kenaf fibre reinforced epoxy unidirectional composites. Sains Malaysiana 48(1):173–181. https://doi.org/10.17576/jsm-2019-4801-20
15. Kumar RS, Muralidharan N, Sathyamurthy R (2022) Optimization of alkali treatment process parameters for kenaf fiber: experiments design. J Nat Fibers 19(11):4276–4285. https://doi.org/10.1080/15440478.2020.1856276
16. Komal UK, Singh I (2022) Sustainable treatments of pineapple leaf fibers for polylactic acid based biocomposites. J Nat Fibers 19(16):13438–13456. https://doi.org/10.1080/15440478.2022.2095550
17. Chaitanya S, Singh I (2018) Ecofriendly treatment of aloe vera fibers for PLA based green composites. Int J Precis Eng Manuf Green Technol 5(1):143–150. https://doi.org/10.1007/s40684-018-0015-8
18. Chaitanya S, Singh I (2018) Sisal fiber-reinforced green composites: effect of ecofriendly fiber treatment. Polym Compos 39(12):4310–4321. https://doi.org/10.1002/pc.24511

Fabrication of Bio-degradable Material for Bone Tissue Engineering Applications

Devaraju Aruri, Vamshi Krishna Thallapelli, and Anil Kumar Sunkari

1 Introduction

Rather than using subtractive manufacturing techniques, three-dimensional (3D) printing, sometimes referred to as AM, or additive manufacturing and rapid development, is a way to join materials to build items from 3D model information, typically layer by layer [1]. As a flexible technological platform for computer-aided design and rapid production, 3D printing has developed. Using AM, or additive manufacturing, instead of traditional development and subtractive manufacturing, which often uses molding and machining, makes it possible to produce custom items out of substances such as metals and polymeric materials [2]. Applications for products made from 3D printers include construction of tissue engineering design, gold design, and many more fields [3]. The need for bone transplant materials is currently on the rise [4]. Years of investigation on different types of bone replacements have concentrated on defect repair [5–7]. The ability to carry cells, stimulate osteoblast growth, enhance biocompatibility, appropriate decomposition, along with osteoblast formation at the region of defect is all crucial factors to take into account when choosing an efficient bone replacement product [8–10]. The process that formerly made it possible to fabricate planar goods from digital files has been developed into AM techniques [11]. Using biomaterials the 3D produced components are constructed step by step [12, 13]. The benefits of 3D produced structures over conventional technique include architectural pore form [14]. Since they can be created and produced in three dimensions biomedical devices are essential for the development of tissues. Unlike traditional approaches for creating 3D porosity frameworks accurate plans based on patient 3D imaging data may be used to create personalized replacements [15–18]. Procedures

D. Aruri · V. K. Thallapelli (✉) · A. K. Sunkari
Department of Mechanical Engineering, Kakatiya Institute of Technology and Science, Warangal, Telangana, India
e-mail: vamshi4477@gmail.com

R. Sahu et al. (eds.), *Advancements in Materials Processing Technology, Volume 1*, Springer Proceedings in Materials 48, https://doi.org/10.1007/978-981-97-4958-4_9

for 3D printers with enough pore connectivity, optimal pore morphology and acceptable porosity offer promise for superior conventional bone substitutes [19]. Some of the substances we used to create our three-dimensional printable framework was polylactic acid (PLA) [20]. However,, have several disadvantages, such as low water solubility and frangibility [21]. Because it is readily availableand polylactic acid has become one of those polymers that decompose and have the widest range of uses.number of tasks in the packaging, electrical, automotive and prosthetic device industries [22, 23]. According to various trends in orthopedic research and practice [24], the use of PLA is harmless and secure to use. According to several studies have been shown that these biomaterials are biocompatible both in vivo and in vitro [25]. PLA and its substitutes have shown suitable for biological adaptability. They are non-toxic, non-inflammatory, and commonly used in bone regeneration [26]. In rats with skull abnormalities, polylactic acid grafts have been demonstrated to increase bone development and encourage reconstruction of the bone [27].

2 Literature Review

The process of tissue engineering makes extensive use of 3D printers to produce scaffolds that mimic the extracellular matrix of real tissues. Because it is simple to produce, compatible with life, and environmentally friendly [1–7].

Numerous investigations have been carried out to examine the application of 3D produced polylactic acid supports in bone tissue engineering process. Utilized a modified fused deposition modeling (FDM) printing system to build PLA frameworks with a hierarchical pore structure. It was demonstrated that the frameworks have high mechanical strength and aided in the proliferation and osteogenic lineage differentiation [12–18].

3D printing and electrospinning to build PLA frameworks with a microfiber-reinforced structure were able to adhere, multiply, to differentiate into osteogenic linkage on the compatible with the body frameworks [13].

Bioprinting technique to produce a PLA framework with a porous structure and incorporated stem cell (hBMSCs) produced from human bone marrow. After being implanted in rat with living cells, the framework enhanced the proliferation to differentiation into osteogenic, leading to positive outcome in bone repair [28].

Other research has looked into using PLA based composite materials to 3D print bone tissue engineering scaffolds. They used a mixture of PLA and graphene oxide to create scaffolds with superior mechanical characteristics and biocompatibility for bone regeneration [13].

All things considered, our results suggest that 3D-printed polylactic acid structures hold great potential for applications in tissue engineering for bone. However, more investigation is required to maximize the mechanical properties, biocompatibility, and clinical performance of framework design and production parameters.

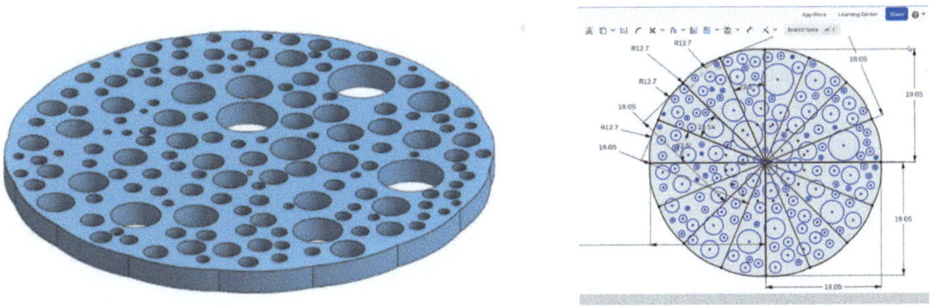

Fig. 1 Scaffold design in onshape (CAD software)

3 Methodology

3.1 Scaffold Design

Onshape, a free cloud-based CAD software, was used to create the scaffold geometry. The scaffold was created with a highly interconnected porous structure that resembled genuine bone tissue as shown in Fig. 1. For 3D printing, the scaffold design was exported as.stl files.

3.2 Material Selection

Polylactic acid (PLA) filament was chosen as the scaffold material because of its biocompatibility, biodegradability, and favorable mechanical qualities. The PLA filament used has a diameter of 1.75 mm and a melting point of 170–230 °C as shown in Fig. 2.

3.3 Scaffold Preparation

The scaffolds were made on a Creality Ender 3 3D printer that had been customized by replacing the original nozzle with a 0.4 mm brass nozzle and modifying the printer settings to optimize the printing parameters for PLA. The scaffolds were printed with varied pore diameters (400–700 m) and porosities (50–70%) as shown in Fig. 3. The following printer settings were used: nozzle temperature 215 °C, bed temperature 60 °C, printing speed 50 mm/s, layer height 0.2 mm. After that, the scaffolds were disinfected with ethylene oxide gas.

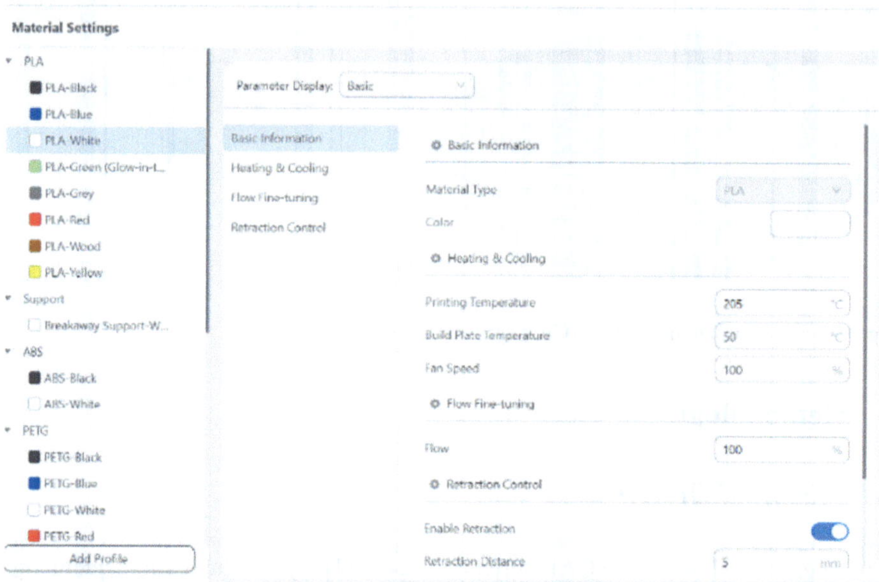

Fig. 2 Adjusting material settings in Snapmaker Luban (slicing software)

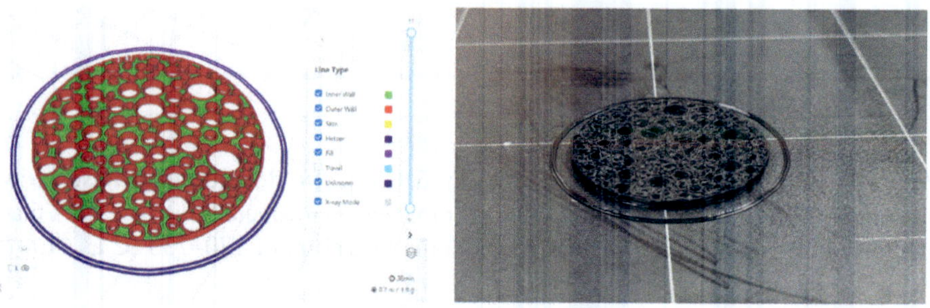

Fig. 3 3D-printed scaffold

3.4 *Scaffold Characterization*

A tensile testing machine (MTS Systems Corporation, Eden Prairie, MN) was used to analyze the mechanical properties of the PLA scaffolds as shown in Fig. 4. The specimens were cut into rectangular forms (20, 5, 2 mm) and tested to failure at a crosshead speed of 5 mm/min. The stress–strain curves were used to calculate the tensile strength, elongation at break, and Young's modulus.

The scaffold's tensile strength shows the maximum stress that the scaffold can withstand before failing. The elongation at break is a measure of the scaffold's ductility and is the deformation at which the scaffold breaks. The Young's modulus,

Fig. 4 Specimen placed in UTM for mechanical testing

which is calculated as the slope of the stress–strain curve in the elastic area, is a measure of the scaffold's stiffness or rigidity.

The overall goal of the scaffold characterization was to evaluate the mechanical properties of the PLA scaffolds utilizing the tensile test. This information was crucial in determining the scaffolds' potential for bone tissue engineering applications.

3.5 Degradation

The deterioration of the scaffold is a critical factor in the success of bone tissue creation. The scaffold must deteriorate at a rate that corresponds to the rate of new bone growth. If the scaffold declines too fast, it may not provide enough mechanical support for new bone growth, whereas gradual deteriorate may inhibit new bone formation.

PLA is a biodegradable polymer that is frequently utilized in 3D-printed bone tissue engineering scaffolds. PLA scaffolds degrade via hydrolysis, which breaks down the polymer into lactic acid monomers. The body's normal metabolic processes then metabolize the lactic acid.

The molecular weight of the polymer, the crystallinity of the scaffold, and the porosity of the scaffold are all aspects that can influence the breakdown rate of PLA scaffolds. In general, the higher the scaffold's molecular weight and crystallinity, the slower the breakdown rate. Conversely, the higher the scaffold porosity, the faster the degradation rate.

4 Experiments and Results

4.1 Mechanical Properties Testing

1. Compressive Strength Testing: The PLA scaffolds' compressive strength was assessed by subjecting the scaffold samples to a uniaxial compressive load until failure. The scaffolds have a compressive strength of 25 MPa, according to the data.
2. Tensile Strength Testing: The PLA scaffolds' tensile strength was determined by subjecting the scaffold samples to a uniaxial tensile load until failure. The scaffolds had a tensile strength of 8 MPa, according to the data.
3. Elastic Modulus Testing: A compressive load test was used to determine the elastic modulus of the PLA scaffolds. The elastic modulus of the scaffolds was determined to be 350 MPa.
4. Overall, these results indicate that 3D-printed PLA scaffolds have mechanical properties that are appropriate for bone tissue engineering applications. The PLA scaffolds' compressive strength and elastic modulus are comparable to that of human cancellous bone, while their tensile strength is slightly lower.

4.2 Degradation Test

The results of the degradation study revealed that the scaffolds exhibited gradual degradation over time. After 12 weeks of immersion in PBS, the scaffolds experienced a weight loss of approximately 10%. This weight loss indicates the degradation and dissolution of the PLA material within the scaffolds.

Furthermore, the pH of the PBS solution decreased slightly during the degradation process. This change in pH suggests the release of acidic degradation byproducts as the PLA scaffolds underwent degradation.

Combining the degradation results with the mechanical properties testing, it can be inferred that the 3D-printed porous PLA scaffolds possess suitable degradation behavior for their potential applications in tissue engineering. The gradual degradation allows for the gradual transfer of load to the newly formed tissue during the regeneration process. However, further research is still needed to optimize the mechanical properties and degradation characteristics of PLA scaffolds to enhance their performance for bone tissue engineering applications [28].

5 Conclusion

In conclusion, using 3D-printed PLA scaffolds for bone tissue engineering is a promising strategy for bone repair and regeneration, with the potential to match the mechanical properties of native bone tissue. PLA scaffolds' mechanical qualities, such as stiffness and strength, can be adjusted by altering their internal architecture and porosity. Furthermore, in vitro degradation investigations have revealed that PLA scaffolds deteriorate gradually over time, with the rate of degradation being dependent on the scaffold features. The degrading behavior of PLA scaffolds must be carefully considered to ensure their long-term usefulness in vivo.

Overall, 3D-printed PLA scaffolds are an appealing solution for bone tissue engineering applications due to their good mechanical properties and regulated degrading behavior. More research is needed to optimize the design and manufacturing parameters of PLA scaffolds in order to improve their mechanical characteristics and degradation behavior, as well as to assess their long-term performance in vivo. 3D-printed PLA scaffolds have the potential to change the area of bone tissue engineering and regenerative medicine as 3D printing technology and materials science develop.

References

1. Ngo TD, Kashani A, Imbalzano G, Nguyen KTQ, Hui D (2018) Additive manufacturing (3D printing): a review of materials, methods, applications and challenges. Compos B Eng 143:172–196
2. Ligon SD, Liska R, Stampfl J, Gurr M, Mülhaupt R (2017) Polymers for 3D printing and customized additive manufacturing. Chem Rev 117(15):10212–10290
3. Altaf K, Qayyum JA, Rani AMA, Ahmad F, MegatYusoff PSM, Baharom M, Aziz ARA, Jahanzaib M, German RM (2018) Performance analysis of enhanced 3D printed polymer molds for metal injection molding process. Metals 8(6):433
4. Henkel J, Woodruff MA, Epari DR, Steck R, Glatt V, Dickinson IC, Choong PF, Schuetz MA, Hutmacher DW (2013) Bone regeneration based on tissue engineering conceptions—a 21st century perspective. Bone Res 1:216–248. https://doi.org/10.4248/BR201303002
5. McAllister BS, Haghighat K (2007) Bone augmentation techniques. J Periodontol 78:377–396. https://doi.org/10.1902/jop.2007.060048
6. Browaeys H, Bouvry P, De Bruyn H (2007) A literature review on biomaterials in sinus augmentation procedures. Clin Implant Dent Relat Res 9:166–177. https://doi.org/10.1111/j.1708-8208.2007.00050.x
7. Aghaloo TL, Moy PK (2007) Which hard tissue augmentation techniques are the most successful in furnishing bony support for implant placement? Int J Oral Maxillofac Implants 22:49–70
8. Moore WR, Graves SE, Bain GI (2001) Synthetic bone graft substitutes. ANZ J Surg 71:354–361. https://doi.org/10.1046/j.1440-1622.2001.02128.x
9. Navarro M, Michiardi A, Castaño O, Planell JA (2008) Biomaterials in orthopaedics. J R Soc Interface 5:1137–1158. https://doi.org/10.1098/rsif.2008.0151
10. Rezwan K, Chen QZ, Blaker JJ, Boccaccini AR (2006) Biodegradable and bioactive porous polymer/inorganic composite scaffolds for bone tissue engineering. Biomaterials 27:3413–3431. https://doi.org/10.1016/j.biomaterials.2006.01.039

11. Travieso-Rodriguez JA, Jerez-Mesa R, Llumà J, Gomez-Gras G, Casadesus O (2021) Comparative study of the flexural properties of ABS, PLA and a PLA–wood composite manufactured through fused filament fabrication. Rapid Prototyp J 27:81–92. https://doi.org/10.1108/RPJ-01-2020-0022
12. Habibovic P, Gbureck U, Doillon CJ, Bassett DC, van Blitterswijk CA, Barralet JE (2008) Osteoconduction and osteoinduction of low-temperature 3D printed bioceramic implants. Biomaterials 29:944–953. https://doi.org/10.1016/j.biomaterials.2007.10.023
13. Luo Y, Wu C, Lode A, Gelinsky M (2013) Hierarchical mesoporous bioactive glass/alginate composite scaffolds fabricated by three dimensional plotting for bone tissue engineering. Biofabrication 5:015005. https://doi.org/10.1088/1758-5082/5/1/015005
14. Yang C, Wang X, Ma B, Zhu H, Huan Z, Ma N, Wu C, Chang J (2017) 3D-printed bioactive Ca_3SiO_5 bone cement scaffolds with nano surface structure for bone regeneration. ACS Appl Mater Interfaces 9:5757–5767. https://doi.org/10.1021/acsami.6b14297
15. Montjovent MO, Mathieu L, Schmoekel H, Mark S, Bourban PE, Zambelli PY, Laurent-Applegate LA, Pioletti DP (2007) Repair of critical size defects in the rat cranium using ceramic-reinforced PLA scaffolds obtained by supercritical gas foaming. J Biomed Mater Res A 83:41–51. https://doi.org/10.1002/jbm.a.31208
16. Abdal-Hay A, Sheikh FA, Lim JK (2013) Air jet spinning of hydroxyapatite/poly(lactic acid) hybrid nanocomposite membrane mats for bone tissue engineering. Colloids Surf B Biointerfaces 102:635–643. https://doi.org/10.1016/j.colsurfb.2012.09.017
17. Kothapalli CR, Shaw MT, Wei M (2005) Biodegradable HA-PLA 3-D porous scaffolds: effect of nano-sized filler content on scaffold properties. Acta Biomater 1:653–662. https://doi.org/10.1016/j.actbio.2005.06.005
18. Jeong SI, Ko EK, Yum J, Jung CH, Lee YM, Shin H (2008) Nanofibrous poly(lactic acid)/hydroxyapatite composite scaffolds for guided tissue regeneration. Macromol Biosci 8:328–338. https://doi.org/10.1002/mabi.200700107
19. Guvendiren M, Molde J, Soares RM, Kohn J (2016) Designing biomaterials for 3D printing. ACS Biomater Sci Eng 2:1679–1693. https://doi.org/10.1021/acsbiomaterials.6b00121
20. Vidakis N, Petousis M, Savvakis K, Maniadi A, Koudoumas E (2019) A comprehensive investigation of the mechanical behavior and the dielectrics of pure polylactic acid (PLA) and PLA with graphene (GnP) in fused deposition modeling (FDM). Int J Plast Technol 23:195–206. https://doi.org/10.1007/s12588-019-09248-1
21. Kumar MP, Ponnusamy S, Reddy Nallamilli MS (2021) The influence of process parameters on the impact resistance of 3D printed PLA specimens under water-absorption and heat-treated conditions. Rapid Prototyp. J. 27:1108–1123. https://doi.org/10.1108/RPJ-02-2020-0037
22. Afonso JA, Alves JL, Caldas G, Gouveia BP, Santana L, Belinha J (2021) Influence of 3D printing process parameters on the mechanical properties and mass of PLA parts and predictive models. Rapid Prototyp J 27:487–495. https://doi.org/10.1108/RPJ-03-2020-0043
23. Von Windheim N, Collinson DW, Lau T, Brinson LC, Gall K (2021) The influence of porosity, crystallinity and interlayer adhesion on the tensile strength of 3D printed polylactic acid (PLA). Rapid Prototyp J 27:1327–1336. https://doi.org/10.1108/RPJ-08-2020-0205
24. Athanasiou KA, Niederauer GG, Agrawal CM (1996) Sterilization, toxicity, biocompatibility and clinical applications of polylactic acid/polyglycolic acid copolymers. Biomaterials 17:93–102. https://doi.org/10.1016/0142-9612(96)85754-1
25. Mikos AG, Sarakinos G, Leite SM, Vacanti JP, Langer R (1993) Laminated three-dimensional biodegradable foams for use in tissue engineering. Biomaterials 14:323–330. https://doi.org/10.1016/0142-9612(93)90049-8

26. Levy FE, Hollinger JO, Szachowicz EH (1994) Effect of a bioresorbable film on regeneration of cranial bone. Plast Reconstr Surg 93:307–311. https://doi.org/10.1097/00006534-199402000-00011
27. Hollinger JO (1983) Preliminary report on the osteogenic potential of a biodegradable copolymer of polyactide (PLA) and polyglycolide (PGA). J Biomed Mater Res 17:71–82. https://doi.org/10.1002/jbm.820170107
28. Zhang H et al (2019) Fabrication and degradation study of 3D printed PLA scaffolds for bone tissue engineering. In: Proceedings of the 5th international conference on biomaterials and tissue engineering, vol 56(2), pp 123–135

Processing Challenges in the Fabrication of In Situ Al-Mg_2Si Composite Sheet Through Continuous Casting Route

Dheeraj Kumar Saini and Pradeep Kumar Jha

1 Introduction

Innovation in composite materials has revolutionized various industries, offering superior properties and unmatched versatility in diverse applications. The selection and implementation of appropriate fabrication methods play a crucial role in optimizing the performance of composite materials. In the context of composite production, continuous production techniques have gained significant attention due to their ability to generate long, uninterrupted lengths of product. Continuous production offers numerous advantages, including improved efficiency, enhanced mechanical properties, reduced material waste, and cost-effectiveness. The production of composite sheets through conventional methods poses challenges, such as converting a composite billet into a sheet through rolling and extrusion, which is highly energy-intensive and prone to cracking. Fabrication of direct sheets from the melt in a single-step process can adequately address these issues. In the quest for this, Saini et al. summarize a need to design a twin-roll continuous casting setup to fabricate composite sheets in a single-step continuous casting process [1]. In the subsequent study, they developed the composite sheet of Al-Mg_2Si using a twin-roll continuous caster and highlighted the processing issues in fabricating particle-reinforced composite sheets [2]. The quality of the composite sheet not only depends on the temperature and roll velocity but the certain material selection, designed constraints and development routes that must be addressed to get the desired product quality. Designed constraints involve tundish selection that should ensure the uniform distribution of particles inside the matrix and should not promote the clustering and agglomeration of reinforced particles. Material selection involves the roll materials

D. K. Saini (✉) · P. K. Jha
Process Modeling and Heat Treatment Laboratory, Department of Mechanical and Industrial Engineering, Indian Institute of Technology Roorkee, Roorkee, India
e-mail: dk_saini@me.iitr.ac.in

R. Sahu et al. (eds.), *Advancements in Materials Processing Technology, Volume 1*,
Springer Proceedings in Materials 48, https://doi.org/10.1007/978-981-97-4958-4_10

and their ability to stick with the cast component, cast material processing temperature, and freezing range. During twin roll continuous casting, rolls are subjected to high thermal stress, wear, and corrosion-oxidation [3]. The coating on the rollers is required to prevent these phenomenon and sticking the cast material to the rolls. To avoid the sticking, literature reported the use of lubricants, which leads to the formation of an interlayer between the melt and rollers. This significantly retard the heat transfer rate from the rollers [4, 5]. Therefore, a lower casting speed is desired while applying the lubricants.

The freezing range refers to the temperature range between the liquidus and solidus temperatures of the alloy, during which the alloy transforms from a fully liquid state to a solid state. The freezing range affects several aspects, such as solidification behavior, casting speed, microstructure heat transfer of the continuous casting process, and sheet quality [6–8]. Continuous casting is limited to alloys with a narrow freezing range [9]. A wider freezing range tends to result in the formation of multiple phases and potential microstructural defects such as macrosegregation. This can affect the cast sheet's mechanical properties and surface quality [8]. The two approaches, ex situ and in situ, are used to develop composite in the conventional casting process [10–12]. While fabricating a composite sheet, it is required to know the selection of an available route to cast composite. Ex situ production leads to inhomogeneity in the microstructure, which requires an additional stirring mechanism. In contrast, the in situ route has better chances of mixing and shows better bonding between matrix and reinforcement. Therefore, the quality of the developed composite sheet depends on the processing challenges involve while continuous operation.

The present study highlights the processing challenges while developing particle-reinforced near-eutectic Al-Mg_2Si composite sheets using a vertical twin-roll continuous casting process. This study aims to address the concerns associated with continuous casting techniques and pave the way for improved sheet production in the industry. This study examines the design of continuous casting setup, selection of processing parameters (such as casting speed and superheat temperature), material selection criteria to cast composite sheet, and processing defects in the cast components. The research employs stereoscope, optical microscope, field emission scanning electron microscope (FESEM), differential scanning calorimetry (DSC), and % thermogravimetric (TG) analysis to characterize the defects, reinforcement distribution, and solidification behavior of the composite sheet.

2 Methodology

2.1 Casting

Commercially pure Al, Mg (99% purity), and Al-12Si were used as a feedstock to cast the Al-Mg_2Si composite. The pure Al and Al-Si alloy was kept in a graphite-coated SiC crucible at a temperature of 650 °C in a muffle furnace. Pure Mg was

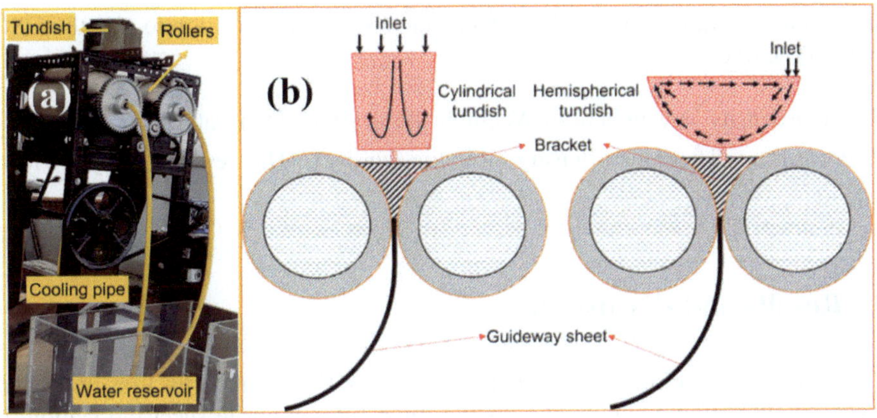

Fig. 1 **a** Continuous casting setup to cast composite sheet, **b** schematic diagram of continuous casting setup having a cylindrical or hemispherical shape of tundish

added lastly in the melts to avoid oxidation losses. The melt was held for one hour after complete melting at 650 °C. A C_2Cl_6 was used as a degasser to prevent the gaseous entrapment in the melts. 0.5% of the C_2Cl_6 was added to the melts and held for 5 min. The melt was cooled to 600 °C to make it suitable for pouring at a continuous casting setup. An in-house developed facility, shown in Fig. 1a, was used to cast the composite sheet. The prepared melt at 600 °C was poured onto the casting setup, while the roller speed was set at 0.072 m/s, ensuring the formation of the desired composite sheet.

2.2 *Metallography, Solidification, and Porosity Measurements*

The metallographic practices outlined in ASTM E3-11 were followed for the metallography. SiC polishing paper followed by the alumina slurry of 0.01 μm size was used for sample polishing. A Keller's reagent was used to etch the specimen. Phase-specific analysis was conducted using a stereoscope, optical microscope and FESEM electron microscope. To understand the solidification behavior of the melt, a differential scanning calorimetry (DSC) and % thermogravimetric analysis (TG) were employed. The DSC measured the heat flow during the process, while %TG analyzed the changes in weight as a function of temperature. Alumina crucible was used to hold 0.2 g of Al-Mg_2Si composite. A 10 °C/min heating rate was used to conduct the experiment.

A relative density measurement technique was used to measure the experimental density of the samples. The percentage porosity was calculated by comparing the experimental and theoretical densities. Origin Pro 2023b learning edition and X-pert, high score software, were used for data analysis and visualization to plot the DSC and %TG curves.

$$\text{Relative density} = \text{weight in air Weight in the air} - \text{Weight in water}$$

$$\%\text{Porosity} = (\text{Theoretical density} - \text{Relative density})/(\text{Theoretical density}) \times 100$$

3 Results and Discussion

This study is categorized into four core dimensions. For instance, the first dimension discusses designing a continuous casting setup for composite sheet fabrication. It involves designing the setup to ensure uniform particle distribution and selection of roller material, roller coating, and bracket material. The second-dimension studies the processing parameters, such as superheat temperature and casting speed. The third dimension discusses the selection of processed material depending on its processing temperature, freezing range, and development route. The fourth-dimension studies the defects occurring during composite sheet processing, such as macrosegregation, solidification shrinkage, gaseous porosity, and melting loss.

3.1 Designing of Continuous Casting Setup for Composite Sheet Fabrication

Ensure Uniform Particle Distribution

While designing the continuous casting setup for the particle-reinforced composite sheet fabrication, knowing the density difference between the reinforcement and the matrix is important. Density difference is one of the material's bulk properties that play a significant role in the even distribution of the reinforcement. The mixture was partly solid during the solidification, as in the compo casting in the continuous operation. Small or insignificant gravity-induced segregation between the matrix and particle occurs if the mixture is at rest, and the matrix's density and reinforcement density are the same. A significant difference in the reinforcement and matrix density leads to the settling or floatation of the reinforcement during solidification. During the processing of the Al-Mg_2Si composite, the density difference was 26%, whereas Mg_2Si has a lesser density (1.99 g/cm^3) than the aluminum matrix (2.7 g/cm^3). Therefore, there was a probability of gravity-induced segregation while processing. To avoid this segregation, the melt pool should have turbulence to avoid gravity-induced segregation. During the continuous casting, tundish above the rolls is used to maintain the specific flow rate of the material based on the casting speed. Therefore, a design promoting turbulence could be preferred for continuous operation. Hence, a hemispherical or cylindrical shape of the tundish can be preferred for continuous casting, as shown by a schematic diagram in Fig. 1b. The present study selected a cylindrical type of tundish with a 0.076 m/s casting speed to ensure reinforcement homogeneity in the microstructure.

Roller Material and Coating

The type of roll material (steel or copper) in continuous casting affects the cooling rate, strip thickness, supercooling, and cast component quality. Copper rolls have higher thermal conductivity than steel rolls, allowing a higher cooling rate, which helps refine the microstructure. However, Cu rolls' superiority over steel rolls is unclear. The copper rolls over steel rolls are selected for higher productivity and rapid solidification. The literature suggests that twin-roll casting can lead to the formation of a weakly pronounced basal texture, and the surface quality of the cast product is a critical issue. The surface roughness of the roll material also impacts the casting surface [13]. The present study used steel rolls with water-cooling arrangements to control the cooling rate by maintaining the water flow rate at 0.083 l/s. The rolls were Ni-coated to prevent oxidation and sticking to the rolls with the cast material. A casting operation in Fig. 2a shows the sticking of Al-Mg_2Si melts with the Ni-coated mild steel rollers.

Bracket Material

To control the width of the sheet and cooling rate, the selection of bracket material is essential. A bracket is used between the rolls to hold the melt between the rollers. Different types of bracket materials, such as Cu, steel, and insulating materials, can be used. Higher conductivity material allows the rapid temperature to drop off the melt pool and does not ensure the continuous feeding of the material due to early solidification. An insulating material was selected to avoid the premature solidification of the melt. The present study used mild steel brackets and Plaster of Paris brackets to cast composite sheets. Mild steel bracket requires higher superheating temperature

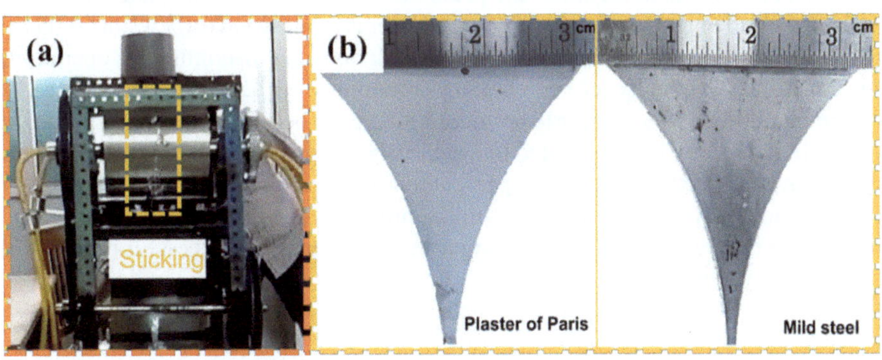

Fig. 2 **a** Continuous casting setup showing the sticking of Al-Mg_2Si melts with roll surface, **b** bracket made up of plaster of Paris and mild steel

and has a higher probability of choking the continuous operation. Figure 2b shows the bracket used in the continuous casting setup made up of steel and plaster of Paris.

3.2 *Processing Parameters*

Superheat Temperature and Casting Speed

The higher superheating temperature leads to a longer heat flux radiation time, necessitating a lower casting speed. Similarly, a higher casting speed requires a lower superheat temperature. However, the simultaneous adjustment of both variables is likely to have a detrimental effect on strip quality.

Casting speed impacts the production rate and heat transfer coefficient (HTC) [2, 14]. During casting, insufficient solidification at the roll nip area interrupts the continuous operation. The surface region of the roller nip cools faster compared to the thickness region of the cast sheet. As the roller speed increases, the temperature at the surface and middle portion of the sheets rises due to reduced heat transfer caused by shorter contact time. This situation can sometimes result in incomplete solidification of the molten material at the roller exit, interrupting continuous operation. If the casting speed is too low, the molten material solidifies at the interface with the roller. The metal in contact with the roller cools faster than the central region, causing increased rolling force and stress concentration at the edges, leading to cracks near the strip's edges. Hence, the selection of casting speed is the critical criterion while fabricating the composite sheet.

During the casting of the Al-Mg_2Si composite sheet, a casting speed ranging from 0.05 to 0.96 m/s and superheat or melt inlet temperature ranging from 603 °C to 657 °C were selected. Higher casting speed (613 °C and 0.096 m/s) does not allow the melt to solidify at a kissing point completely and leads to the insufficient strength of the sheet to withstand metallostatic pressure. As a result, the sheet does not come out from the roll exit, as shown in Fig. 3a. A lower casting speed (603 °C and 0.05 m/s) leads to the premature solidification of the melt pool and discontinuous operation. This condition is typically termed choking or jamming of the roller due to the early melt pool solidification. The early solidified melt pool between the rollers is shown in Fig. 3b. Therefore, it is suggested to use the higher superheating temperature with a lower casting speed and a lower superheat temperature with a higher casting speed to ensure successful composite sheet production. The present study used a 600 °C inlet temperature with 0.072 m/s casting speed to ensure the successful production of composite sheets.

Fig. 3 **a** Continuous operation shows discontinuity due to higher casting speed, **b** premature solidified melt pool during the continuous casting

3.3 Material Selection

Freezing Range

The freezing range has an impact on the casting speed. A narrower freezing range allows for higher casting speed due to rapid solidification of the melt. However, if the freezing range is too narrow, it may lead to cracking or other defects due to thermal stresses. The freezing range also influences the microstructure of the cast sheet. A narrower freezing range generally leads to a finer and more uniform microstructure, which can improve the mechanical properties of the sheet. On the other hand, a wider freezing range may result in coarser and less homogeneous microstructures, potentially affecting the sheet's mechanical properties. Different material compositions widen or narrow the freezing range, as shown in Fig. 4a. To optimize the continuous casting process for composite sheets, it is essential to carefully select material compositions with suitable freezing ranges, control the cooling rates and adjust the process parameters to achieve the desired microstructure and mechanical properties in the cast sheet. A low-freezing range material is desired for continuous casting. The composition of the continuous cast sample should be selected to reach its solidus temperature at the exit of the rolls. Generally, higher freezing range material has greater chances of macrosegregation, commonly seen in the continuous cast product. Therefore, selecting a near-eutectic composition is desirable for continuous casting.

In situ/Ex situ Development

While fabricating a composite sheet using continuous casting, achieving the homogeneous distribution of reinforcement or secondary phases in the matrix is necessary to ensure uniform mechanical properties. The ex situ and in situ routes are used to develop composite in the conventional casting. The main difference between these approaches is how the reinforcing particles are introduced inside the melt. Ex situ

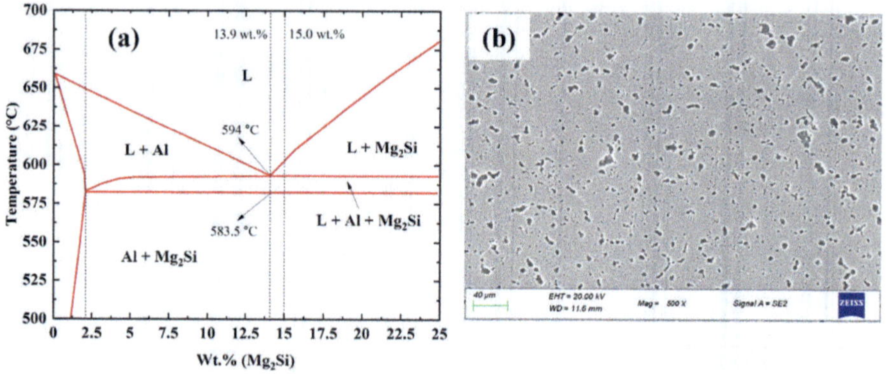

Fig. 4 **a** Phase diagram of Al-Mg_2Si composite, **b** FESEM micrograph of Al-15Mg_2Si shows the uniform distribution of Mg_2Si particles in the Al matrix

composites refer to multiphase materials, where the reinforcement phase is synthesized separately and then incorporated into the matrix through secondary processes such as powder metallurgy and infiltration. In situ composites involve the synthesis of the reinforced phase during the processing itself [1]. An in situ developed composite is better mixed with the matrix with good wettability, whereas the ex situ method required external stirring to ensure homogeneity in the microstructure. The present study used an in situ method to develop Al-Mg_2Si composite sheet. The microstructure in Fig. 4b shows the uniform distribution of the Mg_2Si (black) within the Al matrix (gray).

3.4 *Processing Defects*

Macrosegrigation

Macrosegregation refers to the non-uniform distribution of alloying elements and impurities within a cast product. It primarily occurs due to the differences in solidification behavior and density of various constituents. Figure 5a shows the macrosegregation in the Al-Mg_2Si composite sheet due to variation in chemical composition along the thickness of the sheet, where regions 1 and 2 show the different solute concentration. Figure 5b shows the grain size variation at the composite sheet's surface. During solidification, macrosegregation can lead to variations in the cooling rate and solute concentration in different casting regions. These variations can result in heterogeneous nucleation and growth of grains, leading to a non-uniform grain size distribution. Solidification shrinkage is one of the primary macrosegregation mechanisms. As the molten metal solidifies, it undergoes a volumetric contraction due to the phase change from liquid to solid. This solidification shrinkage can lead to the formation of shrinkage porosity, which acts as a pathway for the transport of

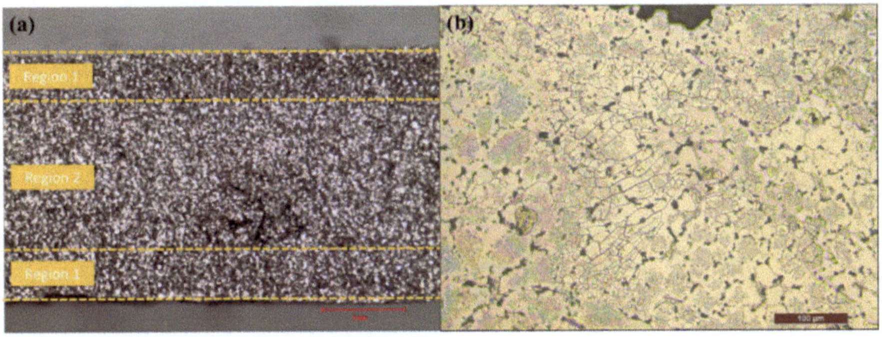

Fig. 5 Variation in **a** composition along the thickness of the continuously cast composite sheet in region 1 and region 2 and **b** grain size due to different nucleation and growth

solute-rich liquid. The solute-rich liquid tends to flow toward regions of lower pressure, accumulating alloying elements and impurities, leading to macrosegregation [15].

Solidification Shrinkage

Shrinkage porosity or solidification shrinkage is a widespread problem in continuous casting that occurs during solidification due to the no longer compensation of the loss of volume during contraction. This leads to defects in the form of porosity in the fabricated sheet. The porosity in the sheet is present in the form of surface sinks, internal microporosity, and microporosity. Figure 6 shows the internal macroporosity shrinkage in the 3 mm ticked sheet. Figure 7 shows the internal microporosity shrinkage in the continuously cast Al-Mg_2Si composite sheet. The porosity in the continuously cast Al-Mg_2Si sheet was reported to be 10%. During continuous casting, shrinkage was mostly seen at the sheet's central portion, the last solidified region that cannot compensate for the volume loss.

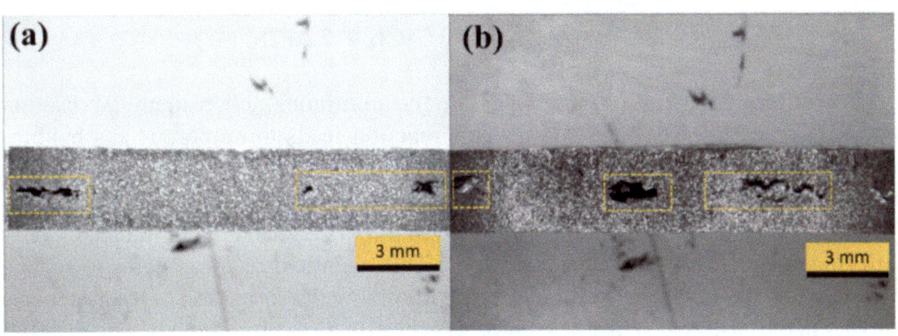

Fig. 6 Stereoscopic images of internal solidification shrinkage in the continuously cast Al-Mg_2Si composite sheet cross section

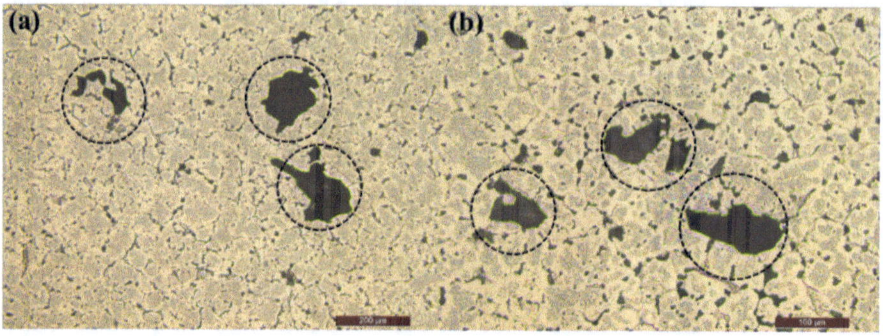

Fig. 7 Optical micrograph shows the internal microporosity shrinkage defect in the continuously cast Al-Mg_2Si composite sheet

Gaseous Porosity

Gaseous porosity is a common defect during the continuous casting of aluminum components. It refers to gas-filled voids or bubbles within the solidified metal structure. Various gases, such as hydrogen, nitrogen, and oxygen, can become entrapped in the liquid metal during casting. These gases may originate from the melting, atmosphere, or reactions within the melt. Certain gases, mainly hydrogen, can dissolve in the liquid metal at high temperatures. When the metal cools and solidifies, these dissolved gases can come out of the solution and form bubbles. To avoid such defects before the continuous casting, degassing was done by mixing the hexachloroethane to remove dissolved gases. Hexachloroethane is primarily employed to prevent the formation of hydrogen gas porosity. It forms aluminum chloride ($AlCl_3$) and gaseous byproducts, such as carbon monoxide (CO) and chlorine gas (Cl_2), shown in the following equation:

$$C_2Cl_6 \rightarrow C_2Cl_4 + Cl_2$$
$$3\,Cl_2 + 2\,Al \rightarrow 2\,AlCl_3$$
$$3\,C_2Cl_6 + 2\,Al \rightarrow 3\,C_2Cl_4 + 2\,AlCl_3$$

Upon entering the hexachloroethane into the aluminum melt, a chemical reaction generates gases like HCl and $AlCl_3$. This reaction leads to numerous gas bubbles within the aluminum melt. The hydrogen atoms in the melt diffuse into these bubbles and are carried away due to the pressure difference between the melt and the bubbles. The gas bubbles that rise to the surface can trap impurities and contaminants, thereby purifying the melt by capturing inclusions. This chemical reaction helps to reduce the hydrogen content in the molten metal, minimizing the formation of hydrogen gas porosity in the castings. While hexachloroethane effectively removes hydrogen gas, it may not be as effective in degassing other types of gases, such as nitrogen or oxygen. Therefore, additional degassing techniques or alternative agents may be necessary to address these specific gas-related issues in aluminum composite casting processes.

During the casting of the Al-Mg_2Si composite sheet, a C_2Cl_6 degasser was used. Gaseous porosity in the cast component is majorly seen in the spherical shape. The microstructure of the Al-Mg_2Si composite sheet does not show the microporosity in Figs. 6 and 7, which are typically spherical. Therefore, using the C_2Cl_6 as an effective degasser is recommended to avoid gaseous porosity in the Al-Mg_2Si composite sheet fabrication.

Melting Loss

Melt oxidation is a common problem for Al and Mg alloys. The fabrication of the Al-Mg_2Si sheet was performed in the open atmosphere, leading to the melt's oxidation. The DSC plot in Fig. 8 shows the oxidation of the melt in the open atmosphere by an exothermic peak during heating. At a temperature of 552.2 °C, materials start dissociating. Dissociation of the solid leads to increased melt oxidation, shown by an increase in % TG from 552.2 °C to 700 °C. The cooling curve of the same shows that if it does not come out of the melt, it solidifies as an inclusion in the melt.

During the oxidation and degassing processes, the alloying system's chemical composition alters due to the loss of elements in $AlCl_3$, Al_2O_3, SiO_2, MgO, etc. Therefore, maintaining a specific mixture composition during the casting poses a challenge in fabricating composite sheets.

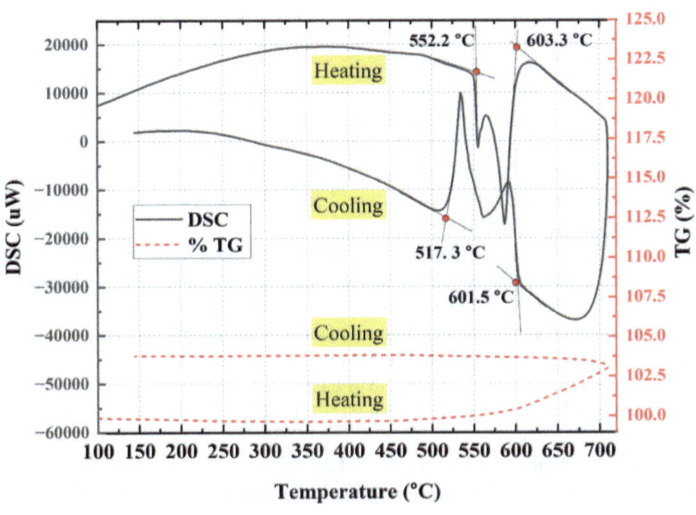

Fig. 8 DSC and %TG plot for the heating and cooling cycle for Al-Mg_2Si composite

4 Conclusions

An experimental study of continuous casting of Al-Mg_2Si composite sheet was carried out and subjected to many processing challenges such as designing continuous casting setup, optimizing processing parameters such as casting speed and superheat temperature, material selection, and defects occurring in the composite sheet. Based on the above study, the following conclusion can be made:

(a) A design promoting turbulence in the tundish and melt pool is preferred to ensure a homogeneous distribution of reinforcement.
(b) Roller coating is suggested for continuous operation to avoid corrosion, oxidation, and high thermal stresses.
(c) Composite material with a shorter freezing range and less density difference in matrix and reinforcement developed through the in situ route can get uniform microstructure without external stirring.
(d) Simultaneous adjustment of both superheating temperature and casting speed has detrimental effects on the quality of the composite sheet.
(e) Macrosegregation, solidification shrinkage, oxidation, and melting losses are the major defects associated with the continuous casting of Al-Mg_2Si composite sheet.

References

1. Saini DK, Jha PK (2023) Fabrication of aluminum metal matrix composite through continuous casting route: a review and future directions. J Manuf Process 96(January):138–160. https://doi.org/10.1016/j.jmapro.2023.04.041
2. Saini D, Jha PK (2023) Numerical modelling and experimental study of solidification behaviour of Al-Mg_2Si composite sheet fabricated using continuous casting route. J Heat Mass Transfer. https://doi.org/10.1115/1.4062758
3. Sanz A (2004) New coatings for continuous casting rolls. Surf Coat Technol 177–178:1–11. https://doi.org/10.1016/j.surfcoat.2003.06.024
4. Haga T, Nishiyama T, Suzuki S (2003) Strip casting of A5182 alloy using a melt drag twin-roll caster. J Mater Process Technol 133(1–2):103–107. https://doi.org/10.1016/S0924-0136(02)00251-0
5. Kundu A, Biswas P, Mallik A, Das S (2022) Electromagnetic twin-roll casting of aluminium alloy sheets: an overview. JOM 74(12):4876–4897. https://doi.org/10.1007/s11837-022-05490-y
6. Hector LG, Kim WS, Richmond O (1996) Freezing range effect on shell growth instability during alloy solidification. Am Soc Mech Eng (Paper) 63(September):594–602
7. Sigworth GK, Wang C (1993) Evolution of porosity in long freezing range alloys. Metall Trans B 24(2):365–377. https://doi.org/10.1007/BF02659139
8. Anyalebechi PN (2004) Technical issues impeding the proliferation of continuous casting processes in the aluminum industry. Proceed TMS Fall Extr Process Conf 2004(December):191–215
9. Jolly M, Katgerman L (2020) Modelling of defects in aluminium cast products. Progress Mater Sci 123:100824. https://doi.org/10.1016/j.pmatsci.2021.100824

10. Mandal A, Chakraborty M, Murty BS (2008) Ageing behaviour of A356 alloy reinforced with in-situ formed TiB_2 particles. Mater Sci Eng A 489(1–2):220–226. https://doi.org/10.1016/j.msea.2008.01.042
11. Ranjan S, Jha PK (2023) Investigation on the thermodynamic stability of phases evolved in Al-based hybrid metal matrix composite fabricated using in-situ stir casting route. J Manuf Process 95(January):14–26. https://doi.org/10.1016/j.jmapro.2023.03.084
12. Gobalakrishnan B, Rajaravi C, Udhayakumar G, Lakshminarayanan PR (2021) Effect of ceramic particulate addition on aluminium based ex-situ and in-situ formed metal matrix composites. Met Mater Int 27(9):3695–3708. https://doi.org/10.1007/s12540-020-00868-6
13. Kuru C, İpek SK, Dağdelen E, Özşahin O, Ulus A (2017) Material surface roughness change in twin roll casting of aluminium as cast sheet product. In: Light metals 2017, the minerals, metals & materials series, pp 833–842.https://doi.org/10.1007/978-3-319-51541-0
14. Guthrie RIL, Isac M, Kim JS, Tavares RP (2000) Measurements, simulations, and analyses of instantaneous heat fluxes from solidifying steels to the surfaces of twin roll casters and of aluminum to plasma-coated metal substrates. Metall Mater Trans B 31(5):1031–1047. https://doi.org/10.1007/s11663-000-0079-y
15. Eskin DG (2014) Mechanisms and control of macrosegregation in DC casting. Light Metals 9781118889:851–860. https://doi.org/10.1002/9781118888438.ch143

Role of Energy and Materials in Industry 4.0—A Pragmatic Deliberation

Hitesh R. Thakare, Nilesh P. Salunke, Sudhir U. Patil, and Ashok D. Parekh

1 Introduction

1.1 *Importance of Energy*

Energy is fundamental for economic and ecological prosperity of any nation. However, the problem associated with energy is its visibility. When an organization wants to improve its efficiency and profit margins, it can focus on man, machine, material and money aspects of the business. The organization can either reduce the manpower required for a particular operation, product or service, or it can improve their productivity by investing into their training and skills upgrading. New and advanced machines can always be deployed. Multiple vendors can be contacted to provide raw materials with cheaper costs. Financial institutions such as banks and non-banking financial corporations (NBFCs) are nowadays highly eager to provide business loans in shorter duration of time, thanks to market competition and increased customer awareness.

All these parameters are evident in nature, thereby immediately addressable when it comes to improving their efficiency. That is not the case with energy. An organization comes to know about its energy expenditure mostly at the end of the month, when it receives its electricity bill from the electricity distribution utility. Summarized statement of fuel bills is always maintained by accounts/finance department in association with maintenance/utilities department of the company. However, such statements are

H. R. Thakare (✉) · N. P. Salunke
SVKM's Institute of Technology, Dhule, Maharashtra 424001, India
e-mail: thakare.hitesh@gmail.com

S. U. Patil
SNJB's College of Engineering, Chandwad, Maharashtra 423101, India

A. D. Parekh
Sardar Vallabhbhai National Institute of Technology, Surat, Gujarat 395007, India

R. Sahu et al. (eds.), *Advancements in Materials Processing Technology, Volume 1*,
Springer Proceedings in Materials 48, https://doi.org/10.1007/978-981-97-4958-4_11

infrequently analysed with productive criticism and aligned with production output. In such opaque scenario, it becomes essential to quantify the energy use, irrespective of energy type, to gain competitive edge in the market.

With the advent of Industry 4.0, there is supposed to be a highly frequent (smooth) connection between cyber-physical systems of manufacturing industry. These connections are expected to extend towards all the elements of industrial supply chain. The flow of information throughout the production system shall enable the manufacturer and the customer to know the status of resource availability. The industrial value chain is expected to be digitalized in its entirety.

However, this extended proficiency of Industry 4.0 is expected to go hand in hand with Sustainable Development Goals (SDGs) set by United Nations in 2015. This become particularly important in a scenario, wherein we have already consumed more than 70% of our fossil fuel reserves in the past 100 years, which have taken millions of years to be created through various natural processes. Inefficiency utilization of energy for the past several decades has already created global environmental disasters such as poor air quality, untimely floods and draughts, melting of ice glaciers, rise in sea levels, deforestation and loss of marine aquatic life.

Coming to the business aspect of it, for every product and service, competition in the market in increasing by day. As an example of automotive sector, a few decades ago, if a person wanted to buy a Maruti 800 car, he/she had to wait for several months. As the time passed, numerous options are available in the market today in the form of Ford, Honda, Toyota, Mahindra, to name a few. Foreign players such as Morris Garages and Kia have made a grand entry in the Indian automobile market and forced the other players to improve their product portfolio in order to survive in the market. The profit can be generated either by increasing the price per unit or reducing the cost per unit. Out of these, first option drives away the customer while second option aids the efficiency improvement. Transition to Industry 4.0 is going to make competitive situation even tougher.

1.2 Steps Taken by Governments

Indian government formulated Energy Conservation Act in 2001 (EC Act 2001) to promote the drive for energy efficiency in the country. This act enables the formation of Bureau of Energy Efficiency (BEE) which subsequently started various programmes such as,

1. Standards and Labelling Programme (S&L).
2. Certification of Energy Managers and Energy Auditors.
3. Bachat Lamp Yojana (BLY).

As observed from Fig. 1, standards and labelling programmes started by various countries have shown tangible energy efficiency improvement and reduced electricity consumption. Such programmes have helped to reduce the electricity consumption up to 16% in the countries like US and EU, where the programmes have been run

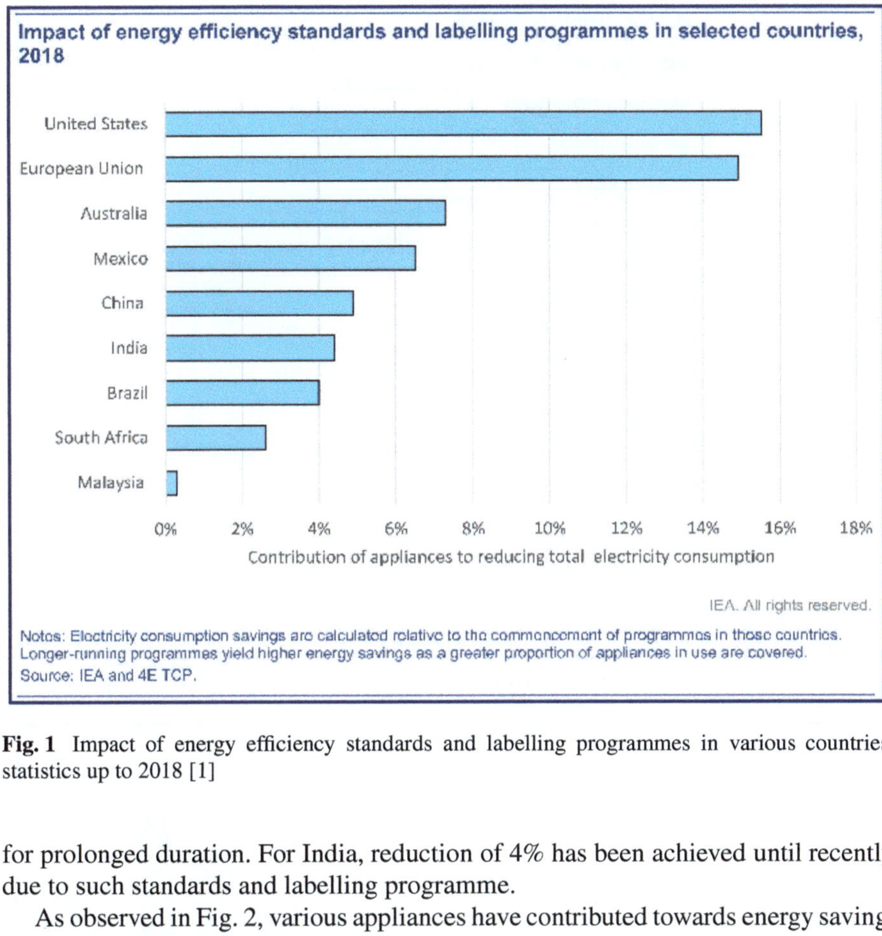

Fig. 1 Impact of energy efficiency standards and labelling programmes in various countries, statistics up to 2018 [1]

for prolonged duration. For India, reduction of 4% has been achieved until recently due to such standards and labelling programme.

As observed in Fig. 2, various appliances have contributed towards energy saving, room air conditioners being the highest contributor. Once again, the programmes running for longest duration have contributed to highest energy saving. Electric motors are yet to achieve a significant potential of energy saving, considering their widespread industrial use.

It is observed from Fig. 3 that there has been sharp rise in the quantum of connected devices such as smart metres, sensors, smart cooking appliances in the past few years. This rapid rise also necessitates the provision for higher energy consumption. Simply speaking, the number of connected devices equipped with automated controls is more than the population of the planet as of now. The growth of these connected devices has been at the rate of 33% per year for the last 5 years duration. In 2021, the number of connected devices crossed the mark of 9 billion, while it was 7 billion in 2020. As per the estimates of Cisco, more than 500 billion devices may get connected by the year 2030 [2].

At the same time, as observed in Fig. 4, these smart devices are becoming more and more energy efficient. General purpose devices such as smart LED lighting and

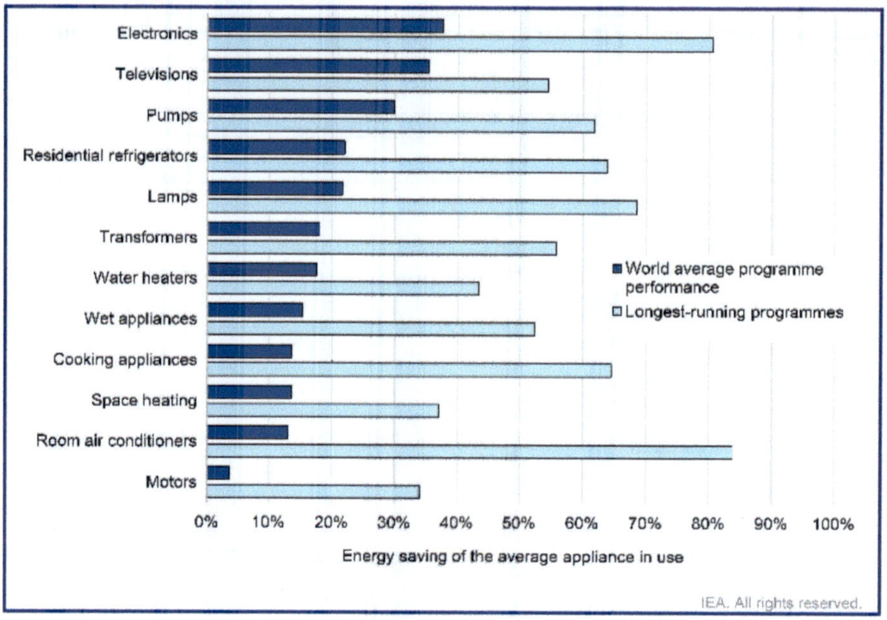

Fig. 2 Contribution of different appliances in achieving energy savings from energy efficiency standards and labels over life of programmes [1]

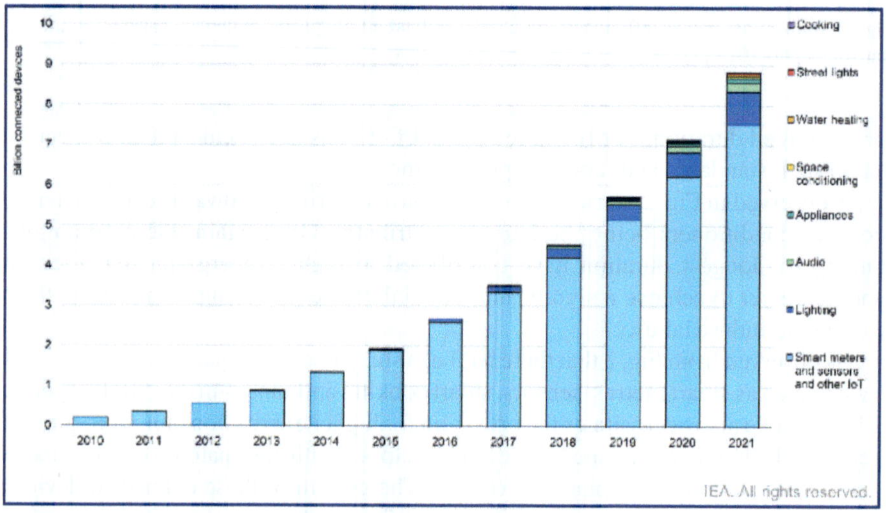

Fig. 3 Rise in number of connected devices in the last 10 years [1]

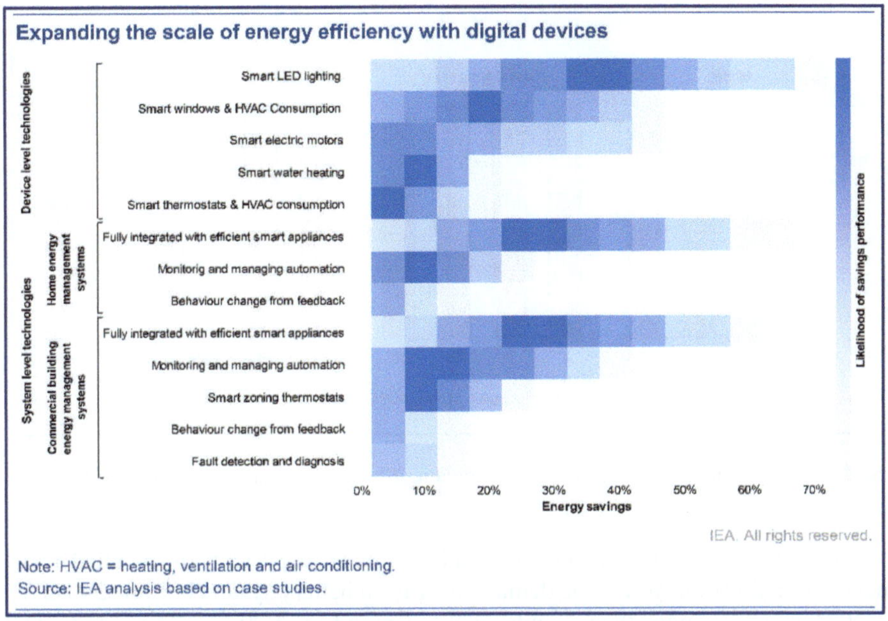

Fig. 4 Energy saving accumulated through various digitally connected devices [1]

HVAC systems have contributed to 40% to almost 60% in energy savings. Both in the case of home and commercial buildings energy management systems, integration of smart appliances has resulted in about 50% energy savings.

2 Pragmatic Concerns Pertaining to Energy Transition in Industry 4.0

2.1 Problems with Resource Demand

The digital devices which are at the core of cyber-physical systems of Industry 4.0 require raw materials for their manufacturing. Because of rapid increase in the deployment of such devices, followed by their accelerated diffusion in the digital ecosystem, there is spike in demand for such raw materials. This is exhibited in Fig. 5.

World is sharply shifting its focus towards e-mobility. Key components required in EV battery manufacture and operation are made out of lithium, nickel, cobalt and graphite. It is interesting to observe from Fig. 5 that China has clear dominance when it comes to mining, material processing as well as battery cells production for EV systems.

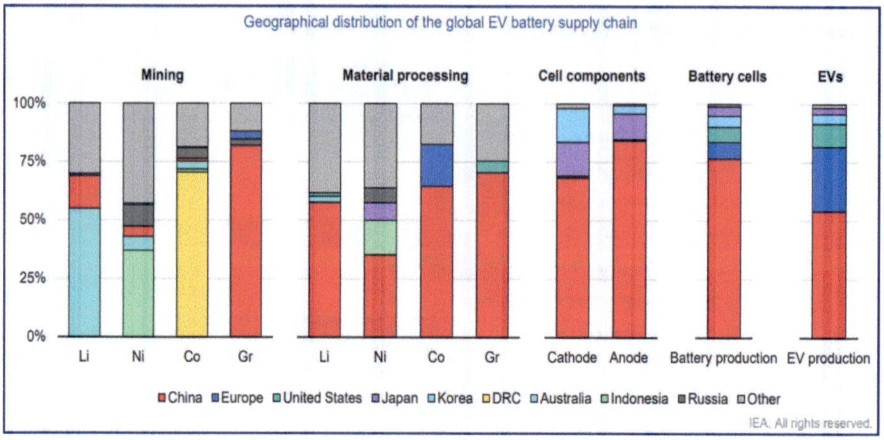

Fig. 5 Contribution of various countries in providing EV system-related resource materials [3]

As observed in Fig. 6, there has been sudden spike in the pricing of so-called heavy metals from 2021 onwards. The demand for Li-ion battery system became 340 GWh in 2021 which is more than double of the demand in 2020 [3]. The government registrations of electric vehicles alone increased by about 120% in this time frame. Out of this demand, 200 GWh was the demand in Chinese markets alone, showing an increment of 140% than that in 2020. Speaking about process chemistry, batteries having higher nickel content in their cathode provide higher energy density. However, the related manufacturing processes are highly controlled and complex in nature [4]. Between the beginning of 2021 and May 2022, there has been twofold increase in cobalt prices, sevenfold increase in lithium prices and twofold increase in nickel prices. This pricing surge can be attributed to not only the rise in battery demand but also to pressure created on supply chains as a result of geo political tensions going on around the world, Russia–Ukraine crisis being primary of them.

2.2 *Overstretching of Governments in Policy Formation*

The speed of device deployment, subsequent digitization and ultimately, execution of Industry 4.0 aspects are highly desired by the players in the economic sphere due to its ability to enhance the efficiency of entire manufacturing chain radically. On the other hand, the policies and frameworks to regulate such developments are usually formed at comparatively slower pace. If this holds true in the future as well, then it will have adverse effects in the domain of data security and privacy, working conditions and terms for future workforce, environmental concerns and sustainability aspects, etc.

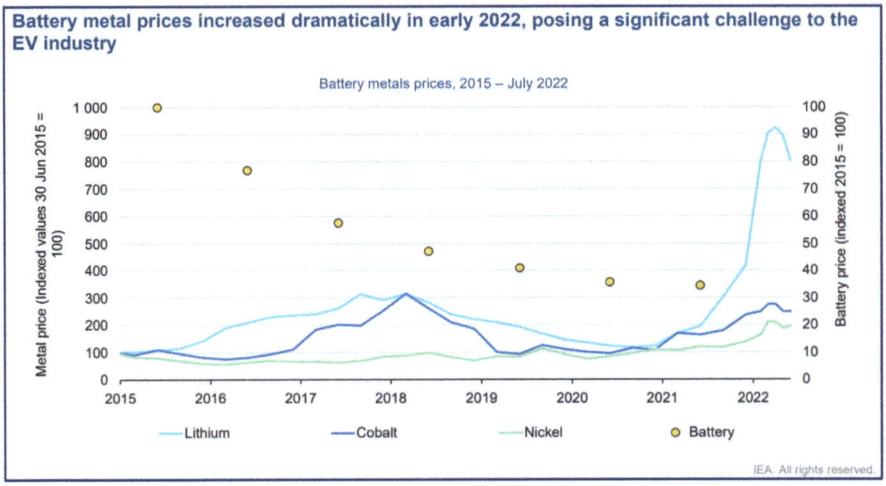

Fig. 6 Spike observed in prices of materials for EV batteries over the past years [3]

2.3 Impact of Energy Transitions on Employment Scenario

Recently, IEA has published an outlook on Indian Energy Scenario. It encompasses several scenarios such as Stated Policies Scenario (STEPS) which considers a balanced assessment India's energy system's direction in future with reference to today's policy situations and restraints along with postulation that COVID-19 pandemic situation will be under control. The India Vision Case (IVC) scenario is somewhat more optimistic and assumed higher speed of economic recovery and growth of the economy in long term. This scenario also assumes that energy policy will be implemented with its full potential. The Sustainable Development Scenario (SDS) however takes into consideration provisions of Paris agreement as well as clear air and energy access goals and examines the course of action needed to achieve these goals. As the country develops, there will be accelerated drive for clean and sustainable energy sources in India. This will also necessitate employment of skilled labour force in the domain of energy. The freshly recruited manpower will be needed to be trained while the existing manpower will need to undergo training to update their awareness and skillsets in understanding the energy aspects related to a just transition. As observed in Fig. 7, there is expected to be a ramp up in the job availability in the domain of electric vehicles, efficient energy utilization and production of end use renewables. A substantial growth is expected particularly in the SDS scenario due to higher employment expectations from EV Industry.

These employment opportunities are also affected due to their location. For example, SDS scenario focuses on substantial reduction of coal-based power plants as well as reduced mining of coal in domestic mines. This is going to severely affect poorer states of the country, wherein approximately 5,00,000 official workers are

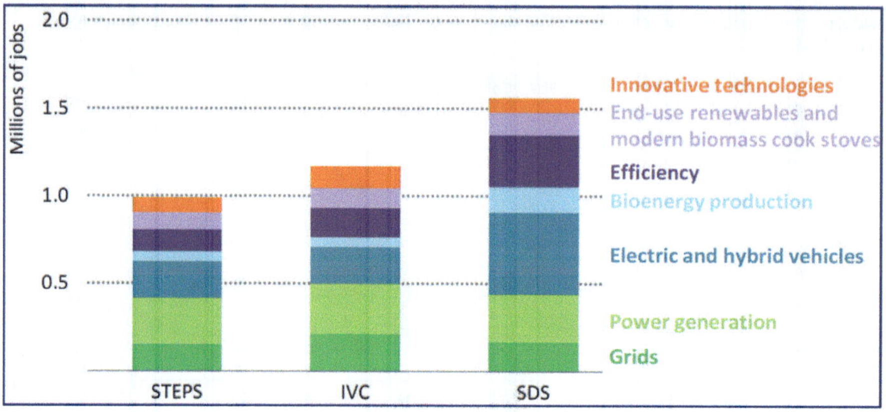

Fig. 7 Possible effect of energy transition on employment scenario in India [5]

dependent upon this industry and even more informal workers. To avoid dire socio-economic impact of this transition, it is necessary to create effective communication channels to enable retraining of such affected workforce in other domains of manufacturing, which can be very well facilitated by skill mapping between the affected workforce and growing occupational segments.

3 Solutions Explored Thus Far

3.1 Digitization of Energy Sector

Energy sector, being highly important for industrial and general growth, has already been going through digitization over the course of past few years. In India, access to mobile services through 4G Internet technology, accelerated smartphone market back by torrent of e-commerce sector has resulted in "never before" opportunities and business models which can transform energy sector. Blockchain technology is also going to play key role in revolutionizing the energy scenario.

The foremost expectation from digitalization is the integration of generated electricity from renewable energy into local power grids to achieve an efficient grid management. The entity called as "Smart Grid" may serve the purpose which utilizes the potential of information and communication technology so as to manage generation, distribution and device level consumption of electricity procured from various sources. Preferably, such sources shall be decentralized in nature and can be effectively utilized to address the problem of variable electricity demand.

3.2 Role of Energy Exchanges

Establishment of energy exchanges has enabled as well as accelerated the digitization of energy sector, at least for industrial and commercial organizations. In an efficient Industry 4.0 scenario, there is expected to be efficient flow of information regarding availability of manufacturing resources. Furthermore, dynamism of interaction between cyber-physical manufacturing systems is going to be very high. In such dynamic scenario, a manufacturing industry will not only rely on long-term energy contract but it will also need to procure energy from time to time through short-term energy contracts. Organizations such as Indian Energy Exchange (IEX), Power Exchange India Ltd. (PXIL) and Hindustan Power Exchange (HPX) are digital marketplaces, which deal with nationwide procurement and delivery of electricity from conventional sources as well as renewables along with energy saving certificates. These digital platforms play an important role as an aggregator wherein they have developed an ecosystem consisting of power generators and distribution utilities. For example, IEX has an ecosystem consisting of 29 states and 05 union territories. It encompasses the transactions between 55+ distribution utility companies and 500+ power manufacturers. At present, 4000+ consumers are provided energy procurement services through the portal of IEX [6].

3.3 Mini Grids–Alternative to Barriers of Grid Connections in Low Population Density Areas

It is well known that electrical supply lines should have higher voltages and lower current values to ensure least amount of energy loss in current carrying conductor. Our field work in rural areas of Nashik District, Maharashtra state, India, indicated that for regions located beyond 25–30 km away from substations of utility, the voltage in supply lines is significantly lower than prescribed. We measured voltage and current values in several transformers in this region. We also observed temperatures as high as 191 °C in these transformers as a result of increased current values.

Mini grids can function effectively in such distant and hilly areas. Frequently during monsoon, such regions experience power failure and substantial efforts, both in terms of manpower and finances, are necessary to resume the services post monsoon devastation. Such mini grids are also convenient in areas having low population density, distant location away from the main grid and rough terrain difficulties. These constraints also result in prohibitively high cost of expanding the main grid to the end user of such areas. As the cost of renewable energy technologies goes on declining, mini grids could be set up with economic viability and clarity about tariff structures.

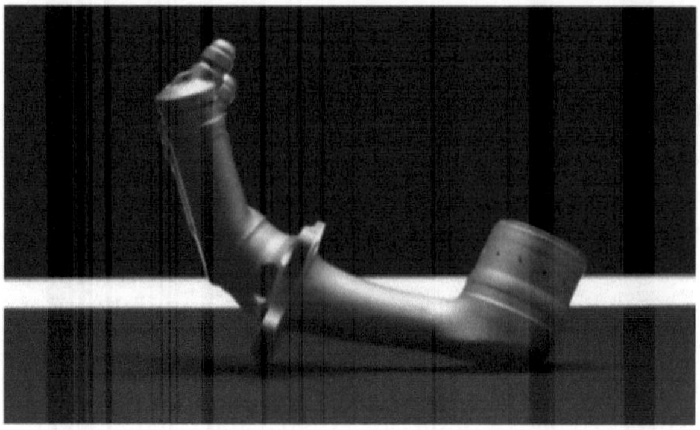

Fig. 8 Fuel nozzle of LEAP engine of General Electric, manufactured by 3D printing [7]

3.4 Application of Rapid Prototyping

Rapid prototyping is a synonym for integration of several complementary technologies such as Computer Aided Design (CAD) and 3D printing, which is a type of additive manufacturing. This combination is highly capable to rapidly manufacture prototypes of a product and inspect them subsequently, as opposed to conventional subtractive manufacturing and lengthy design and optimization process. As an example, by the application of 3D printing technology, number of subparts required for fuel nozzles of LEAP engine of General Electric have reduced down from 18 to only 1. This innovation resulted in weight reduction of the component by 25%. Furthermore, geometry of the component was optimized to achieve gain in combustion efficiency. This enhanced efficiency will enable the engine to achieve fuel saving for its operating life with lower CO_2 emissions. The 3D printing technology was able to achieve intricate shape of tip's interior geometry, which GE's engineers had already attempted to cast 8 times and failed. This component is exhibited in Fig. 8.

3.5 Digitally Enable Business Models—Virtual Power Plants (VPPs)

Digitalization of energy sector and diffusion of digital technologies has created immense scope for new business models which are looking forward to tap the expanded energy market through new means of delivering value to energy systems operators as well as end users. Simply speaking, VPP is digitally enabled network that helps to link together, aggregate and centrally manage the DERs such as rooftop solar PV plants, battery storage units with utility power plants along with monitoring the variations on demand side [1]. The share of DERs in the capacity addition for VPPs is expected to grow from 51% in 2020 to about 83% in 2029.

As an example, Tesla is developing the facility for aggregation of energy from rooftop solar PV panel and battery storage units of about 3000 households. This plant has electricity generation capacity of 20 MW and energy storage capacity of 54 MWh. In near future, this framework is expected to expand by including 50,000 units in its operation.

3.6 Education and Awareness

Several researchers have taken initiative in preparing research works for Industry 4.0 technologies such as triboelectric sensors as well as wearable devices equipped with nanogenerators [8–10]. Energy system forecasting techniques such as use of neural networks and machine learning models are also gaining momentum [11, 12].

4 Conclusion

It goes without saying that prudent implementation of Industry 4.0 philosophy at the grassroot level is going to transform industrial manufacturing, supply chain, information flow and resource allocation, in essence, the entire business landscape, forever. With progressive flow of information, it becomes imperative to create awareness about efficient usage of energy. There is still huge potential for energy savings, particularly in rural and disconnected areas of society. At the same time, procuring the energy from sustainable sources needs to be accelerated. With manufacturing and operational costs going down, rapid prototyping technologies have huge potential to help in energy conservation, an aspect which is usually overlooked.

Acknowledgements First and second authors are thankful to the Hon. Management of SVKM's Institute of Technology, Dhule, for providing necessary resources during preparation of this manuscript.

References

1. IEA Report on Energy Efficiency (2021), pp 17–80
2. IEA report on energy applications within IoT and digitalization strategies (2020), p 1
3. IEA report on global supply chains of EV batteries (2022), pp 6–7
4. Li W, Erickson EM, Manthiram A (2020) High-nickel layered oxide cathodes for lithium-based automotive batteries. Nat Energy 5:26–34. https://doi.org/10.1038/s41560-019-0513-0
5. Birol F (2021) Indian energy outlook 2021, report of international energy agency 2021. IEA Publications
6. https://www.iexindia.com/Aboutus.aspx?id=Gy9kTd80D98%3d&mid=Gy9kTd80D98%3d. Last accessed on 25 Jan 2023
7. https://www.ge.com/news/reports/transformation-3d-walnut-sized-part-changed-way-ge-avi ation-builds-jet-engines. Last accessed on 22 Jan 2023

8. Bhagwan N, Evans M (2023) A review of industry 4.0 technologies used in the production of energy in China, Germany and South Africa. Renew Sustain Energy Rev 173:113075. https://doi.org/10.1016/j.rser.2022.113075
9. Choy TH, Zhou F, Xu W, Wong MC, Ye T, Hao J, Chai Y (2018) Enhanced output power of a freestanding ball-based triboelectric generator through the electrophorus effect. J Mater Chem A 6:18518–18524. https://doi.org/10.1039/C8TA05198D
10. Chen J, Wang Z (2017) Reviving vibration energy harvesting and self-powered sensing by a triboelectric nanogenerator. Joule 1(3):480–521. https://doi.org/10.1016/j.joule.2017.09.004
11. Wang Y, Zou R, Liu F, Zhang L, Liu Q (2021) A review of wind speed and wind power forecasting with deep neural networks. Appl Energy 304:117766. https://doi.org/10.1016/j.apenergy.2021.117766
12. Wu Z, Chu W (2021) Sampling strategy analysis of machine learning models for energy consumption prediction. In: IEEE 9th international conference on smart energy grid engineering (SEGE), 11–13 Aug 2021. https://doi.org/10.1109/SEGE52446.2021.9534987

Computational Analysis of Various Fin Configurations—A Comprehensive Assessment

Hitesh R. Thakare, Nilesh P. Salunke, Bhavesh Deore, Kunal Karankal, Pranav Gujar, and Rohit Patil

1 Introduction

1.1 Importance of Fins as Extended Surfaces for Heat Transfer

It is well known from the science of heat transfer that fins are helpful to enhance the effective area available for heat interaction and thereby improve the heat transfer coefficient as well as heat transfer rate. Fins are mainly beneficial because they do not require any active source of energy to establish the fluid motion. This mechanism particularly addresses the needs of devices which are located outside such as high-power LED street lamps as well as thin client computers [1]. However, observation of temperature distribution profile of a fin indicates exponential decrement of temperature towards the fin tip, which results in reduced temperature differential between fin and surrounding fluid, thus ultimately reducing the heat transfer rate. Due to high applicability of fins in enhancement of gas side heat transfer coefficient, there has been immense interest in the research pertaining to fin performance improvement. Some notable works which involve study of input power required for array of LEDs [2], fins spacing for large fin arrays [3], providing holes in the fin base of a fin array [4], to name a few.

H. R. Thakare (✉) · N. P. Salunke · B. Deore · K. Karankal · P. Gujar · R. Patil
Department of Mechanical Engineering, SVKM's Institute of Technology, Dhule, Maharashtra 424001, India
e-mail: thakare.hitesh@gmail.com

R. Sahu et al. (eds.), *Advancements in Materials Processing Technology, Volume 1*, Springer Proceedings in Materials 48, https://doi.org/10.1007/978-981-97-4958-4_12

1.2 Objective of Present Work

Present work is an effort to bring forward details of few but distinguishing research works carried out in the domain of fin performance enhancement. The geometries, thermos-physical conditions and target applications of discussed research works are diverse, hence thorough attention is given to individual work. This is intended to help future researchers by providing them a comprehensive picture of research work.

2 Review of Published Literature on Computational Analysis of Fins

Huang et al. [1] provided perforations in the base surface of the fins so as to improve air flow below the fin base. Authors also carried out numerical investigation regarding air flow pattern and heat transfer enhancement for various fin array configurations, as shown in Fig. 1.

Authors considered the assumption of three-dimensional laminar flow with incompressible fluid in unsteady mode. Fins were constructed of aluminium with thickness of 1 mm. The flow physics was modelled using commercial package Fluent 12.0 using pressure-based implicit solver. Unsteady simulation was accelerated using pressure implicit with splitting of operators (PISO) scheme. The convergence criteria for mass conservation were 10^{-4}, while it was 10^{-6} for energy and momentum.

Results of local heat flux indicated that in the absence of perforations, effective heat transfer took place in the fin tip region and inefficient heat transfer in the internal region of the fin, as observed in Fig. 2a. This was mainly attributed to flow structure which consisted of recirculating zone formed due to rise of hot air and downward flow of cold air as seen in Fig. 3a. Such recirculating zone did not exist in case of perforated fin array, which improved the air ventilation, as observed in Fig. 3b. Such

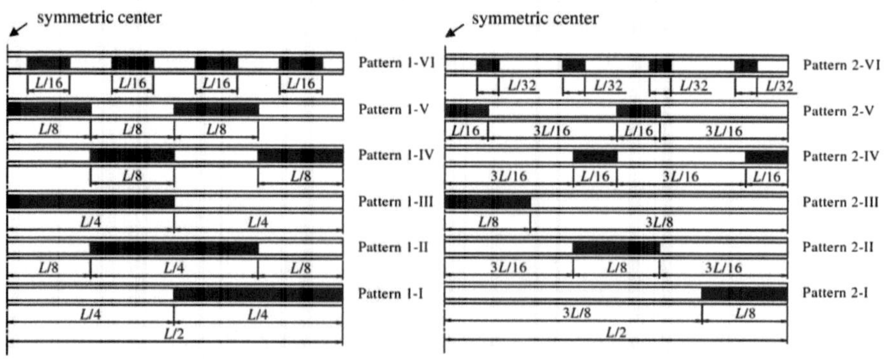

Fig. 1 Various patterns of perforations created at fin base [1]

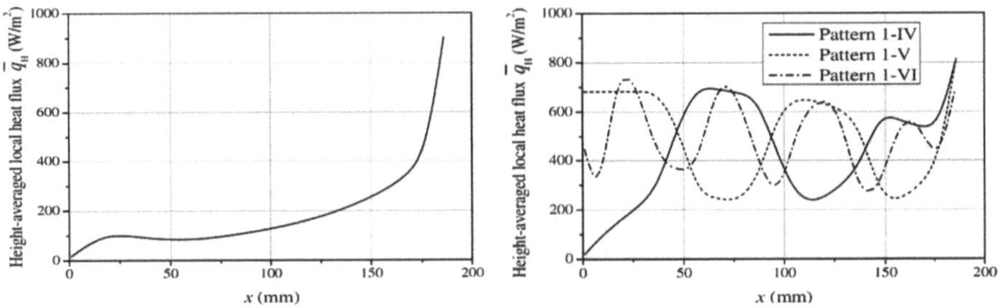

Fig. 2 Distribution of local heat flux in fins **a** without perforations **b** with perforations [1]

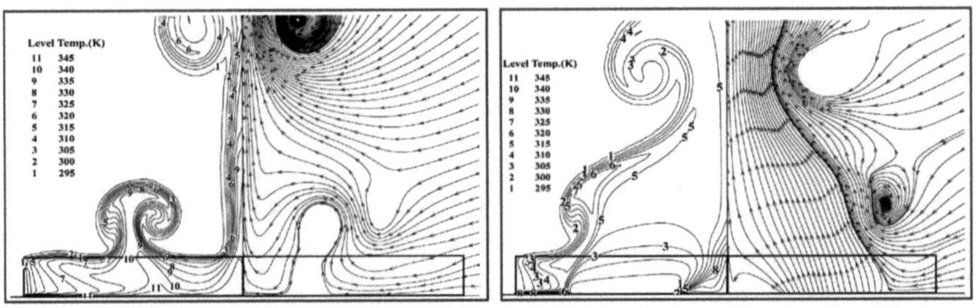

Fig. 3 Temperature and flow profile for fins configuration **a** without perforations **b** with perforation [1]

air flow pattern also improved the local heat flux along the length of the fin, as shown in Fig. 2b.

Park et al. [5] performed experimental analysis to study the natural convection when fins mounted on a vertical cylinder are branched fins, as shown in Fig. 4. Author's focus of research was on variation of angle of fin branch (10–40°), number of fins (9–36) and temperature of base surface. Through experimentation, authors also suggested an empirical correlation for Nusselt number. Experimental results indicated that thermal resistance of branched fin heat sink was about 20% lower than that of plane fin heat sink, as observed from Fig. 5a. Authors postulated that if fin spacing is small, then boundary layer developed during natural convection flow shall merge, thereby, heat transfer from one fin is blocked due to the presence of second fin. For a large branch angle, such obstruction of flow does not happen. This was also evident as during experimentation and mathematical correlation, as shown in Fig. 5b, wherein lower thermal resistance was observed for higher branch angle.

The contour map of thermal resistance for different values of fin thickness and number of fins, as shown in Fig. 6, indicated that minimum thermal resistance was observed for certain number of fins and fin thickness. However, with increase in the

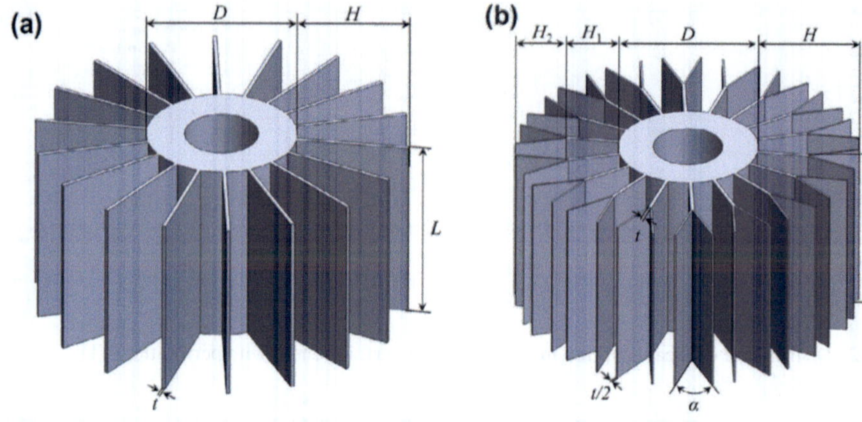

Fig. 4 Vertical cylinder heat sink with **a** plain plate fins **b** branched fins [5]

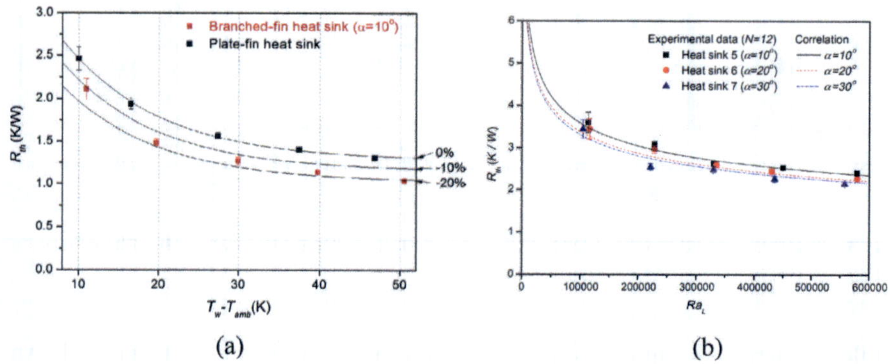

Fig. 5 Variation of thermal resistance **a** between heat sink with plate fin and branched fin **b** for different angle of branch fins [5]

values of these parameters, the previously discussed boundary layer overlap results in decrease of heat transfer coefficient.

Chingulpitak et al. [6] studied the effect of diameter and number of circular perforations on the performance of finned heat sink, wherein fin area was kept constant at 1875 mm^2 (75 × 25 mm) with uniform distribution of perforations. Authors used Shear Stress Transport (SST) k–ω model in software package ANSYS CFX 17.0, due to its ability to incorporate the advantages of standard k–ε model and standard k–ω model. The residual error criteria for all the parameters were set at 10^{-5}. Grid independence study indicated that grid size between 9,394,572 and 12,193,439 was suitable with marginal loss of accuracy. Numerical results were validated through comparison with experimental results of Shaeri and Bonner [7] in terms of pressure drop and thermal resistance for different values of Reynolds number, as observed in Fig. 7. Flow visualization studies indicated that air flow consisted of recirculation

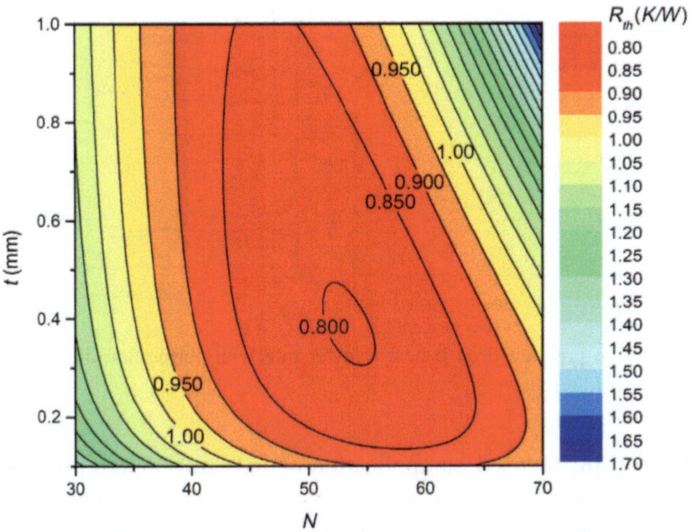

Fig. 6 Contour of thermal resistance for various values of fin thickness and number of fins [5]

and mixing zones. The span of recirculation zone was higher towards the centre of the perforation. Perforation helped to increase the flow turbulence. The distance between the perforations also played an important role. With shorter distance between perforations, the size of recirculation zone increased.

With increase in perforation diameter, the total pressure drop magnitude also increased at any given Reynolds number, even though frictional pressure drop decreased due to reduction in available surface area of the fins, as observed in Fig. 8. When the diameter of perforations was increased from 1 to 10 mm, pressure drop increased by 96.3%. Heat transfer rate is higher for perforation diameter 3 mm, and thereafter, it decreased for higher perforation diameter.

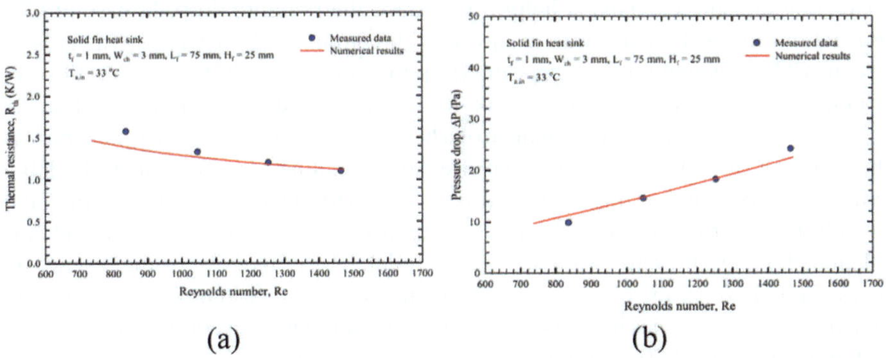

Fig. 7 Validation of numerical results with experimental data for **a** pressure drop **b** thermal resistance [6]

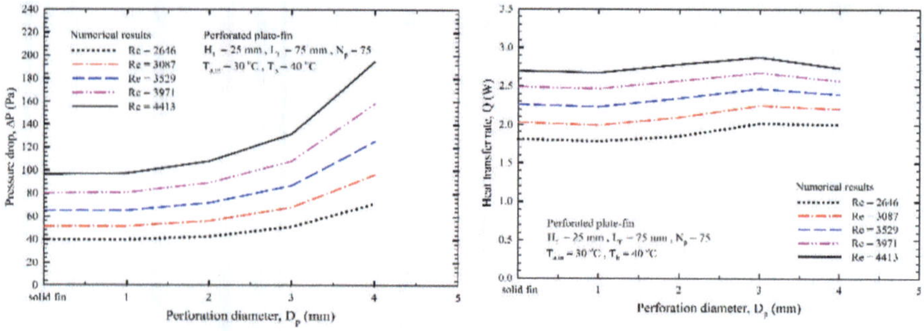

Fig. 8 Variation of **a** pressure drop **b** heat transfer rate with perforation diameter at different Reynolds numbers [6]

Hithaish et al. [8] focused on application of triangular micro-pin fin heat sink to aid the heat transfer from electronic systems. Authors postulated that triangular fins shape can improve flow separation. Differentiating factor of this study was consideration of previously unexplored various fin parameters like orientation, height and placement of fins in different directions. Authors used finite volume method through ANSYS Fluent 14.5. Convergence criteria were set at 10^{-6} for all parameters. Authors proposed a Thermal Performance Index (TPI) which indicates improvement in thermal performance of fin relative to pressure drop increase due to as compared to unfinned arrangement of heat sink. Results of numerical investigation were validated using experimental data of Liu et al. [9] and numerical data of Hasan [10]. Authors reported substantial increase in pressure drop with increase in fin height. The pressure drop for Hf 0.50 was 110% higher and that with Hf 0.75 was 353% higher as compared to Hf 0.25. This is also observed in contours presented in Fig. 9a which shows vigorous mixing of fluid layers and no development of boundary layer.

Higher pressure drop was observed for clockwise (CW) than counter clockwise (CCW) orientation for constant fin height of 0.75. Changing the fin orientation created uneven gap which creates passage for some of the fluid to travel forward without contacting the surface. This is helpful to reduce the pressure drop. In case of CW orientation, the wake zone was in upwards direction, while it was downwards in case of CCW orientation. The pressure drop magnitude was 3.9% less for 30° CW orientation, while it was 5.3% higher for 40° CCW orientation as observed in Fig. 9b.

Song et al. [11] studied performance of a heat sink equipped with perforated fins, as shown in Fig. 10a. Authors studied the effect of fin inclination, fin number, Rayleigh number and orientation angle on heat transfer performance through experimental as well as numerical analysis. Radiation heat transfer in the numerical model was calculated using Discrete Ordinates (DO) model. The emissivity of the test set-up was set as 0.4, as per the observations of Sundar et al. [12]. Grid independence study indicated that a mesh having 1,447,512 cells was sufficiently accurate with variation of temperature by 0.5%. Authors used two non-dimensional parameters, namely finning factor and porosity factor to express the thermal resistance performance of heat sink. As observed in Fig. 10b, numerical results agreed with experimental data

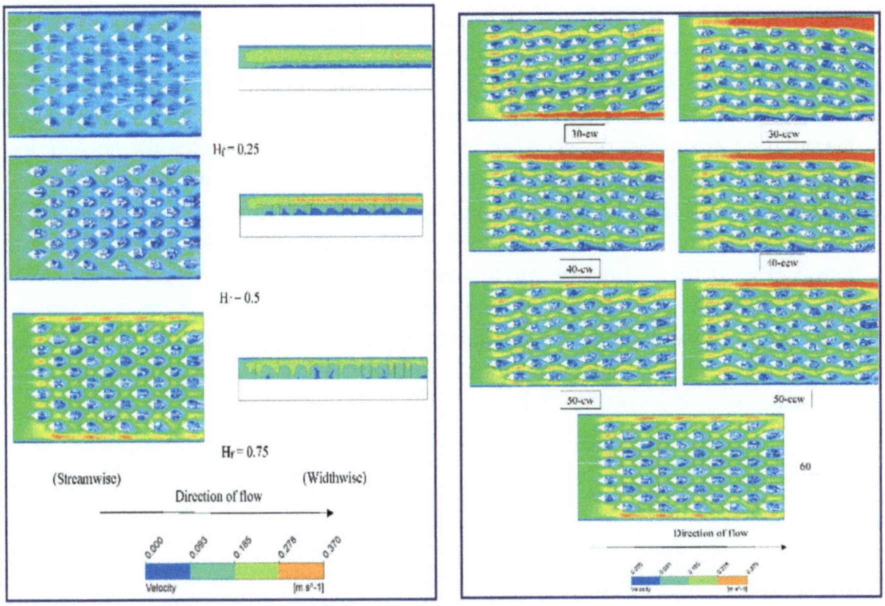

(a) (b)

Fig. 9 Streamline contours for **a** different fin heights **b** different fin orientations at Re = 500 [8]

with deviation of about 2.7%. It also shows that with increase in orientation angle, thermal resistance increased.

The heat flux supplied had substantial effect of thermal resistance of heat sink with respect to number of perforations per fin as well as radius of perforation. With the

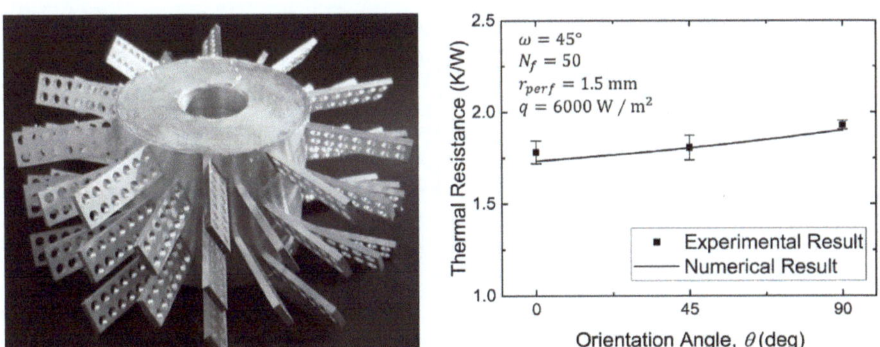

(a) (b)

Fig. 10 **a** Perforated fins used for experimental set-up **b** validation of numerical results with experimental results [11] increase in number of perforations, thermal resistance decreased. This was attributed to weakening of air blocking effect with increase in number of perforations, as per the observations of Sundar et al. [12]. Lower thermal resistance was observed for lower radius of perforation and higher number of perforations per fin. There was a substantial deterioration observed in heat dissipation performance for perforation radius less than 1.5 mm. For all the values of Rayleigh number studied, thermal resistance of heat sink reduced when number of fins was increased, as observed in Fig. 11a. also, lowest thermal resistance was observed when heat sink orientation angle was 0° and fin inclination angle was 45° as observed in Fig. 11b. This was attributed to higher flow resistance created in case of higher heat sink orientation angle. In case of lower heat sink orientation angle, the flow passage of fluid becomes longer and enables higher heat migration, as observed from streamline structure and temperature distribution profile in Fig. 12 [11].

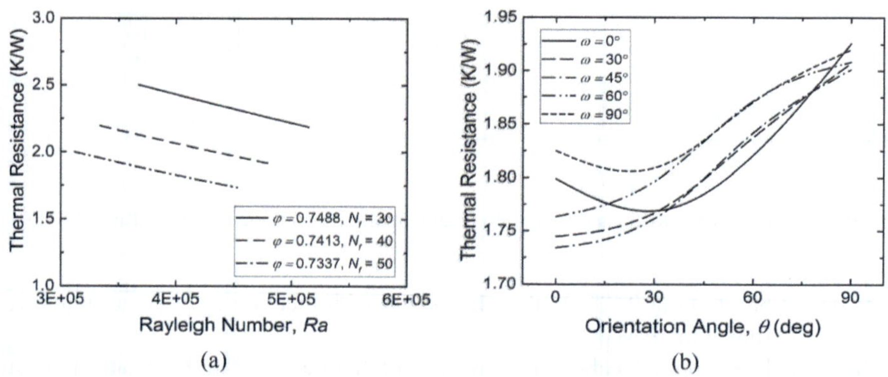

Fig. 11 Trend of thermal resistance of heat sink **a** for different Rayleigh number and porosity factor **b** for different orientation angles [11]

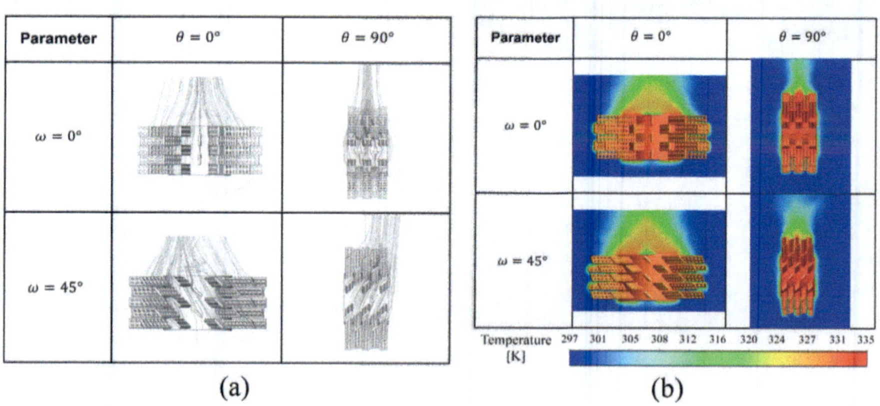

Fig. 12 Comparison of **a** streamlines **b** temperature profile for different heat sink orientation and fin inclination [11]

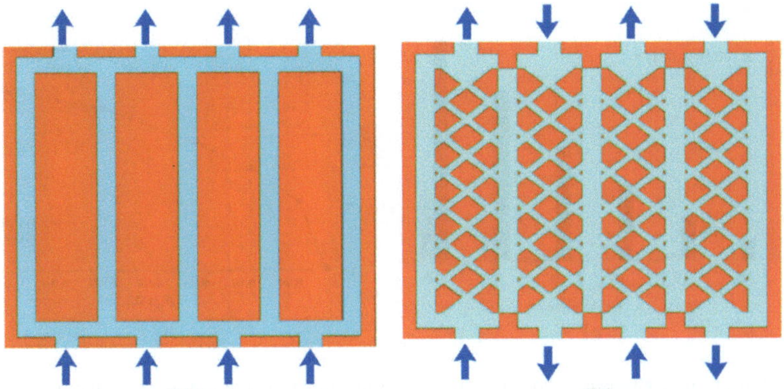

Fig. 13 Schematic of mini-channel heat sink **a** with plate fins **b** with tessellated fins [13]

Cao et al. [13] proposed a novel mini-channel heat sink with tessellated fins, as shown in Fig. 13. Such heat sinks are highly essential for miniaturization of cooling technology necessary for electronic equipment. The tessellated fins were created by interrupting the parallel channels to create secondary channels, thus providing more heat transfer area. Flow was considered to be compressible during numerical simulation carried out on ANSYS Fluent 15.0, with radiation and gravity effects neglected. Water was used as coolant flowing through the heat sink and its properties were treated constant. Flow rate of water was 7.48–30 ml/s, which generated Reynolds number of 573–2295.

As observed from Fig. 14a, with increase in water flow rate, pressure drop increases for both plate fin and tessellated fin heat sink. However, pressure drop is more in plate fin heat sink than in tessellated fin heat sink due to availability of secondary channels in case of tessellated fins. The Nusselt number was observed to be higher for tessellated fins than plate fins and it increased with increase in pumping power, as shown in Fig. 14b. At flow rate of 30 ml/s, the improvement in Nusselt number was about 11.9% using tessellated fins as a result of improved flow structure. The tessellated structure helps to break the momentum boundary layer and reduces boundary layer thickness to promote the heat transfer rate. The temperature contours indicate that non-uniform temperature distribution was observed for plate fins from inlet towards outlet due to improper flow mixing. While, only one hot spot was observed in case of tessellated fins due to enhanced flow of fluid in the crevices. The temperature difference of 45.8 K and 21.6 K was observed in case of plate fin and tessellated fin, respectively, which indicates better heat transfer in case of tessellated fins.

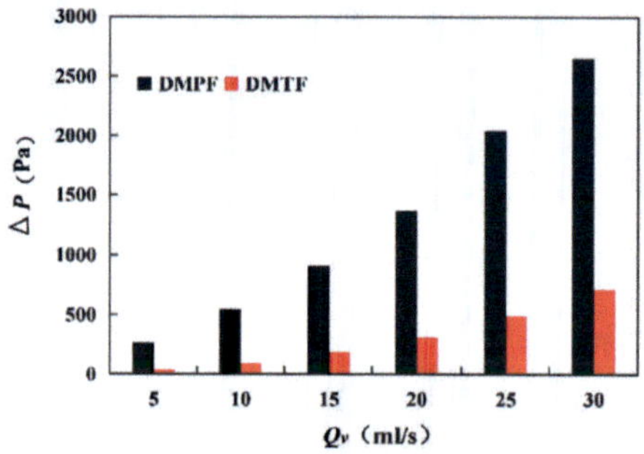

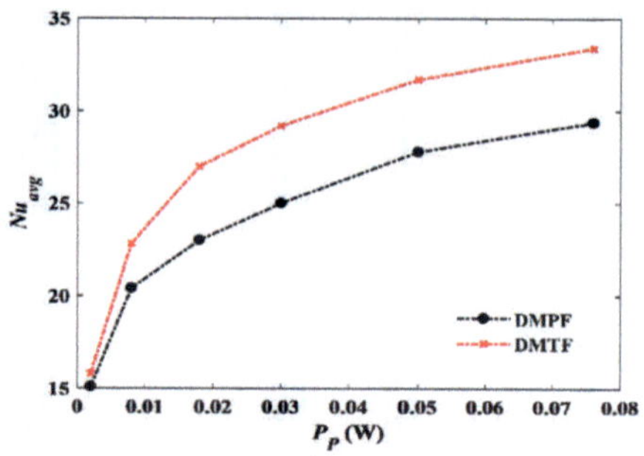

Fig. 14 Variation of **a** pressure drop for different flow rates **b** Nusselt number with pumping power for plate fin and tessellated fin heat sinks [13]

3 Conclusion

Analysis of heat transfer performance of fins and its subsequent improvement is a very important topic for research, considering varied applications of fins in the domain of passive cooling. Flow of cooling fluid through fins is affected by several parameters, such as fin spacing, fin length, orientation of fins and heat source, to name a few. Some researchers have proposed innovative designs and geometrical variations which have shown substantial improvement in heat transfer performance. Present work has put forth essential details of various notable research works in this area which will prove useful to the future researchers.

References

1. Huang GJ, Wong SC, Lin CP (2014) Enhancement of natural convection heat transfer from horizontal rectangular fin arrays with perforations in fin base. Int J Therm Sci 84:164–174. https://doi.org/10.1016/j.ijthermalsci.2014.05.017
2. Christensen A, Graham S (2009) Thermal effects in packaging high power light emitting diode arrays. Appl Therm Eng 29:364–371. https://doi.org/10.1016/j.applthermaleng.2008.03.019
3. Wong SC, Huang GJ (2013) Parametric study on the dynamic behavior of natural convection from horizontal rectangular fin arrays. Int J Heat Mass Transf 60:334–342. https://doi.org/10.1016/j.ijheatmasstransfer.2013.01.019
4. Meng XR, Ma XL, Lu JF, Wei XL (2012) A study on improving in natural convection heat transfer for heat sink of high power LEDs. Adv Mater Res 383–390:6834–6839. https://doi.org/10.4028/www.scientific.net/AMR.383-390.6834
5. Park KT, Kim HJ, Kim DK (2014) Experimental study of natural convection from vertical cylinders with branched fins. Exp Therm Fluid Sci 54:29–37. https://doi.org/10.1016/j.expthermflusci.2014.01.018
6. Chingulpitak S, Seon H, Godson L, Wongwises S (2019) Fluid flow and heat transfer characteristics of heat sinks with laterally perforated plate fins. Int J Heat Mass Transf 138:293–303. https://doi.org/10.1016/j.ijheatmasstransfer.2019.04.027

7. Shaeri MR, Bonner R (2017) Heat transfer and pressure drop in laterally perforated-finned heat sinks across different flow regimes. Int Commun Heat Mass Transf 87:220–227. https://doi.org/10.1016/j.icheatmasstransfer.2017.07.022
8. Hithaish D, Saravanan V, Umesh CK, Seetharamu KN (2022) Thermal management of electronics: numerical investigation of triangular finned heat sink. Therm Sci Eng Prog 30:101246. https://doi.org/10.1016/j.tsep.2022.101246
9. Liu M, Liu D, Xu S, Chen Y (2011) Experimental study on liquid flow and heat transfer in micro square pin fin heat sink. Int J Heat Mass Transf 54:5602–5611. https://doi.org/10.1016/j.ijheatmasstransfer.2011.07.013
10. Hasan MI (2014) Investigation of flow and heat transfer characteristics in micro pin fin heat sink with nanofluid. Appl Therm Eng 63:598–607. https://doi.org/10.1016/j.applthermaleng.2013.11.059
11. Song G, Kim D, Song D, Sung J, Yook S (2021) Heat-dissipation performance of cylindrical heat sink with perforated fins. Int J Therm Sci 170:107132. https://doi.org/10.1016/j.ijthermalsci.2021.107132
12. Sundar S, Song G, Zahir MZ, Jayakumar JS, Yook SJ (2019) Performance investigation of radial heat sink with circular base and perforated staggered fins. Int J Heat Mass Transf 143:118526. https://doi.org/10.1016/j.ijheatmasstransfer.2019.118526
13. Cao X, Liu H, Shao X, Shi H (2022) Experimental and numerical investigation on the heat transfer enhancement for Mini-channel heat sinks with tessellated fins. Appl Therm Eng 211:118353. https://doi.org/10.1016/j.applthermaleng.2022.118353

Experimental Investigations of CO_2 Laser Cutting of Glass Fiber-Reinforced Unsaturated Polyester Composites

K. Anil, R. Sai Sanjana, Md. Sameer, and P. Srikanth

1 Introduction

The glass fiber-reinforced unsaturated polyester (GFRUP) thermoset composite has attracted significant interest in the area of manufacturing engineering due to its versatile properties and wide range of applications [1]. The composite material indicated above has exceptional mechanical strength, corrosion resistance, and dimensional stability, making it suitable for many sectors such as automotive, aerospace, construction, and marine [2]. The laser cutting operations may be affected by the complex structural characteristics of GFRUP, which include a three-dimensional arrangement of monomers and carbon atom connections inside the glass fiber network. The use of laser cutting has emerged as a very effective and precise method for manipulating glass fiber-reinforced unsaturated polyester (GFRUP) materials. The procedure involves the use of a focused laser beam to specifically increase the temperature of the designated substance, resulting in its vaporization and the creation of a precise and uncontaminated cut. In comparison with conventional cutting techniques such as mechanical cutting or sawing, laser cutting has many advantages, including reduced material waste, little heat damage, and the ability to achieve elaborate and complicated patterns [3]. Moreover, laser cutting is a non-contact method that eliminates the need for direct touch between the cutting tool and the material, thereby reducing the risk of tool deterioration and environmental contamination. While laser cutting does provide significant advantages, the complex structure of GFRUP poses challenges in the tuning of laser cutting settings.

The use of fiber-reinforced polymer composites is widespread owing to its advantageous attributes, including their lightweight nature and high-strength properties. The use of CO_2 laser cutting has been recognized as an effective technique for precise

K. Anil · R. Sai Sanjana · Md. Sameer (✉) · P. Srikanth
Kakatiya Institute of Technology and Science, Warangal, India
e-mail: mds.me@kitsw.ac.in

R. Sahu et al. (eds.), *Advancements in Materials Processing Technology, Volume 1*,
Springer Proceedings in Materials 48, https://doi.org/10.1007/978-981-97-4958-4_13

manipulation of materials. This literature review offers a comprehensive examination of the significant results documented in several research on the process of CO_2 laser cutting of fiber-reinforced polymer composites. The evaluation places significant emphasis on the examination of the kerf profile. The study done by Fatimah and colleagues [4] investigates the use of CO_2 laser cutting on composites composed of glass fiber-reinforced polymers. This research critically studies the kerf profile and investigates the influence of cutting factors on the cut quality. In their research, Löhr et al. [5] have presented a model that utilizes neural network technology for the purpose of cutting PMMA sheets with varying thickness. The findings of their study have shown the efficacy of the model in accurately predicting kerf features. Choudhury and Shirley [6] performed an empirical investigation pertaining to the laser cutting process of polymeric materials. The study primarily centered on the examination of the impact of laser settings on the quality of cut edges and the features of the kerf. The research undertaken by Riveiro et al. [7] examines the CO_2 laser cutting method applied to carbon fiber-reinforced plastic composites. The study conducted by the authors focuses on the evaluation of the effects of laser power, cutting speed, and focal location on the resultant edge quality. The research undertaken by Najjar et al. [8] focuses on the prediction of kerf quality characteristics in the laser cutting procedure of basalt fiber-reinforced polymer composites. The researchers use a hybrid approach, integrating neural network and chimpanzee optimization methodologies, in order to develop a model and enhance the optimization process. The use of this particular approach has shown to be quite advantageous in the acquisition of significant observations pertaining to the anticipation and regulation of kerf quality. The works described above aim to improve our understanding of the process of CO_2 laser cutting of fiber-reinforced polymer composites. These studies include several aspects, including the study of kerf profiles, modeling techniques, and experimental investigations. The findings provide valuable insights into improving the efficacy of the laser cutting process. Multiple research [9–11] have shown significant advancements in the domain of laser cutting of GFRP composites.

Tabar et al. [12] studied the three different characteristics, such as specific point energy, material removal rate, and cutting volume, of GFRP and CGFRP using two different CO_2 laser machines. They found that the cutting volume efficiency declines as the specific point energy increases. The material removal rate increases when the sheet thickness decreases and the specific point energy is reduced for a given power density. Kechagias et al. [13] investigated stand-off distance on kerf geometry using a CO_2 laser on 3D-printed PET-G components. They concluded that the chosen ANN architecture accurately predicted experimental results for laser cutting of PET-G material, specifically in terms of KA and Ra. Mahesh et al. [14] conducted a study on optimizing the process parameters of CO_2 laser drilling on a composite material consisting of hybrid S-glass, Al_2O_3, and perlite powder. The researchers investigated the control of hole delamination and hole tapering through the optimization of parameters using response surface methodology (RSM) and multi-objective genetic algorithm (MOGA) techniques. Singh et al. [15] carried out a study on optimizing the kerf geometry of a coir hybrid composite using a pulsed CO_2 laser. From the above literature, there is not sufficient work done on GFRUP composites using three parameters such as laser power, speed, and focal length.

Therefore, the authors have made an effort to determine the optimal parameter settings for the machining of GFRUP composites, which have extensive utilization in diverse applications, through the application of Taguchi analysis.

2 Materials and Methods

The GFRUP composite material was manufactured using a manual lay-up technique, which included glass fiber and vinyl ester resin, as seen in Fig. 1. The methodology included the implementation of a tri-layered woven glass fiber arrangement, where each layer had a mass of 20 g, in order to produce a laminate with a thickness of 4 mm. A constant ratio of 1:2.5 (fiber to epoxy) was maintained, whereby each layer of glass fiber was accompanied by 50 g of epoxy resin. The use of promoter and accelerator was employed to enhance the curing process and improve the mechanical properties of the final product.

2.1 Laser Cutting

The research used a carbon dioxide (CO_2) laser cutting device with a power output of 130 kilowatts (kW). The production of smoke during the laser cutting of GFRUP sheets is a significant observation, which highlights the need for the installation of an air ventilation system. Figure 2 illustrates the cutting device used for laser cutting a sheet of 4 mm in thickness. The continuous wave mode was used to concentrate the laser beam onto the top surface of the sheet.

Fig. 1 GFRUP composite preparation

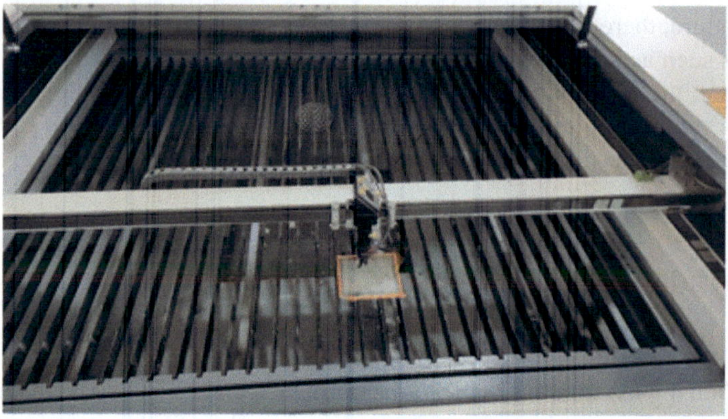

Fig. 2 Laser cutting of GFRUP sheet

2.2 Kerf Width Measurement

The kerf width was measured using a digital vernier caliper. The data was gathered at a designated site where the cutting tool's velocity was maintained at a consistent level. To assure accuracy, the specific data point underwent three measurements, and the resulting arithmetic mean of these measurements was afterward recorded as the final output. The determination of the kerf size was derived from the measurements of the aperture. The determination of the kerf taper angle (°) included acquiring measurements for the diameter of the aperture at the point of entrance (Dentry), the diameter of the aperture at the point of exit (Dexit), and the thickness (T). The values were acquired using a digital vernier caliper, as seen in Fig. 3. The taper angle was calculated using Eq. 1.

$$\text{Kerf taper angle}(\circ) = \tan^{-1}\left(\frac{D_{\text{entry}} - D_{\text{exit}}}{2T}\right) \quad (1)$$

2.3 Design of Experiments

The investigation includes the implementation of laser cutting procedures using the L9 Orthogonal Array. The research considered the parameters and their respective levels, as shown in Table 1.

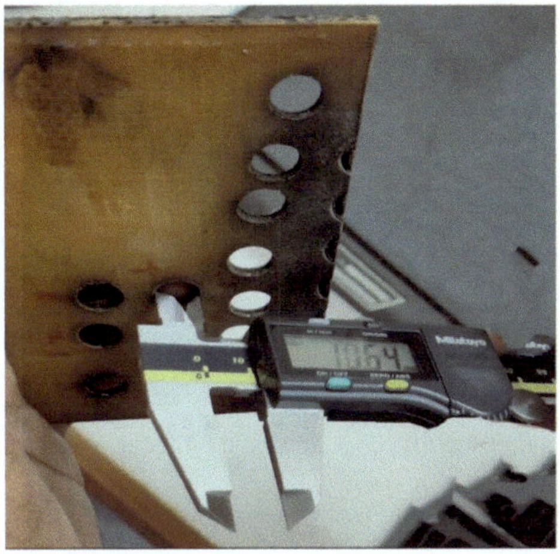

Fig. 3 Kerf width measurements

Table 1 Design of experiments

Exp No	Power (*W*)	Speed (mm/s)	Focal positions (mm)
1	60	5	6
2	60	7	8
3	60	9	10
4	65	5	8
5	65	7	10
6	65	9	6
7	70	5	10
8	70	7	6
9	70	9	8

3 Results and Discussion

The tests were done using the methodology of design of experiments, and the calculation of taper angles was performed using Eq. (1) and shown in Table 2. The S/N ratio value for taper angle is shown in Table 3 subsequent to the acquisition of experimental data. Furthermore, the response table for the means of Taper angle is shown in Table 4. The calculation of the signal-to-noise ratio was performed for the taper angle, using the smaller-the-better criteria and Eq. (2).

$$\frac{S}{N} = -\log_{10}\left(\frac{1}{n}\sum_{j=1}^{n} y_{ij}^2\right) \tag{2}$$

Based on the data provided in Table 4, it is evident that the variables of speed, focal position and power have been awarded rankings 1, 2, and 3, respectively. This

Table 2 Results of the performed experiments

Exp No	Power (W)	Speed (mm/s)	Focal positions (mm)	D_{entry}avg	D_{exit}avg	Taper angle (°)
1	60	5	6	10.5	10.36	1.002
2	60	7	8	10.6	10.48	1.210
3	60	9	10	10.63	10.5	0.780
4	65	5	8	10.59	10.5	0.860
5	65	7	10	10.58	10.51	0.780
6	65	9	6	10.57	10.5	1.140
7	70	5	10	10.7	10.53	1.630
8	70	7	6	10.47	10.37	1.430
9	70	9	8	10.47	10.51	0.210

Table 3 S/N ratios of taper angles of different experiments

Exp No	Power (W)	Speed (mm/s)	Focal positions (mm)	Taper angle (°)	S/N ratio of taper angle
1	60	5	6	1.002	− 0.0174
2	60	7	8	1.210	− 1.6557
3	60	9	10	0.780	2.1581
4	65	5	8	0.860	1.3100
5	65	7	10	0.780	2.1581
6	65	9	6	1.140	− 1.1381
7	70	5	10	1.630	− 4.2438
8	70	7	6	1.430	− 3.1067
9	70	9	8	0.210	13.5556

Table 4 Response table for means of taper angle

Level	Power	Speed	Focal positions
1	0.9973	1.1640	1.1907
2	0.9267	1.1400	0.7600
3	1.0900	0.7100	1.0633
Delta	0.1633	0.4540	0.4307
Rank	3	1	2

ranking has been generated based on the higher magnitudes of delta. The primary determinant of taper angle is speed, as shown by its top ranking. Subsequently, Rank 2 and Rank 3 represent diminishing levels of contributing factor in relation to taper angle, with power being the least influential element.

Based on the data shown in Table 4, it can be seen that the cutting speed parameter has the most pronounced level of dominance in comparison with the other cutting parameters. The findings derived from Fig. 4 demonstrate that increasing the laser power to 70 W resulted in a significant decrease in the kerf taper angle. When the laser beam is operated at a high power level, it exhibits the ability to effectively distribute thermal energy over the whole thickness of the profile. This results in a wider hole while yet preserving a minimal decrease in hole width. Consequently, the removal rate of material undergoes an increase during high power operation, resulting in a decrease in the creation of taper. Furthermore, it was noted that when the cutting speed was increased to 9 mm/s, there was a decrease in the taper angle to 0.21. This behavior may be attributed to the reduction in contact time between the laser and the composite material. Hence, as stated in citation [16], the diminishment of the material on the upper surface of the composite specimen was negligible. The increase of laser beam penetration throughout the full thickness of a material generates heat dispersion in several directions, hence causing a decrease in energy intensity and ultimately resulting in a reduced kerf width. The reduction in taper angle has been recorded in prior scholarly investigations [17]. The correlation between focal length and kerf angle demonstrates that an initial increase in focal length leads to a subsequent reduction in kerf angle. Nevertheless, successive increments in focal length do not provide any additional impact on the kerf angle. Based on the aforementioned findings, it can be deduced that the cutting speed has a substantial effect on the taper angle, whereas the influence of power on the taper angle is rather insignificant.

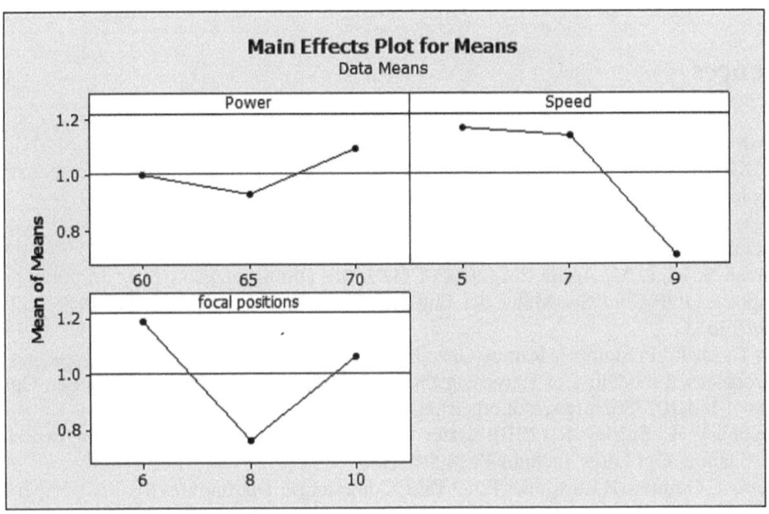

Fig. 4 Mean effects plot for means of taper angle

4 Conclusions

To summarize, the study conducted an experiment to examine the CO_2 laser cutting process of a 4-mm-thick sheet of glass fiber-reinforced unsaturated polyester (GFRUP). Based on the findings and analysis, the following conclusions are drawn.

- The CO_2 laser is capable of cutting GFURP composite sheets and can achieve improved kerf quality on cut surfaces by optimizing parameter settings.
- The Taguchi approach is used to determine the optimal values of laser cutting parameters, which have been identified as 65 W of laser power, 9 mm/s of cutting speed, and a focal position of 8 mm.
- Cutting speed is the dominant parameter among all the cutting parameters.
- Kerf taper angle is significantly reduced with increase in laser power and it is found that as the cutting speed increased to 9 mm/s taper angle decreased to 0.21.
- Increasing focal length decreases kerf angle initially but does not affect when further increased.

Acknowledgements We authors would also like to acknowledge the facilities provided by KITSW-AICTE IDEA Lab to carry CO_2 laser cutting operations and Composites Materials Lab at Department of Mechanical Engineering at KITSW for the support in carrying this research work.

References

1. Crawford RJ (1998) Plastics engineering, 3rd edn
2. Yao XF, Bijlaard FS, Kolstein MH et al (2005) Tensile strength and fracture behaviors of complex GFRP composites with a central hole. J Compos Mater 39:1247–1260
3. Davim JP, Barricas N, Conceição M, Oliveira C (2008) Some experimental studies on CO2 laser cutting quality of polymeric materials. J Mater Process Technol 198(1–3):99–104
4. Fatimah S, Ishak M, Aqida SN (2012) CO2 Laser cutting of glass fiber reinforce polymer composite. IOP Conf Ser Mater Sci Eng 36(1):012033. https://doi.org/10.1088/1757-899X/36/1/012033
5. Löhr C, La Fé-Perdomo I, Ramos-Grez JA, Calvo J (2021) Kerf profile analysis and neural network-based modeling of increasing thickness PMMA sheets cut by CO2 laser. Opt Laser Technol 144:107386. https://doi.org/10.1016/j.optlastec.2021.107386
6. Choudhury A, Shirley S (2010) Laser cutting of polymeric materials: an experimental investigation. Opt Laser Technol 42(3):503–508
7. Riveiro A, Quintero F, Lusquiños F, del Val J, Comesaña R, Boutinguiza M (2012) Experimental study on the CO2 laser cutting of carbon fiber reinforced plastic composite. Compos A Appl Sci Manuf 43(8):1400–1409
8. Najjar MR, Sadoun AM, AbdElaziz M, Abdallah AW, Fathy A, Elsheikh AH (2022) Predicting kerf quality characteristics in laser cutting of basalt fibers reinforced polymer composites using neural network and chimp optimization. Alex Eng J 61(12):11005–11018
9. Yao XF, Bijlaard FS, Kolstein MH et al (2005) Tensile strength and behaviors of complex GFRP composites with a central hole. J Compos Mater 39:1247–1260
10. El-Hassan H, El Maaddawy T (2019) Microstructure characteristics of GFRP reinforcing bars in harsh environment. Adv Mater Sci Eng 2019:1–19

11. Tagliaferri V, Di Ilio A, Visconti C (1985) Laser cutting of fibre reinforced polyesters. Composites 16(4):317–325
12. Tabar MY, Hashemzadeh M, Rejani AR (2023) CO2 laser cutting of reinforced polyester (CGFRP and GFRP) sheets: an experimental investigation into specific point energy, cutting volume efficiency and material removal rate (MRR). J Manuf Process 97:137–147
13. Kechagias JD, Fountas NA, Ninikas K, Petousis M, Vidakis N, Vaxevanidis N (2022) Surface characteristics investigation of 3D-printed PET-G plates during CO2 laser cutting. Mater Manuf Processes 37(11):1347–1357. https://doi.org/10.1080/10426914.2021.1981933
14. Guru Mahesh G, Kandasamy J (2023) Optimization of CO2 laser drilling process parameters of GFRP/Al2O3/perlite composites. Mater Today Commun 35:105962
15. Singh Y, Singh J, Sharma S et al (2021) Multi-objective optimization of kerf-taper and surface-roughness quality characteristics for cutting-operation on coir and carbon fibre reinforced epoxy hybrid polymeric composites during CO_2-pulsed laser-cutting using RSM. Lasers Manuf Mater Process 8:157–182
16. Moradi M, KaramiMoghadam M, Shamsborhan M, MalekshahiBeiranvand Z, Rasouli A, Vahdati M, Bakhtiari A, Bodaghi M (2021) Simulation, statistical modeling, and optimization of CO2 laser cutting process of polycarbonate sheets. Optik 225:164932. https://doi.org/10.1016/j.ijleo.2020.164932
17. Solati A, Hamedi M, Safarabadi M (2019) Comprehensive investigation of surface quality and mechanical properties in CO_2 laser drilling of GFRP composites. Int J Adv Manuf Technol 102:791–808. https://doi.org/10.1007/s00170-018-3164-6

[illegible]

Mathematical Modeling to Predict Angular Distortion in Duplex Stainless Steel 2101 Plates

Kashish Badwal, Sneha Rabha, and Pradeep Khanna

1 Introduction

MIG welding is a process in which an arc is established between a workpiece and a consumable filler wire [1]. The arc and the molten puddle are protected from atmospheric contamination using an inert gas. This process can run semi-automatically or automatically [2, 3]. Due to its various advantages like higher welding speeds, greater deposition rates, easy automation, and good welding quality, this process is widely used in industry [4, 5]. The current study focuses on duplex stainless steel plates, which combine the formability of ferritic stainless steel with the corrosion resistance features of austenitic stainless steel [6]. It is widely used in petrochemical, mining, pharmaceutical, seawater desalination, and nuclear power industry [7, 8]. While duplex stainless steel accounts for a small portion of the global stainless steel market, it offers a number of advantages such as high strength, toughness, ductility, and corrosion resistance [9, 10]. The microstructure of the material is composed of a combination of austenite and ferrite phases [11, 12]. As a result, duplex stainless steels have properties that are similar to those of austenitic and ferritic stainless steels.

In this study, MIG welding was used to weld 80 mm × 100 mm × 4 mm duplex SS plates, with the objective of developing a mathematical model for angular distortion (AD). The input parameters studied were (WS), (WFR), and (V), and a three-factor, three-level, face-centered design was used to conduct the experimental runs for the model development. The adequacy of the mathematical model was evaluated using the ANOVA technique, and RSM was utilized to graphically analyze the results. After the model was generated, the parameters were optimized to minimize the angular distortion.

K. Badwal (✉) · S. Rabha · P. Khanna
Netaji Subhas University of Technology, New Delhi, India
e-mail: kashishbadwal@gmail.com

R. Sahu et al. (eds.), *Advancements in Materials Processing Technology, Volume 1*,
Springer Proceedings in Materials 48, https://doi.org/10.1007/978-981-97-4958-4_14

2 Materials and Methodology

A welding power supply with a rated capacity of 400 amps and a flat V–I characteristic was utilized. To ensure a consistent predetermined WS, a mechanical welding manipulator was used during each run. The angular distortion (AD) was measured using a surface plate, a vernier height gauge, and counterweights. The welding setup is illustrated in Fig. 1.

The input parameter working ranges for this study are presented in Table 1. A filler rod material of SS 308L with a 1.2 mm diameter and an Argon shielding gas with a constant flow rate of 15 lpm were employed.

A total of 20 experimental runs were carried out using the central composite design statistical technique in the Minitab 19 software. The welding was executed according to the standard runs created by the software.

An image of the welded specimens as per design matrix can be seen in Fig. 2. Subsequent welding, vernier height gauge setup as shown in Fig. 3 was used to determine the angular displacement of the weld plates. The design matrix along with the values of AD corresponding to each experimental run can also be seen in Table 2.

Fig. 1 The welding setup (Welding Lab, Netaji Subhas University of Technology)

Table 1 Input variables and their ranges of variation

S. no.	Input parameter	Symbol	Unit		Values	
1	WS	A	cm/min	35	40	45
2	WFR	B	m/min	5	6	7
3	V	C	Volts	18	20	22

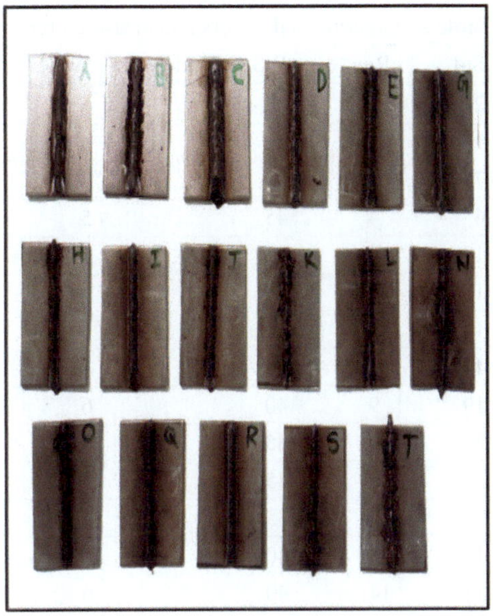

Fig. 2 Welded plates

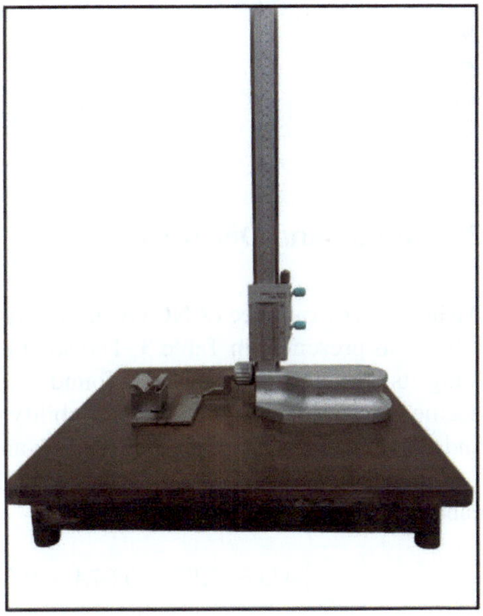

Fig. 3 Vernier height gauge (Metrology Lab, Netaji Subhas University of Technology)

Table 2 Experimental factors and measured responses (Minitab 19)

Std	Run	WS (cm/min)	WFR (m/min)	V (Volts)	AD (Degrees)
20	1	40	6	20	1.5
14	2	40	6	22	1.972
7	3	35	7	22	1.555
6	4	45	5	22	1.915
12	5	40	7	20	2.045
16	6	40	6	20	1.715
18	7	40	6	20	1.895
4	8	45	7	18	1.120
19	9	40	6	20	1.850
3	10	35	7	18	1.290
2	11	45	5	18	0.315
9	12	35	6	20	0.755
15	13	40	6	20	1.445
13	14	40	6	18	1.475
10	15	45	6	20	0.835
17	16	40	6	20	1.650
8	17	45	7	22	2.100
5	18	35	5	22	1.705
11	19	40	5	20	0.825
1	20	35	5	18	0.200

3 Results and Discussions

An analysis of variance (ANOVA) table was generated using Design of Experiments (DOE), as presented in Table 3. The adequacy of the model was assessed by examining its sufficiency, which was found to be significant, while the lack of fit was deemed not significant. The model's ability to explain the relationship between input and output parameters was confirmed by an R^2 score of 87.18%.

The mathematical model developed by the software after dropping the insignificant terms is given below.

$$\begin{aligned} \text{AD} = 1.58 + 0.07A + 0.31B + 0.48C + 0.10AC \\ - 0.23BC - 0.63A^2 + 0.29C^2 \end{aligned} \quad (1)$$

Table 3 ANOVA table (Minitab 19)

Source	Sum of squares	*df*	Mean square	*F*-value	*P*-value	
Model	5.24	9	0.5826	7.55	0.0020	Significant
A—Speed	0.0608	1	0.0608	0.7889	0.3953	
B—Wire feed rate	0.9923	1	0.9923	12.87	0.0050	
C—Voltage	2.35	1	2.35	30.46	0.0003	
AB	0.0003	1	0.0003	0.0041	0.9505	
AC	0.0820	1	0.0820	1.06	0.3267	
BC	0.4325	1	0.4325	5.61	0.0394	
A^2	1.11	1	1.11	14.43	0.0035	
B^2	0.0000	1	0.0000	0.0006	0.9816	
C^2	0.2352	1	0.2352	3.05	0.1113	
Residual	0.7712	10	0.0771			
Lack of fit	0.6064	5	0.1213	3.68		Not significant
Pure error	0.1648	5	0.0330			
Cor total	6.01	19				

3.1 *Direct Effects*

WS affects AD in a variety of ways, as Fig. 4a demonstrates, the initial rise in AD was most likely caused by the interactions of the other factors, while the subsequent fall in AD with increasing WS was most likely caused by the weld's heat input per unit length rapidly declining as WS grew after a certain point. There is a drop in AD as WS is elevated because as heat is lowered, thermally induced stresses are also reduced.

Figure 4b clearly illustrates the relationship between WFR and the magnitude of AD. This relationship may be explained as when WFR rises, the heat input to weld likewise rises, allowing the heat to penetrate the weld more deeply and raising AD.

Figure 4c makes it very evident that as V rises, current magnitude rises as well. This can be explained by the fact that an abrupt rise in current leads to rise in heat input to weld, which raises the welds' AD. Furthermore, as V increases, the arc expands, distributing heat over a broader area and increasing the AD.

3.2 *Interactive Effects*

From Fig. 5a–c, it can be inferred that WFR has a positive impact on AD because when WFR increases, the heat input to weld also grows, causing the effect to spread farther into the weld and increasing AD. It has been found that WS increases AD at first and later shows a negative effect. The most probable reason for an increase in

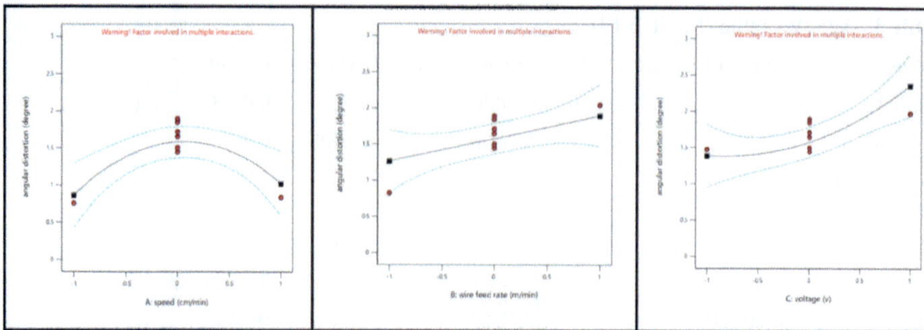

(a) Direct effect of WS (Minitab 19) | **(b)** Direct effect of WFR (Minitab 19) | **(c)** Direct effect of V (Minitab 19)

Fig. 4 **a** Direct effect of WS (Minitab 19), **b** direct effect of WFR (Minitab 19), **c** direct effect of V (Minitab 19)

AD is the dominating influence of WFR, the reason for subsequent decrease in AD with larger WS is that welding speed reduces the heat input per unit length of the weld. There is a drop in AD as WS is elevated because as heat is lowered, thermally induced stresses are also reduced. Due to the fact that a rise in current increases the amount of heat input, increasing the AD of the welds, it can be stated that V has a positive effect on AD. WS has been discovered to affect AD both positively and negatively. At last, it can be inferred that both V and WFR have a positive effect on AD, the justification for which has been given above.

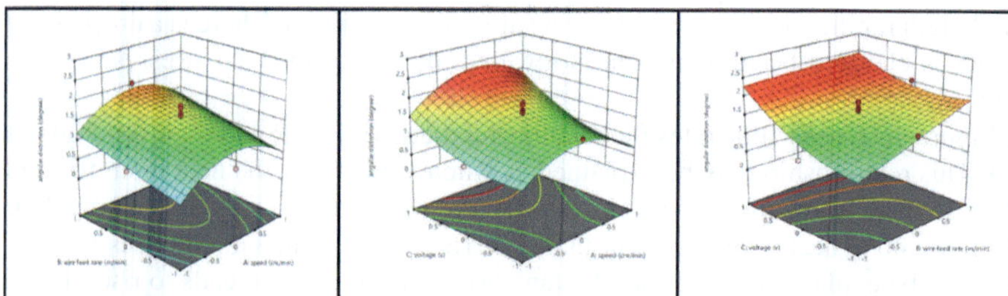

(a) Combined influence of WFR and WS on AD (Minitab 19) | **(b)** Combined influence of V and WS on AD (Minitab 19) | **(c)** Combined influence of V and WFR on AD (Minitab 19)

Fig. 5 **a** Combined influence of WFR and WS on AD (Minitab 19), **b** combined influence of V and WS on AD (Minitab 19), **c** combined influence of V and WFR on AD (Minitab 19)

4 Conclusion

The investigation led to the following conclusions:

1. In this study, the face-centered composite design statistical technique was successfully applied to study the effect of input parameters on angular distortion.
2. The magnitude of angular distortion is negatively impacted by WS and positively impacted by both WFR and V.
3. The minimum and maximum angular distortion of about 0.2° and 2.1° is attained at lowest magnitudes of WS, WFR, and V and at the highest values of WS, WFR, and V, respectively.

References

1. Jha RK, Kumar A, Chakraborty A, Khanna P (2018) Effects of process parameters on angular distortion of GMAW Welded C45 plates. Int J Res Eng Technol 7(07):168–174
2. Mohamat SA, Ibrahim IA, Amir A, Ghalib A (2012) The effect of flux core arc welding (FCAW) processes on different parameters. Proc Eng 41:1497–1501. ISSN 1877-7058
3. Davies CM, Ahn J, Tsunori M, Dye D, Nikbin KM (2014) The influence of pre-existing deformation on GMA welding distortion in thin steel plates. J Mater Eng Perform Europe
4. Gery D, Longb H, Maropoulos P (2005) Effects of welding speed, energy input and heat source distribution on temperature variations in butt joint welding. J Mater Process Technol 167:393–401
5. Fouda AS, Al-Sarawy AA, Ahmed FS, El-Abbasy HM (2009) Corrosion inhibition of aluminum 6063 using some pharmaceutical compounds. Protect Metals Phys Chem Surf 45. ISSN 2070-2051
6. Hooda A, Dhingra A, Sharma S (2012) Optimization of MIG welding process parameters to predict maximum yield strength in AISI 1040. Int J Mech Eng Robot 1:203–213. ISSN 2278-0149
7. Kashish, Singh C, Khanna P (2021) Effect of welding parameters on angular distortion of submerged arc welded low carbon steel plates. IOP Mater Sci Eng 1:1–8. ISSN 1126-012015
8. Wang J, Lu M, Zhang L, Chang W, Xu L, Hu L (2012) Effect of welding process on the microstructure and properties of dissimilar weld joints between low alloy steel and duplex stainless steel. Springer Int J Miner Metall Mater 19:518–524
9. Acar I, Cevik B, Gulenc B (2023) Weldability of dissimilar stainless steels by MIG welding with different gas combinations, vol 48. Springer
10. Ghosh N, Rudrapati R, Pal PK, Nandi G (2017) Parametric optimization of gas metal arc welding process by using Taguchi method on ferritic stainless steel AISI409. Mater Today Proc 4:2213–2221
11. Reisgen U, Oechsner M, Sharma R, Ellermeier J, Andersohn G, Engler T, Zokoll E, Heider B, Gonzalez Olivares E (2020) Influence of preheating on lamellar gray cast iron for surface layer welding applications with plasma-transferred arc powder and metal inert Ga welding processes with duplex steel as filler material, vol 29. Springer, pp 724–740
12. Devendranath Ramkumar K, Singh A, Raghuvanshi S, Bajpai A, Solanki T, Arivarasu M, Arivazhagan N, Narayanan S (2015) Metallurgical and mechanical characterization of dissimilar welds of austenitic stainless steel and super-duplex stainless steel—a comparative study. J Manuf Process 19:212–232

Conclusion

Analysis of a Hybrid Composite Material for a Relation Between Ply Angle, Strength and Fibre Volume Fraction

Mahatma Patel, Manjeet Keshav, and Diya Patel

1 Introduction

For any material, mechanical strength plays pivotal role, and composite materials are being considered. They are the perfect replacement of conventional metallic materials. They consist of two or more materials with varying properties that are combined macroscopically and are insoluble in each other. The resulting material especially fibre-reinforced composites exhibits desirable characteristics such as a high strength-to-weight ratio, high tensile strength, low coefficient of thermal expansion, good corrosion resistance, dimensional stability, non-conductivity, non-magnetism, ease of manufacturing and light weightiness, which are absent in typical metals [1, 2]. The properties of the composite materials can be tailored according to the specific requirements of the application. Along with all the proposed benefits, one can further enhance the properties by selecting proper materials, orientation of their fibres, manufacturing method, etc. [3, 4]. This is where the concept of hybrid composites comes into scenario. Primary goal of diving into hybrid composites is that one can attain the whole range of properties of all the constituent materials, which cannot be obtained with single reinforcement. Here, one obtains benefits of each used reinforcement [5, 6].

Hybridization strengthens and improves the performance of composite materials. Hybrid composites can be classified into two types such as matrix hybrid and fibre hybrid composites. During curing if the additional fibres dissolve into the matrix, then it is called matrix hybrid and if they remain intact, they are called fibre hybrid [7]. Hybrid composite brings forth additional advantages like higher impact energy, impact and fatigue resistance, increased fracture toughness along with all the

M. Patel · M. Keshav (✉)
Pandit Deendayal Energy University, Gandhinagar, Gujarat 382007, India
e-mail: manjeetkshv@gmail.com

D. Patel
Hindustan Institute of Technology and Science, Chennai, Tamil Nadu 603103, India

R. Sahu et al. (eds.), *Advancements in Materials Processing Technology, Volume 1*,
Springer Proceedings in Materials 48, https://doi.org/10.1007/978-981-97-4958-4_15

composite material properties [8]. Other configurations of hybrid composites include intra-ply, inter-ply, sandwich or intimately mixed [5]. Several researches have been carried out on hybridization and the testing of hybrid materials.

Dong and Davis [9] in their study of hybrid epoxy composites concluded that on substituting carbon fibre by glass fibre on the compressive side of test specimen, it improved the flexural strength. Je Hoon Oh et al. [10] for the hybrid composites found that when using $[0C/\pm 45G/\pm 45C/90C]_s$ and $[0G/\pm 45C/90G]_s$ as stacking sequences, if ±45 plies were distributed in even manner, it resulted in increase in bearing strength and it was entirely independent of carbon–epoxy to glass–epoxy ratio. Additionally, even when there was increase in volume fraction of carbon–epoxy plies with ±45° orientation, the bearing strength was found to increase. Ramalingam et al. [11] in their work blended vinyl ester and polyurethane resin with E-glass fibre, and the material was tested experimentally and numerically for mechanical properties like flexural stress. Mamalis et al. [12] studied that there is variation in energy absorption based on ply orientation and speed of crushing. A 45° ply angle was found to be critical for this parameter. Nayak et al. [13] observed that a carbon–glass fibre hybrid composite with carbon plies at the end and the glass ones at the core possesses maximum stiffness and can withstand maximum load. Nayak et al. (2014) [14] fabricated a glass fibre composite material where the epoxy was modified with compounds like Al_2O_3, SiO_2 and their effect on mechanical properties was determined. It was found that SiO_2 modified epoxy had more flexural strength and with alumina the hardness and impact energy was higher. Caminero et al. [15] compared performances of composite materials based on stacking sequence and ply angles. It was observed that $[0/90/(\pm 45)_2)]_s$ laminate inherited good performance under impact similar to [0°] and [±45°] composites, in terms of absorbed energy. $[\pm 45]_{3s}$ gave best impact performance but reduced flexural strength. Aisyah et al. [16] base their study on fabric counts and weave. They determined that plain fabric gives higher impact and tensile strength and better adhesion properties than satin weave. Fabric count of 5 × 5 gave composites with higher flexural modulus than 6 × 6. Dixit and Padhee [17] conducted FEA of composites made with hybridization of natural and synthetic fibres. They found that carbon–glass fibre hybrid composites undergo minimum amount of deflection and has higher directional strength and bonding with epoxy. The laminate with alternating carbon and glass laminas $[0/45/0/-45/0/90/0/45/0/-45]$ gave minimum deformation. Srivastava et al. [18] designed a composite by coating carbon fibres with graphene nano platelets, which resulted in improved flexural strength and a continuous interphase between fibre and epoxy. Hameed and Daway [19] in their study of E-glass and carbon fibre composites individually as well as hybrid concluded that the hybrids of the same gave higher ultimate tensile strength. Pandya [20] in their study of hybrid materials pointed out that when carbon fibres are at inner layers and glass fibres at outer ones, it gives higher ballistic limit velocity. Kim et al. [21] determined that compared to metallic bumper beams, glass–carbon fibre hybrid beams gave better performance under impact and the weight is 33% lesser. Hung et al. [22] observed that in a hybrid composite where carbon layers were at the surface and the glass fibre layers at the core, they show lesser damages from impact compared to a vice versa arrangement.

Moreover, they also highlighted that the flexural strength is controlled by bottom layers. Study by Zhang et al. [23] supports their findings on flexural strength. Attia et al. [24] saw that stacking patterns affect tensile strength but has no effect on tensile modulus and Poisson's ratio. Carbon fibres at the exterior gave higher tensile strength. Along with this, they combined glass fibre with polypropylene fibres for higher flexural strength. Selmy et al. [25] manufactured hybrid composites and found that glass fibres in the exterior and the polyamide fibres in the interior gave higher tensile and flexural properties but reduced shear properties. The resulting laminate eliminates the higher cost and weight of polyamide composite and lesser ductility of glass fibre composite.

Several researches have been carried out which focus on mechanical properties of hybrid composites but there has been negligible focus on establishing a relation between strength, ply angles and volume fraction. This paper focuses on that relation and for the same, a composite laminate was fabricated and a few tests were carried out to evaluate mechanical properties.

2 Methodology

2.1 Materials

For fabrication of hybrid composite, three different materials were used as reinforcements; Unidirectional Carbon fabric with 400 gsm, Bidirectional Carbon fabric with 240 gsm and Unidirectional Fibre E-Glass fabric with 900 gsm. LY556 Epoxy Resin in ratio 10:1 with HY951 hardener was used for matrix material. The reinforcement materials were purchased from Urja Products Pvt Ltd, India, and the matrix materials from Herenba Instruments and Engineers, India.

2.2 Preparation of Hybrid Laminate

Referring to the previous studies, it was decided to use carbon plies at the exterior layer for higher strength. Carbon UD fabric was positioned at the top and bottom, followed by glass fibre UD towards the core and at the centre carbon BD fabric was placed. The material was prepared using the vacuum bagging method to avoid the qualitative variation among specimens. Four different specimens were made by keeping the stacking material sequence same and varying the ply angles of all five layers. The ply angles were selected though permutations and combinations, and four of them were fabricated for further testing and analysis.

Out of various possible combinations, the selected stacking order was as follows (Fig. 1).

1. [$0CU/30GU/0CB/90GU/45CU$].

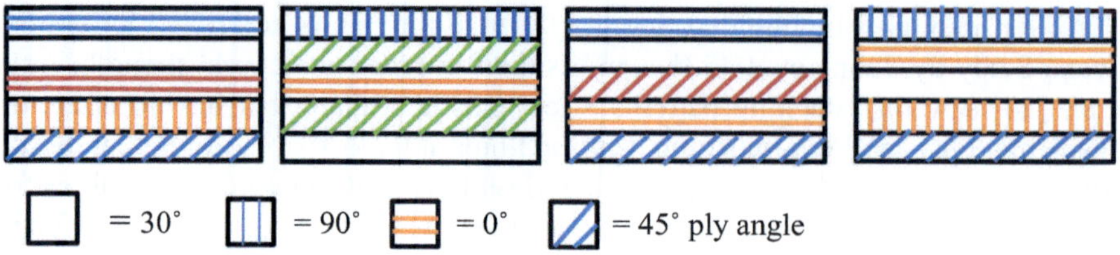

Fig. 1 Stacking order of materials and ply angles for the fabricated specimens [12]

2. [90*CU*/45*GU*/0*CB*/45*GU*/30*CU*].
3. [0*CU*/30*GU*/45*CB*/0*GU*/45*CU*].
4. [90*CU*/0*GU*/30*CB*/90*GU*/45*CU*].

CU = Carbon fibre unidirectional, CB = Carbon fibre bidirectional, GU = Glass fibre unidirectional.

2.3 *Testing*

The prepared specimens were cut accordingly based on the ASTM of a particular test, they were put to test for various mechanical properties, and the comparative analysis was carried out.

Tensile Test

Tensile test is a means to measure the amount of force needed to break a composite material and the amount of elongation it undergoes. It helps gather information about tensile and yield strength, % elongation, tensile strain and modulus, etc. From the previous studies, we already know that for a composite material, tensile strength is highest parallel to fibre orientation [26]. The specimens were placed between the fixed and movable jaws of the universal testing machine, with the load range adjusted to its maximum capacity. The test speed was set at 2 mm/min, and an extensometer/strain gauge was used to determine properties. The specimens were gradually loaded until they reached their ultimate load, and the breaking load was noted.

Bending Test

Bending test is used for measuring bending as well as deflection of composite materials [26]. Three-point bending test was conducted for all the specimens to determine their bending strength. Rectangular specimens (150 × 25 × 3 mm) were utilized for the tests, and all tests were performed at room temperature using a UTM machine. The dimensions of the specimens, including original thickness and gauge length, were measured per ASTM E 370-21. Similar to tensile test, the test speed was kept 2 mm/min and at the end breaking load was noted.

Impact Test

Impact test allows to quantify the response of a composite material to a sudden force/stress. It helps with the data of impact strength and how tough or brittle a material is [27]. One notable benefit of using the drop weight impact test over pendulum impact test methods is that the specimen usually does not require clamping, based on the testing setup. The impact tests were carried out on rectangular specimens (55 × 10 × 3.3 mm) using a Drop Charpy Impact Testing Machine at room temperature, following ASTM A 370-21 standards. An Impactor of appropriate weight was selected, and the height of the Impactor was adjusted accordingly. The Impactor was allowed to drop at the centre of the specimen. The energy released during the impact test was calculated obtained. All the above three tests were conducted at Hertz Testing and Training Centre, Ahmedabad.

Volume Fraction Test

Volume fraction was calculated using Muffle Furnace at CRE Lab, Pandit Deendayal Energy University, Gandhinagar. Initially, the crucible was heated to 600 °C for 30 min. Mass of the crucible and the crucible along with specimen were measured every time. Then, the crucible and the specimen were placed in the furnace and heated to 600 °C till the entire matrix disappeared. After cooling, the post burnout mass was measured. After that, using the following formulas, the volume fraction was calculated as follows:

$$w_{\mathrm{m}} = w_{\mathrm{c}} - w_{\mathrm{f}} \tag{1}$$

$$v_{\mathrm{f}} = \frac{w_{\mathrm{f}}}{\rho_{\mathrm{f}}} \tag{2}$$

$$v_{\mathrm{m}} = \frac{w_{\mathrm{m}}}{\rho_{\mathrm{m}}} \tag{3}$$

$$V_{\mathrm{f}} = \frac{v_{\mathrm{f}}}{v_{\mathrm{f}} + v_{\mathrm{m}}} \tag{4}$$

3 Results and Discussion

The results for tensile test of all the four specimens are shown in Table 1. From the table, it is clearly observed that specimen 3 has the highest tensile strength and possesses highest yield stress; followed by specimens 1 and 2 and the least tensile strength is observed for specimen 4. To conclude for tensile strength, 45° carbon bidirectional fibre at core, 0° and 45° carbon unidirectional lamina at the ends and 30° and 0° glass fibre unidirectional in the middle of CU and CB gives highest tensile strength compared to other arrangements.

Table 1 Experimental results of tensile test

Specimen	1	2	3	4	Unit
Load at yield	14.197	10.636	16.972	7.775	kN
Elongation at yield	7.813	5.957	7.268	6.953	mm
Yield stress	334.47	257.889	371.651	197.033	N/mm^2
Load at peak	17.806	13.298	21.263	9.750	kN
Elongation at peak	9.311	6.938	8.496	8.161	mm
Load at break	0.105	0.106	1.320	0.028	kN
Elongation at break	9.319	6.968	8.653	8.441	mm
Tensile strength	419.417	322.434	465.615	247.084	N/mm^2
% Elongation	4.36	2.86	14.92	10.52	%

Figure 2 shows the test specimens after they underwent the tensile test till they reach the breaking load. Figure 3 shows the graph of load applied on specimen versus cross head travel. For all the four specimens, it can be seen that with the increase in displacement, there is increase in load, and it approaches failure or neck formation. After the point of failure, once can observe a steep decrease in load with increase in displacement, till the generation of fracture in the material.

Bending test results (Table 2) shows that specimen 3 can bear maximum load, before it breaks. Thus, test specimen 3 has the highest bending strength followed by specimens 1, 2 and 4, respectively. The order of test pieces for bending strength can be observed to be same as tensile strength.

For impact test, test was conducted on a bigger sample, compared to the ideal size and then the values were converted on the basis of standard test piece size. The length and width were 55 mm and 10 mm, respectively. On analysing the final test values (Table 3), it was found that specimen 4 with 30° carbon bidirectional fibre at core, 90° and 45° carbon unidirectional lamina at the ends and 0° and 90° glass fibre

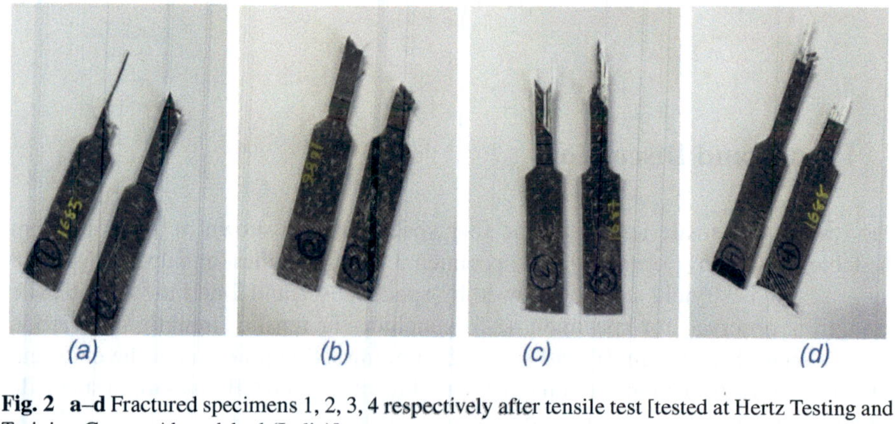

Fig. 2 **a–d** Fractured specimens 1, 2, 3, 4 respectively after tensile test [tested at Hertz Testing and Training Centre, Ahmedabad (India)]

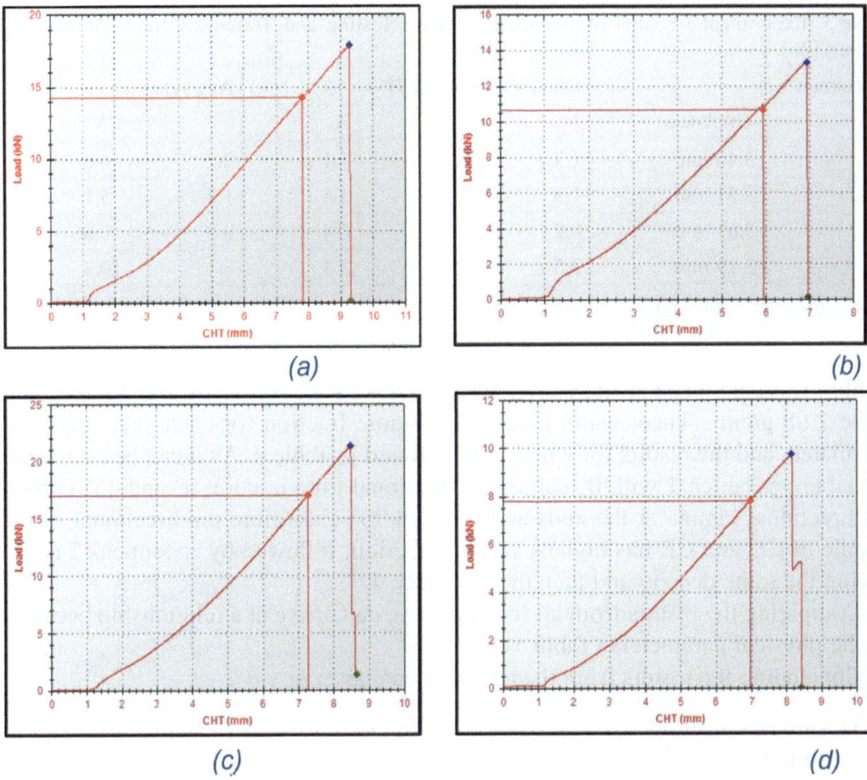

Fig. 3 a–d Plot of load versus displacement for specimens 1, 2, 3 and 4, respectively

Table 2 Experimental results of bending test [Hertz Testing and Training Centre, Ahmedabad (India)] [26]

Specimen	1	2	3	4
Width (in mm)	25	25	25	25
Thickness (in mm)	3.46	3.51	3.61	3.19
Former diameter for bend (in kN)	10	10	10	10
Load at break (in kN)	2.24	1.59	2.39	1.37

unidirectional in the middle of CU and CB has highest impact strength; followed by specimens 1, 2 and 3 in order. It was determined that the material with highest impact strength has the lowest tensile and bending strength and vice versa.

The volume fraction has been calculated by involving density of composite using the mass and dimensions of the material, and the average density of the composite was found to be 1.327 g/cm^3. From the industrial data, it was found that density of carbon fibre UD, carbon fibre BD and glass fibre UD was 1.8, 1.8 and 2.5 g/cm^3 respectively. From these data and using Rule of Mixtures, the cumulative density

Table 3 Experimental results of impact test [Hertz Testing and Training Centre, Ahmedabad (India)] [26]

Specimen		Obtained value (in J)			Avg value (in J)	
No	Thickness	1	2	3		CV
1	3.46 mm	1.7	2.7	2.1	2.2	6.6
2	3.51 mm	1.5	2.1	1.5	1.7	5.1
3	3.61 mm	0.2	0.2	0.4	0.3	0.86
4	3.19 mm	3.5	2.5	3.3	3.1	10.3

CV—Converted value (10 × 10 × 55), Avg—average

of fibre was calculated as there are three different fabrics involved and it was found to be 2.08 g/cm^3. Then, using Eqs. 1–4, volume fraction for each specimen was calculated, and the results for which can be found in Table 4. As it can be seen from the table, specimen 1 with 0° carbon bidirectional fibre at core, 0° and 45° carbon unidirectional lamina at the ends and 30° and 90° glass fibre unidirectional in the middle of CU and CB has highest volume fraction; followed by specimens 2 and 3 having the same density and then the test piece 4.

Comparing the results from all four tests, we can arrive at a relationship between all the physical parameters (Table 5).

Concluding the results from the tests one can observe:

Tensile strength $3 > 1 > 2 > 4$.
Volume fraction $1 > 2 = 3 > 4$.

Table 4 Experimental results of volume fraction burnout test [Hertz Testing and Training Centre, Ahmedabad (India)]

Specimen	1	2	3	4
Mass before burnout test	7.72 g	9.64 g	10.03 g	9.15 g
Mass after burnout test	4.16 g	4.94 g	4.80 g	4.52 g
Volume fraction	40.24%	36.58%	36.58%	36.02%

Table 5 Comparison of the results of all tests [Hertz Testing and Training Centre, Ahmedabad (India)] [25]

Specimen	1	2	3	4
Tensile strength (in MPa)	419.417	322.434	465.615	247.084
Yield stress (in MPa)	334.47	257.889	371.651	197.033
Elongation	4.36%	2.86%	14.92%	10.52%
Impact strength	6.6 J	5.1 J	0.86 J	10.3 J
Bending strength	2.24 kN	1.59 kN	2.39 kN	1.37 kN
Volume fraction	40.24%	36.58%	36.58%	36.02%

Bending strength 3 > 1 > 2 > 4.
Impact strength 4 > 1 > 2 > 3.

where 1, 2, 3, 4 are test specimen numbers. Establishing the relation between ply angle, volume fraction and strength: When the carbon bidirectional and one of the carbon unidirectional fabrics are in the same orientations, tensile, bending and yield strength is found to be higher. From the results, it can be inferred that when looking for higher impact strength or impact absorption, one can take up the combination where carbon weave fabric is at 30° orientation, the glass fibre combination is at 0° and 90° and the carbon unidirectional fibre combination at 90° and 45°. When looking for higher bending strength, one can go with the third combination with carbon bidirectional and carbon unidirectional pair to be at the same orientation and other carbon unidirectional and glass unidirectional fibre to be at the same orientation. But the impact absorption of this material is very less. One can also observe that, when the first carbon UD ply and the fourth glass UD ply are at the same orientation of 90°, impact tolerance can be found higher, whereas when the angle for both is 0°, impact tolerance is about 12 times lesser. No exact relation can be established between volume fraction and other mechanical properties, but one can imply that volume fraction and impact strength are inversely proportional and the bending strength is directly proportional. Even tensile strength is directly proportional to volume fraction. Based on the type of application, one can trade properties and the manufactured hybrid composite materials can be used for better performance. For instance, the places requiring higher impact strength but can allow lower bending/tensile strength, one can choose specimen 4 for that application. Hence, an efficient material has been fabricated which can be used based on requirement as the properties are tailored accordingly and as each laminate gives different benefits, one can choose independently.

4 Conclusion

With the aim to manufacture a material which has higher impact absorption and strength along with it being lightweight compared to the current conventional materials, a composite laminate has been fabricated where carbon fibre was combined with glass fibre and a hybrid composite material was manufactured. Four different test pieces were made with different arrangements of materials and their ply angles. The test pieces made were arranged like carbon fibre unidirectional, glass fibre unidirectional, carbon fibre bidirectional, glass fibre unidirectional and carbon fibre unidirectional in order. The ply angles were then varied and four different test specimens were made. After that, precise experimental testing was conducted; the tests include the tensile test, bending test, impact test and volume fraction test. It has been concluded that, when the top ply of carbon UD and carbon BD are at the same orientation, the volume fraction is higher and when one of the carbon unidirectional and glass fibre at a distance have the same ply angles, the volume fraction is comparatively lower. It was also concluded that volume fraction is directly proportional to strength and inversely proportional to impact tolerance. In the results, a weird elongation pattern was obtained comparatively. Thorough microanalysis of this result is one of the milestones for future work in this regard. Moreover, more analysis can be carried out on volume fraction as well as by changing arrangements and ply angles. Thus, to conclude, superior quality material is made and shows promising results in terms of impact strength and yield strength, although extended research can still be carried out by trying other combinations of ply angles as well as the materials.

Acknowledgements The authors extend gratitude to Dr. Nirav Patel of Pandit Deendayal Energy University for access to the composite fabrication facility and to the Hertz Testing and Training Centre in Ahmedabad for their testing facilities.

References

1. Mallick PK (1988) Fibre reinforced composites. Marcel Dekker, NewYork
2. Schwartz MM (1984) Composite Materials handbook. Mc Graw Hill, New York
3. Ravishankar B, Nayak SK, Kader MA (2019) Hybrid composites for automotive applications—a review. J Reinforced Plast Compos 38
4. Kaw A (2006) Mechanics of composite materials. Taylor and Francis
5. Kretis G (1987) A review on tensile, compressive, flexural and shear properties of hybrid fibre-reinforced plastics. Composites
6. Atiqah A, Maleque MA, Jawaid M, Iqbal M (2014) Development of Kenaf-glass reinforced unsaturated polyester hybrid composite for structural applications. Composites 56
7. Safri S, Sultan M, Jawaid M, Jayakrishna K (2017) Impact behaviour of hybrid composites for structural applications: a review. Composites
8. Hancox NL (1981) Fibre composite hybrid materials. Applied Science, London
9. Chensong D, Ian D (2014) Flexural strength of bidirectional hybrid epoxy composites reinforced by E glass and T700S carbon fibres. Composites
10. Oh JH, Kim YG, Lee DG (1997) Optimum bolted joints for hybrid composite materials. Compos Struct 38
11. Ramalingam PS, Mayandi K, Srinivasan T, Leno IJ, Ravi R, Suresh G (2020) A study on E-glass fiber reinforced interpenetrating polymer network (vinylester/polyurethane) laminate's flexural analysis. Materials Today
12. Mamalis AG, Robinson M, Manolakos DE, Demosthenous GA, Ioannidis MB, Carruthers J (1997) Crashworthy capability of composite material structures. Compos Struct
13. Nayak S, Sadarang J, Panigrahi I, Nayak RK (2018) Development of carbon/glass fibre reinforcement polymer hybrid composite through modeling and simulation. in Mater Today
14. Nayak RK, Dash A, Ray BC (2014) Effect of epoxy modifiers (Al2O3/SiO2/TiO2) on mechanical performance of epoxy/glass fiber hybrid composites. Mater Sci
15. Caminero MA, Rodriguez GP, Muñoz V (2015) Effect of stacking sequence on Charpy impact and flexural damage. Compos Struct
16. Aisyah MA, Paridah MT, Khalina A, Sapuan SM, Wahab MS, Berkalp O, Lee CH, Lee SH (2018) Effects of fabric counts and weave designs on the properties of laminated Woven Kenaf/carbon fibre reinforced epoxy hybrid composites. Polymers
17. Dixit S, Padhee SS (2019) Finite element analysis of fiber reinforced hybrid composites. Materials Today
18. Srivastava AK, Gupta V, Yerramalli CS, Singh A (2019) Flexural strength enhancement in carbon-fiber epoxy composites through graphene nano-platelets coating on fibers. Composites

19. Hameed DM, Daway EG (2014) Mechanism of hybrid reinforcement and its effect on some properties of binary polymer blend. Eng Technol
20. Pandya KS, Pothnis JJ, Ravikumar G, Naik NK (2013) Ballistic impact behavious of hybrid composites. Mater Design 44
21. Kim D-H, Kim H-G, Kim H-S (2015) Design optimisation and manufacture of hybridglass/carbon fibre reinforced composite bumper beam for automobile vehicle. Compos Struct 131
22. Hung P, Lau K, Chang L, Leng J, Hui D (2017) Impact response of hybrid carbon/glass fibre reinforced polymer composites designed for engineering applications. Composites
23. Zhang J, Chaisombat K, He S, Wang CH (2012) Hybrid composite laminates reinforced with glass/carbon woven fabrics for lightweight load bearing structures. Mater Design 36
24. Attia M, El-Baky MA, Alshorbagy A (2016) Mechanical performance of intraply and inter-intraply hybrid composites based on e-glass and polypropylene unidirectional fibers. J Compos Mater
25. Selmy A, El-baky MA, Hegazy D (2018) Mechanical properties of inter-ply hybrid composites reinforced with glass and polyamide fibers. J Thermoplast Compos Mater
26. Saba N, Jawaid M, Sultan M (2019) An overview of mechanical and physical testing of composite materials. In: Mechanical and physical testing of biocomposites, fibre-reinforced composites and hybrid composites. Woodhead Publishing
27. Saba N, Paridah MT, Abdan K, Ibrahim NA (2016) Effect of oil palm nano filler on mechanical and morphological properties of kenaf reinforced epoxy composites. Constr Build Mater 123

Investigating Microwave Joining as a Solution for Metal Plate/Pipe Connections: A Critical Analysis of Current Techniques and Applications

Manjunath R. Rawal and Duradundi S. Badkar

1 Introduction

The joining process, which involves coalescing materials through external heat sources with or without filler material. Two main methods, conventional and non-traditional, have been extensively researched, but challenges persist in achieving defect-free joints. To overcome these limitations, advanced non-traditional methods like laser welding, friction welding, stir welding, microwave welding, and ultrasonic welding have been developed, chosen based on specific requirements and materials. The paper emphasizes the recent advancements in microwave joining, highlighting its ability to create fast and defect-free joints. Microwave joining is a relatively new method of joining materials, and the development of this process has occurred primarily in the last few decades. In the 1980s, with the development of new materials and the availability of high-power microwave sources, research on microwave joining began to accelerate. Researchers started looking at whether it would be possible to combine different materials like metals, polymers, and ceramics using microwave energy [1]. The first commercially viable microwave joining equipment developed in the 1990s, and research on the subject continued to emphasis on the enhancement of process variables and the creation of new applications. Recently, microwave joining is well-liked approach for joining dissimilar materials, as it allows for a strong bond to be formed without the need for a filler material [2]. The process of heating and sintering a porous, powder metal compact in a microwave field was initially described by Roy et al. in 1999. Previously, this came as a surprise because metals were considered to deflect microwave radiation. Since most metals barely penetrate

M. R. Rawal (✉)
Research Scholar, Shivaji University Kolhapur and DKTE's TEI, Ichalkaranji, Kolhapur, Maharashtra, India
e-mail: manjunathrrawal61@gmail.com

D. S. Badkar
Fabtech Technical Campus, Sangola, Maharashtra 413307, India

R. Sahu et al. (eds.), *Advancements in Materials Processing Technology, Volume 1*, Springer Proceedings in Materials 48, https://doi.org/10.1007/978-981-97-4958-4_16

a few millimeters deep, exposure to microwaves simply heats the surface of the metal [1, 3]. This is known as "skin effect." However, employing powder of metal having grain sizes smaller than the depth of penetration causes volumetric heating. This is because the smaller particle size allows for more efficient energy absorption, resulting in deeper penetration of the microwaves.

Energy from electromagnetic fields is transformed into thermal energy in a relatively fine covering of metals and the most of conductors and calculated as skin depth (d) by Eq. (1) [4],

$$d = \sqrt{\left(\frac{\rho}{\pi f \mu 0 \mu\prime}\right)} \tag{1}$$

where

- ρ material electrical resistivity.
- f alternative magnetic field frequency.
- $\mu 0$ permeability of free space.
- μ' permeability.

For metal powders, the particle size (s) is an important factor. The depth of penetration increases as particle size reduces ($s < d$). Therefore, the ability to achieve volumetric heating in metal powders using microwaves is calculated by the particle size of the powders, the frequency of the microwave energy, the properties of the powders, and the processing parameters. Domestic ovens are working on the principle of microwaves frequency of 2.45 GHz and 900 W power. When it comes to using microwave hybrid heating to combine various materials, the majority of the advancements were made in the nineteenth century (MHH) [1, 4, 5]. As it is effective, defect-free, economical, and fast process as compared to conventional joining processes [3], there are certain limitations while using microwaves for the joining of ceramics and metals, because microwaves are travel through the ceramics and get reflected from the surface of metals [6, 7]. The coalescence of materials and ceramic is possible when microwaves are get absorbed by them and due to this intermolecular energy gets excited, results in a temperature rise, which subsequently produces bonding between them. To overcome this difficulty, susceptor material has to be used. Majorly charcoal, graphite, silicon carbide, aluminum plate, and ceramic crucible are used as a susceptor [8], which is heated as a result of absorbing the microwaves. That heat is then carried through conduction, convection, and radiation mode to the base workpiece to be joined. It is possible to combine two materials that are similar or distinct with or without an interacting powder particle. Generally, the base material powder is used as an interfacing material, which form coalescence easily and form a joint of greater strength [9]. The heat is externally supplied in the case of the traditional joining procedure in order to melt the base material. But in microwave heating, the energy in form of the microwave is get absorbed by materials that lead to form heat within the specimen itself. This phenomenon is shown Fig. 1a. Setup used for MHH is shown in Fig. 1b.

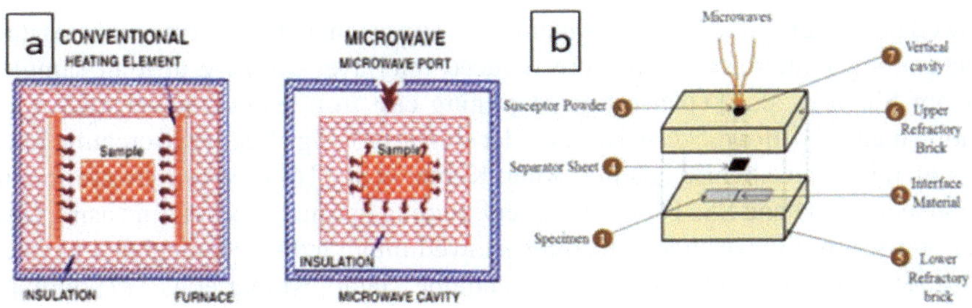

Fig. 1 **a** Difference between conventional and microwave heating [10], **b** microwave hybrid heating (MHH) setup [10]

Microwave heating differs from conventional heating by simultaneously generating heat within materials through microwave energy absorption, enabling bidirectional heating and promoting metallurgical bonding at faying surfaces due to localized excess heat.

2 Literature Review

2.1 Reviews Related to MHH of Plates Joining

Singh et al. [11] studied on the development and characterization of CI joint, which results in defect-free joint having porosity range 1.57–1.88% only. The ultimate tensile strength is 90% of those base metals. Amit Bansal et al. [12] investigated the joints between Inconel 718 and SS-316L with an Inconel 718 interfacing layer. Inclusion of Inconel 718 powder improved mechanical properties: average microhardness 230.5 HV, porosity 0.94%, ultimate T-S 517.5 MPa, and 18.18% elongation. Inconel 718 powder was observed to mix at the joint interface during the MHH process. Badiger et al. [4, 5] used Inconel-625 and nickel-based powder for microwave joining, improving mechanical properties. Nickel powder increased microhardness and tensile strength. XRD showed chromium carbide at the interface, EDS found niobium-rich Laves phase. Inconel-625 filler enhanced joint properties. Bagha et al. [13] investigated how different particle sizes of nickel interface material affect SS304-SS304 welded joint properties. Smaller nickel particles improved hardness and strength but decreased ductility. The research emphasizes the impact of particle size on welded joint characteristics. Tamang et al. [14] explored joining AZ31B and Al6061-T6 using an active braze alloy interlayer by MHH. TiCuSil paste interlayer was applied, and graphite crucible with graphite powder was found to create the fastest joint. Active braze alloy and graphite setup improved joint properties in microwave joining.

Tamang et al. [15] used FEM simulation to investigate microwave welding of dissimilar metals (OFC and SS304) with a Cu powder interlayer. COMSOL software showed interlayer and susceptor temperatures, consistent with experimental measurements. Longer microwave exposure reduced interlayer porosity in joint samples. Satnam Singh et al. [11] investigated joining cast iron using a novel microwave heating method with a nickel-based powder slurry, proving effective in creating joints through microwave hybrid heating. Anup Kumar et al. [16] used microwave joining to connect SS304 and SS316 stainless steel, finding optimal process time using charcoal powder for quicker heating. Joints showed improved mechanical and physical properties compared to the suggested technique. Sanjiv Kumar et al. [17] outline a process for butt joints between SS304 and SS316 using microwave hybrid heating. SS316 powder filler, activated charcoal as susceptor, and microwave exposure create joints. Characterization involved microscopy and hardness tests. Abhishek Sharma et al. [18] examines process parameters for defect-free Inconel-625 joint melting using microwave hybrid heating (MHH). Trials analyze factors like charcoal feeder hole size, graphite sheet thickness, and insulation brick properties. The research aims to enhance understanding and standardization of microwave joining for commercial applicability. Virinder Kumar et al. [19] studied microwave joining of duplex stainless steel (SS2205). Effective joining was achieved with selective microwave hybrid heating in 300 s. Charcoal feeding and powder susceptor concentrated heating for optimal welding time, resulting in increased hardness from heat affected to joining zones. Table 1 shows the literature of some research papers with their findings.

2.2 Reviews Related to MHH of Pipe Joining

Dhirendra Gamit et al. [25–28] explored Microwave Hybrid Heating (MHH) to join copper, stainless steel 316L, and mild steel pipes using specific metal powders and charcoal as a susceptor. Successful connections were achieved within exposure times of 840–900 s for copper, unspecified for stainless steel, and up to 480 s for mild steel with nickel powder. The technique offers potential for eco-friendly, energy-efficient joints with improved properties. Somit Saxena et al. [29] examines microwave heating to join highly conductive metals like copper, highlighting benefits of speed, low temperatures, and improved properties. Experiments using microwave hybrid heating at 2.45 GHz and 900 W detail joining bulk copper pipes with various materials and powder sizes. Dhirendra Gamit et al. [30] used a 3D COMSOL model simulated microwave pipe joining, mirroring experiments. High temperatures at joint zone led to melting for good fusion at 480 s and 900 W. At 540 s, distortion occurred. Susceptor generated energy, raising joint zone temperature, while heat transfer caused cooler pipe ends.

Table 1 Exposure time for different susceptor and base material with interface powder size

Author	Material	Material thickness	Interface material/ blumer	Size of powder	Susceptor	Characterization	Results	Exposure time (isec)	Joint type
Srinath et al. [20]	Stainless steel-316	6 mm	Ni-based powder	40 μm	Charcoal	Microhardness, SEM, XRD, tensile strength, porosity	The joints ultimate tensile strength was determined to be on the order of 309 MPa, with an elongation of 11.50%	300–390	Butt joint (plate)
Srinath et al. [31]	MS-SS316	6 mm	Ni-powder	40 μm	Charcoal	Microhardness, FESEM, X-ray	An elongation of 13.58% and a 346.6 MPa ultimate T-S were found	450	Butt joint (plate)
Singh et al. [21]	Sisal fiber-reinforced polylactic acid (PLA)	–	Charcoal	30–35 μm	Charcoal	NA	No joining was achieved without an acceleration or addition	60–480	Butt joint (plate)
Sharma Gupta [10]	Ceramic–Metal	–	Ni-powder/ 1450XX	30–40 μm	Charcoal	BSE image (back scattered electron)	Produced metal–ceramic composite coated on a metallic substrate by microwave energy	420	Butt joint (plate)
Bansal et al. [11]	MS	5 mm	Ni-powder	40 μm	Charcoal	SEM, XRD, micro-identification	T-S results showed an average joint strength of 250 MPa and 6% elongation	900	Butt joint (plate)

(continued)

Table 1 (continued)

Author	Material	Material thickness	Interface material/ blumer	Size of powder	Susceptor	Characterization	Results	Exposure time (isec)	Joint type
Bajpai et al. [22]	Natural fibers	–	NA	30 μm	Charcoal	Tensile testing, finite element software	By using microwave joining, a stronger joint was obtained in comparison to adhesively bonded joints' strength	200	Butt joint (plate)
Bansal et al. [23]	MS–MS	5 mm	Ni-powder/ 1450XX	40 μm	Aluminum plate	SEM, XRD, EPMA, Vicker's microhardness	(1) The powder and substrate material were well-bonded metallurgically (2) 50% more base material strength was gained by the joint's T-S	600	Butt joint (plate)
Gupta and Kumar [24]	SS	–	Ni-based EWAC102ET	30 μm	Charcoal	Microstructural analysis, X-ray, tensile strength, elongation and microhardness	T-S of the joint increased with longer exposure times and higher Ni-powder concentrations	300–900	Butt joint (plate)

(continued)

Table 1 (continued)

Author	Material	Material thickness	Interface material/ blumer	Size of powder	Susceptor	Characterization	Results	Exposure time (isec)	Joint type
Bansal et al. [12]	Inconel718–SS316L	5 mm	Inconel 718	30 μm	NA	X-ray, FESEM, microhardness	1)The joint interface's average microhardness was found to be 230 ± 5 2)Microwave welds seen an average ultimate tensile strength of 517.5 MPa	600	Butt joint (plate)
Singh et al. [13]	Aluminum alloy	–	Aluminum powder/ 1550XX	40 μm	Silicon carbide	X-ray, XRD, microhardness	An interface and sandwich layer with metallurgical bonding were identified	600	Butt joint (plate)
Badiger et al. [5, 4]	Inconel-625	–	Inconel-625/ 1450XX	40 μm	Graphite sheet	Microstructural examination, X-ray, tensile strength,	Joints' maximum estimated flexural and tensile strengths were 377 and 375 Mpa, respectively	1260	Butt joint (plate)

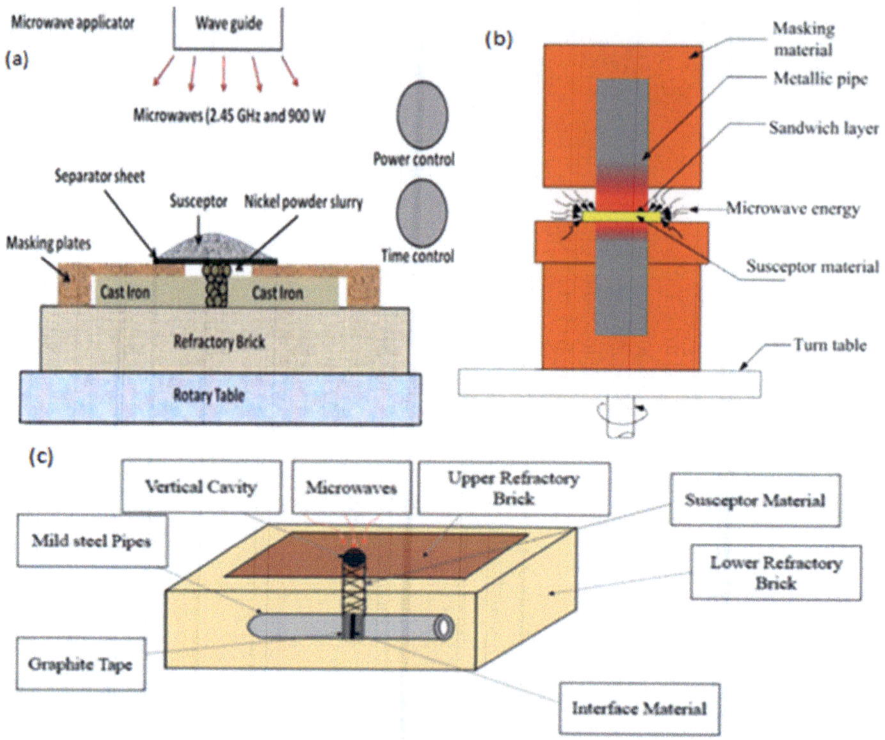

Fig. 2 Different designs used by researchers **a** horizontal butt joining, **b** vertical position pipe joining, **c** horizontal pipe joining

3 Different Types of Setups Used During Experimentation

Microwave joining of plates and pipes requires specific molds to focus waves for even heating. Susceptor materials absorb microwave radiation, melting powder slurry at the joint. Separator sheet prevents direct contact, all placed in a refractory brick mold. Figure 2a–c highlights different mold designs for joining plates and pipes in both horizontal and vertical positions.

4 Characterization Techniques

4.1 Microstructural Properties

Figure 3 shows, EDX (energy-dispersive X-ray spectroscopy) is used with SEM (scanning electron microscope) to analyze specimen composition. SEM scans surfaces with an electron beam to generate images for microstructural analysis. EDX

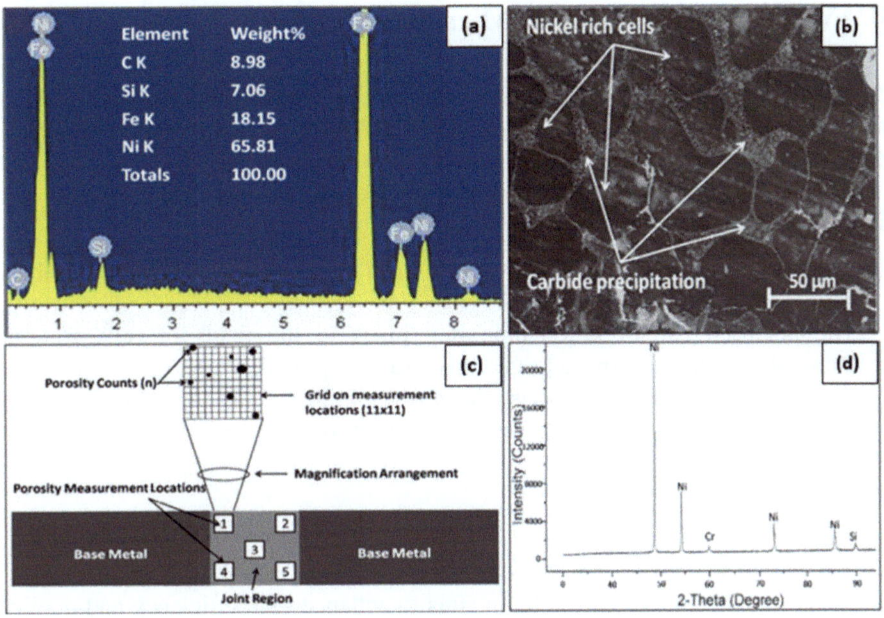

Fig. 3 Characterization results **a** EDX image **b** SEM image **c** porosity measurement **d** XRD Graph [11]

detects emitted X-rays, revealing elemental composition and chemical state. Porosity measurement quantifies voids in materials, essential for assessing material properties. XRD (X-ray diffraction) charts display crystal diffraction patterns, identifying atomic distances and crystal type for material analysis. These advanced techniques provide valuable insights into material composition, structure, and properties, facilitating scientific research, engineering, and industrial applications.

4.2 Mechanical Properties

Tensile testing evaluates weld joint strength, ductility, and toughness. Hardness testing measures weld joint hardness and assesses microstructure. Fatigue testing examines fatigue behavior and crack propagation in the weld. Non-destructive testing (NDT) methods, such as ultrasonic inspection, radiographic evaluation, and magnetic particle analysis, are used for non-destructive assessment of welds. These techniques play vital roles in ensuring weld quality, structural integrity, and safety in various industries. They provide essential data to validate weld performance and identify potential defects without causing damage to the welded joints. This comprehensive approach aids in maintaining high standards of construction, manufacturing, and engineering projects while preventing potential failures.

5 Future Opportunities in Microwave Joining

The existing literature highlights extensive research on microwave joining of various metals, alloys, ceramics, and natural fibers for different applications. Opportunities remain to explore lesser studied materials and assess the technique's potential for joining larger metal components, potentially impacting industries like construction and manufacturing.

1. Little research has been done on the use of MHH to join metal/non-metal pipes that are similar/different.
2. The potential use of aluminum alloy pipes in the fields of refrigeration, automobiles, and aerospace has not been fully explored.
3. Real-time temperature monitoring during the joining process can help to optimize the joint parameters and improve the quality of joint.

6 Conclusions

Recent progress in microwave joining aims to enhance joint efficiency and quality across various materials. Key advancements in MHH include: improved modeling, novel fillers, precise parameter control, expanded applications, non-conventional materials exploration, and hybrid joining.

The key findings from this literature review are as follows:

- Exposure time for microwave joining depends on the material, susceptor type, and interface powder size.
- For 40 μm powder size, the weld bond formed is strong and homogeneous.
- The exposure time for different susceptor materials varies between the range for charcoal 300–600 s (MS, SS, ceramics), for silicon carbide up to 600 s (Aluminum), and for graphite sheet up to 1250 s (Inconel)

References

1. Roy R, Agrawal D, Cheng J, Gedevanishvili S. Full sintering of powdered-metal bodies in a microwave field. Nature 399:668–670
2. Bansal A, Sharma AK, Kumar P (2012) Application of electromagnetic energy for joining Inconel 718 Plates, i-manager's. J Mech Eng 2(4)
3. Bansal A, Ansari S, Sharma S (2017) 3D electromagnetic field simulation of microwave joining of Inconel-718". Int J Mech Eng Technol (IJMET) 8(7):1231–1237
4. Badiger RI, Narendranath S, Srinath MS (2015) Joining of Inconel-625 alloy through microwave hybrid heating and its characterization. J Manuf Process 18:117–123
5. Badiger RI, Narendranath S, Srinath MS (2018) Microstructure and mechanical properties of Inconel-625 welded joint developed through microwave hybrid heating. Proc Inst Mech Eng Part B J Eng Manuf 232(14):2462–2477

6. Bansal A, Sharma AK, Kumar P, Das S (2015) Structure property correlations in microwave joining of Inconel 718. J Minerals Metals Mater Soc (TMS) JOM 67(9):2087–2098
7. Bansal A et al (2017) Galvanic corrosion behavior of microwave welded and post-weld heat-treated Inconel-718 joints. JMEPEG 26:2322–2330
8. Badiger RI, Narendranath S, Srinath MS (2019) Optimization of process parameters by Taguchi grey relational analysis in joining Inconel-625 through microwave hybrid heating. Metallogr Microstruct Anal 8(1):92–108
9. Bansal A, Sharma AK, Kumar P, Das S (2012) Joining of mild steel plates using microwave energy. Adv Mater Res 585:465–469
10. Sharma AK, Gupta D (2012) On microstructure and flexural strength of metal ceramic composite cladding developed through microwave heating. Appl Surf Sci 258:5583–5592
11. Singh S, Singh P, Gupta D, Jain V, Kumar R (2019) Development and characterization of microwave processed cast iron joint. Eng Sci Technol Int J 22(2):569–577
12. Bansal A, Sharma AK, Das S, Kumar P (2016) On microstructure and strength properties of microwave welded Inconel 718/ stainless steel (SS-316L). Proc Inst Mech Eng Part L J Mater Des Appl 230(5):939–948
13. Bagha L, Sehgal S, Thakur A, Kumar H (2019) Low cost joining of SS304-SS304 through microwave hybrid heating without filler-powder. Eng Res Express 1(2):025035
14. Tamang S (2017) An investigation on joining of Al6061-T6 to AZ31B by microwave hybrid heating using active braze alloy as an interlayer. J Manuf Process 28:94–100
15. Tamang S, Aravindan S. Joining of dissimilar metals by microwave hybrid heating: 3D numerical simulation and experiment. Int J Therm Sci
16. Kumar A, Sehgal S, Singh S. Joining of SS304-SS316 through novel microwave hybrid heating technique without filler material. Mater Today Proc
17. Kumar S, Sehgal S, Singh S. Investigations on material characterization of joints produced using microwave hybrid heating. Mater Today Proc
18. Sharma A, Sehgal S, Goyal D. Effects of process parameters in joining of Inconel-625 alloy through microwave hybrid heating. Mater Today Proc
19. Kumar V, Sehgal S. Joining of duplex stainless steel through selective microwave hybrid heating technique without using filler material. Mater Today Proc
20. Srinath MS, Sharma AK, Kumar P (2011) A novel route for joining of austenitic stainless steel (SS-316) using microwave energy. Proc Inst Mech Eng Part B J Eng Manuf 1–11
21. Singh I et al (2011) Feasibility study on microwave joining of green composites, vol 1
22. Bajpai PK et al (2012) Joining of natural fiber reinforced composites using microwave energy: experimental and finite element study. Mater Des 35:596–602
23. Bansal A, Sharma AK, Das S (2013) Metallurgical and mechanical characterization of mild steel-mild steel joint formed by microwave hybrid heating process. Sadhana Acad Proc Eng Sci 38(4):679–686
24. Gupta P, Kumar S (2014) Investigation of stainless-steel joint fabricated through microwave energy. Mater Manuf Process 29:910–915
25. Gamit DN (2020) Size-effect in microwave processing of engineering materials—a review. J Mech Eng Sci (JMES) 14(2):6770–6788
26. Gamit DN, Chudasama MK (2020) Challenges in joining of metallic pipes using microwave hybrid heating. IOP Conf Ser Mater Sci Eng 1–8
27. Gamit D, Mishra RR, Sharma AK (2017) Joining of mild steel pipes using microwave hybrid heating at 2.45 GHz and joint characterization. J Manuf Process 27:158–168
28. Gamit DN (2017) Effect of susceptor heating on joining of metallic pipes using microwave energy. Int J Emerg Technol Eng Res 5:49–52
29. Saxena S (2014) Joining of bulk metallic pipes by microwave hybrid heating process under parametrical regulations. IOSR J Mech Civ Eng 116:62–69
30. Gamit DN. Modeling and simulation of microwave metallic pipe joining process. Int J Res Analy Rev 5(3)
31. Srinath MS, Sharma AK (2011) Investigation on microstructural and mechanical properties of microwave processed dissimilar joints. J Manuf Process 13(2):141–146

Analysis of Various Electrical Tests Performed on Contact Tips for Molded Case Circuit Breaker (MCCB)

Mukesh Kumar Nag, Abhishek Shrivastava, and Binod Kumar Singh

1 Introduction

Electrical circuit breakers are vital switching devices for protection and regulation in electrical circuits. They are reusable, not disposable like fuses [1, 2]. The study of contact materials in MCCBs presents promising research due to their direct exposure to arcing and wear.

Electrical contacts facilitate current flow when closed and insulation when open [3–5]. Ideal materials exhibit high conductivity, wear and corrosion resistance, and environmental friendliness [6].

Resistance brazing offers quick and targeted heating as long as electrodes provide the pressure needed to create electrical contact, according to Timo Mutzel [7]. In this investigation, the movable contacts are Ag–WC (60% silver, 40% tungsten carbide), and the base is ETP copper. Resistance brazes the Ag–WC shim to the ETP copper base. At 650 °C, a brazing filler of 80% Cu, 15% Ag, and 5% P is used. The brazing shim melts lower with phosphorus. Silver–tungsten carbide (Ag–WC) contacts are used for stationary contacts in circuit breakers because they resist arcing erosion [1, 8]. For the stationary connections, an Ag–C shim is furnace brazed onto the copper base.

Mechanical wear benefits processes with minimal material loss. Moveable contacts (Ag–WC) feature tungsten carbide with specific properties, while fixed contacts (Ag–C) use softer silver–graphite [9]. Certification is crucial for dependability, with failure values differing from rated values due to safety measures [10]. A 6000-cycle endurance test requires a lower claimed number due to preventive measures.

M. K. Nag (✉) · B. K. Singh
National Institute of Technology Jamshedpur, Jharkhand 831014, India
e-mail: nag0859@gmail.com

A. Shrivastava
VIT Bhopal University, Bhopal-Indore Highway, Sehore, Madhya Pradesh 466114, India

R. Sahu et al. (eds.), *Advancements in Materials Processing Technology, Volume 1*,
Springer Proceedings in Materials 48, https://doi.org/10.1007/978-981-97-4958-4_17

Contact tips endure arcing and impact loads. Stationary contacts (Ag–WC) are resistance brazed with Cu–Ag–P filler [11]. Silver–graphite contacts (Ag–C) are furnace brazed with 95–5 wt. percent ratio onto copper using copper–phosphorus filler due to cost considerations. Adding silver or tin lowers the copper–phosphorus filler's high melting point. A ternary phase diagram by P. L. Zhang and H. Yayu Yan shows the brazing temperature range around 650 °C [12, 13].

This paper focuses on the metallurgical changes that occur after a specific number of operating cycles. It also includes a comparison between the use of Ag–W (silver–tungsten) in movable contact tips and Ag–C (silver–graphite) in stationary contact tips. The investigation of contact tip failures is conducted based on microstructural and mechanical changes resulting from various electrical tests, including electrical endurance, mechanical endurance, and short-circuit tests. The primary objective of this study is to examine the metallurgical changes in electrical contacts following these electrical tests, while also establishing a relationship between the electrical input and contact tip wear.

2 Materials and Electrical Testing Methods

2.1 *Electrical Contacts*

Circuit breakers use fixed and movable contacts, engaging under pressure when closed. Trip coils activate on faults, opening the circuit and separating contacts, causing an arc. These contacts endure harsh conditions, needing insulation in various electrical equipment. The systematic diagram of molded case circuit breaker with exposed contacts is shown in Fig. 1.

2.2 *Electrical Tests*

Electrical components must be certified for reliability. Ratings and failure values are crucial distinctions. The study covers contact tip short-circuit, mechanical, and electrical endurance testing.

2.2.1 Short-Circuit Test

Short-circuit tests verify circuit breaker's current-limiting capacity. Each tested breaker undergoes "O" and "CO" activities, creating three arcs and energy (I^2t). "I" is circuit current flow, and "t" is total current flow time ($t = 0$ to $t =$ extinguishment, 24–25 ms). High short-circuit currents measured: 42, 65, 80 kA.

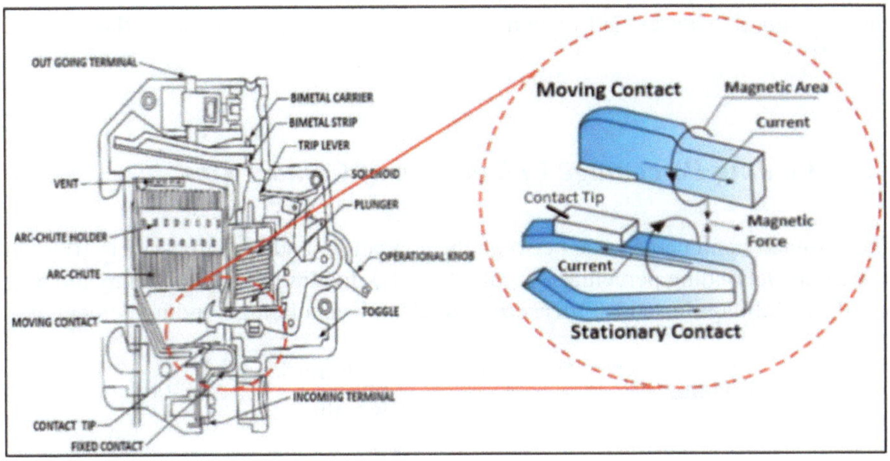

Fig. 1 Molded case circuit breaker with exposed contacts [1]

2.2.2 Mechanical Endurance Test

Mechanical endurance tests compress contact tips and copper bases, subjecting the stationary contact to over 15,000 cycles without current flow. Repetitive impacts cold work the copper, causing wear and evaluating the durability of contact tips, brazed assemblies, and overall circuit breaker components.

2.2.3 Electrical Endurance Test

The electrical endurance test evaluates the prolonged performance of devices like circuit breakers or contactors by subjecting them to repeated opening and closing of contacts under load conditions. The assembly's reliability is verified by repeated impacts between the movable and stationary contacts. 1000 amps and 415 V are used in this project.

3 Methodology

3.1 Sample Preparation

Microstructural investigation (metallography) examines the chemical and atomic structure of metallic alloys, using optical, scanning electron, and transmission electron microscopy. Samples undergo sequential surface finishing, mounting, grinding, and polishing with specific parameters and diamond disks. The goal is to reveal microstructural features and examine brazing bonding percentage in the samples [14–17].

3.2 Bonding Percentage Calculation of Contact Tips

The assumed bond percentage between the contact tip and base material is 100% of its length. Debonding percentage is calculated by dividing the tip length by the pore length, with ImageJ used for measuring pores and calculating bonding percentage.

$$\text{Bonding (\%)} = \frac{\text{Sample length} - \sum \text{Pores length}}{\text{Sample length}} \times 100$$

3.3 Microhardness Testing

ASTM E384 uses Vickers hardness number (HV) with parameters: 500 g-force, 10-s dwell. Contact tips are hardness tested from below the brazed surface to the copper base. Testing sequence was chosen to avoid surface microstructure changes. Figure 2 shows the first indentation near the braze line and subsequent indentations 100 μm downwards with a 100 μm horizontal shift to prevent overlapping.

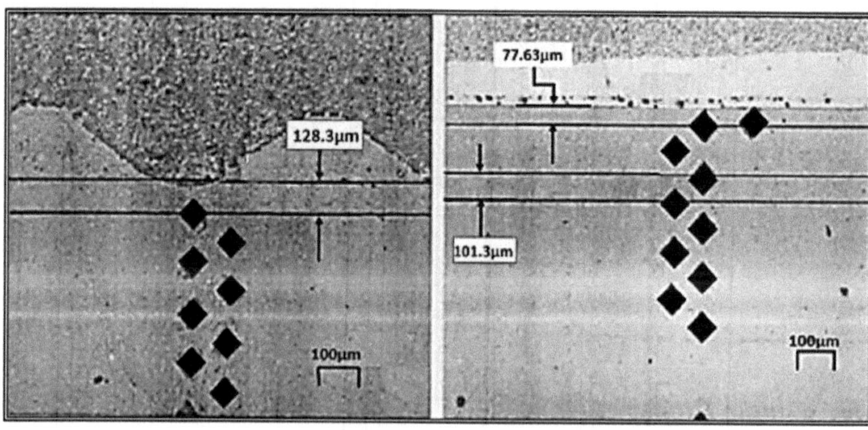

Fig. 2 Indentation marks on the copper base below braze interface (at 50X)

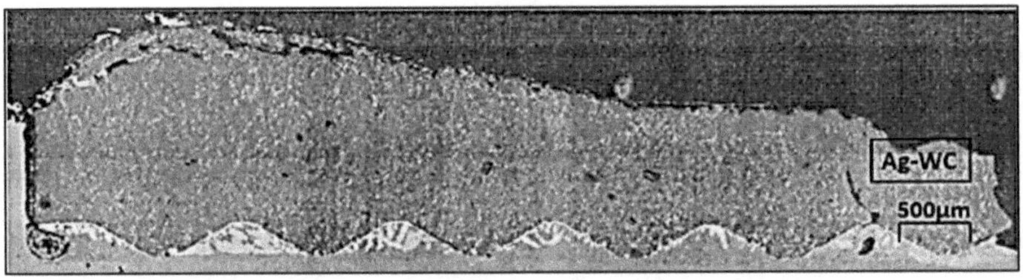

Fig. 3 Unetched micrograph of movable contact tip after short-circuit test (at 50X)

4 Result and Discussion

4.1 *Metallurgical Changes (After Short-Circuit Test)*

Electrical tests often cause wear, tear, and material loss, depending on working conditions. These changes can affect contact tip appearance, electrical resistance, and lifespan. Short-circuit, mechanical, and electrical endurance tests are performed on contact tips to investigate these effects.

4.1.1 Movable Contact Tip

After a short-circuit test, Fig. 3 shows a 50× unetched micrograph of the moveable contact tip. The micrographs show fine fractures and surface degradation from a short-circuit test. Disconnection of the contact tip from the copper base has also occurred. In Fig. 4b, fissures propagate in WC-rich locations. Cracks result from thermal strains in the brazed and direct arcing contact [18]. Figure 4 shows that stationary contact deposits resolidified copper during short-circuit operations (A). The operation raised temperature significantly.

4.1.2 Stationary Contact Tip

Figure 5 illustrates significant material loss, surface wear, and burning. Its softer nature caused more damage to the stationary contact tip [19]. Three OCOCO rounds at 65 kA destroyed contact tip and copper. Stationary contact bonding % calculation is useless since the tip erodes entirely. An electric arc from each circuit-breaking shot disappears in microseconds at the arc extinguisher. Arc-extinguishing energy transmission damages contact tips more.

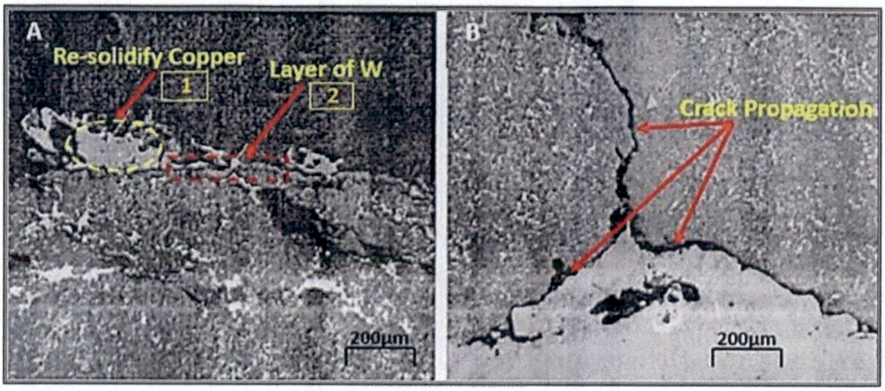

Fig. 4 Micrograph of movable contact tip after short-circuit test: **a** resolidified copper, **b** crack propagation in WC-rich area (at 200X)

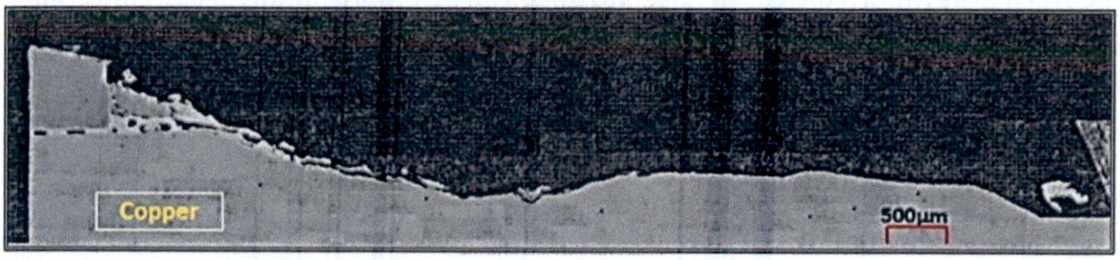

Fig. 5 Unetched microscopic view of stationary contact tip after short-circuit test (at 50X)

4.1.3 Area Measurement After Short-Circuit Test

After the short-circuit test, a digital video microscope with Olympus software measures contact tip area. The design specifies a 37.93 mm^2 fresh contact tip with a coating clearance of $\pm$ 100 μ. Figure 6a shows the calculated fresh contact tip area after the short-circuit test, which is 37.82 mm^2, close to the design specification. Figure 6b shows the remaining contact tip area as 29.79 mm^2. Subtracting this value from the contact tip area in the drawing yields 8.14 mm^2.

4.2 Metallurgical Changes (After Mechanical Endurance Test)

4.2.1 Movable Contact Tip

In Fig. 7, the cracks originate from the brazing interface and propagate into the copper base due to repetitive impact loads during mechanical endurance testing. Cold working reinforces the copper base from impact loading. Copper hardness

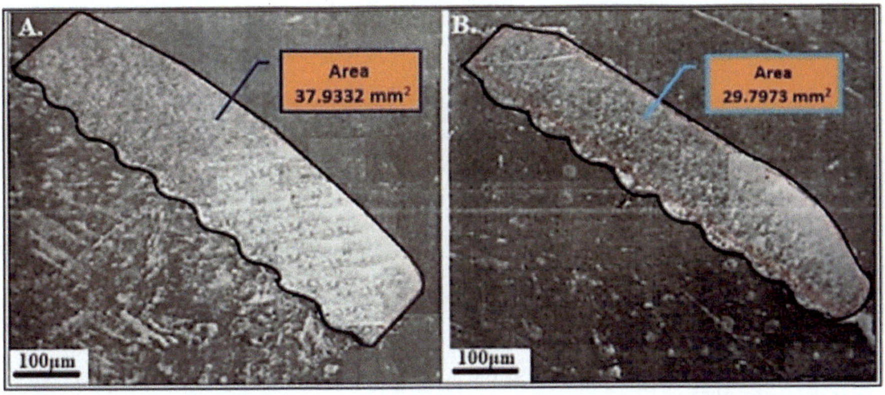

Fig. 6 Image of contact tip **a** before and **b** after short-circuit test measure (at 50X)

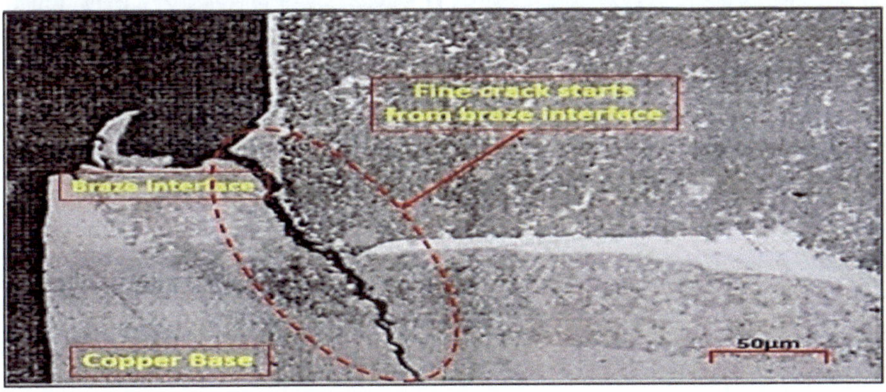

Fig. 7 Crack initiation at brazing interface and propagation into Cu base (at 200X)

increases with cold working, and the transition distance from the brazing interface becomes more apparent on the hardness graphs, which are discussed and presented in further Fig. 11a, b.

4.2.2 Stationary Contact Tip

Mechanical endurance tests showed minimal material loss after 20,000 operations. The movable contact was unaffected by testing. The erosion of the Ag–C surface, being softer than Ag–WC, was expected. The development of cracks in Ag–WC is clearly visible in Fig. 8a, b. Additionally, a few copper dust particles were observed to be entrapped between the Ag–C layers.

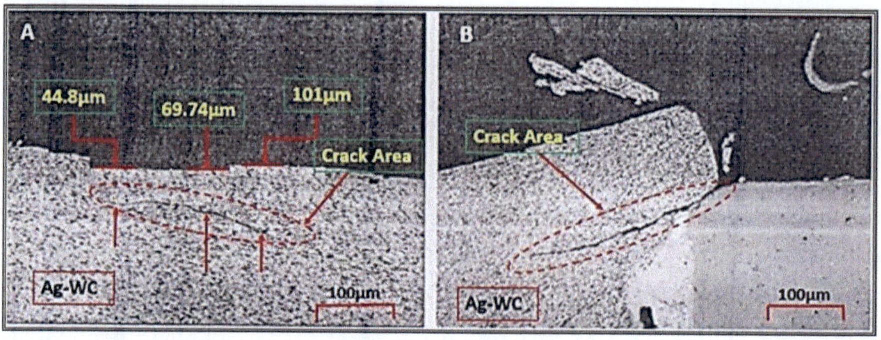

Fig. 8 Stationary contact tip **a** crack observed near the top surface of the tip, **b** crack observed on side of the tip (at 200X)

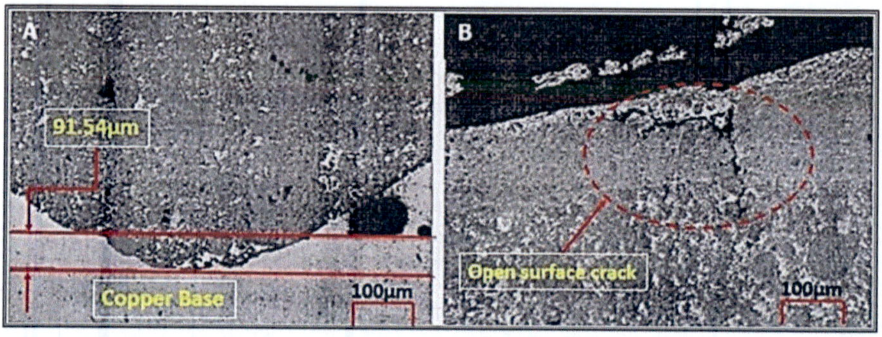

Fig. 9 Movable contact tips (at 200X): **a** contact tip embedding into the copper base due to softening of copper, **b** open to surface cracks on the contact tip surfaces

4.3 Metallurgical Changes (After Electrical Endurance Test)

4.3.1 Movable Contact Tip

Upon further inspection of the micrograph at higher magnification, the following observations can be made, as depicted in Fig. 9.

4.3.2 Stationary Contact Tip

The stationary contact tip deposits a layer of material onto the movable contact tip, increasing contact resistance. Compared to their untested counterparts, the porous silver–copper–tungsten layer and the layer beneath it deform. Electrical endurance tests are mechanical endurance tests with a current, generating more heat. Heat softens copper, and repeated impacts further alter its properties. The silver–graphite

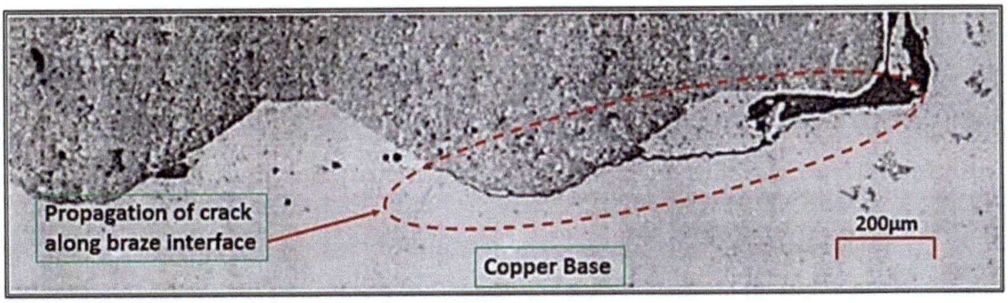

Fig. 10 Propagation of crack along the braze interface line (at 100X)

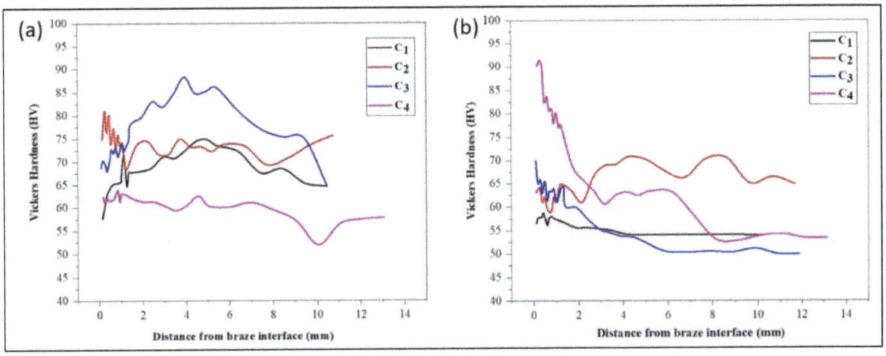

Fig. 11 Hardness profile after different tests **a** movable contacts, **b** stationary contacts

mixture covers surface cracks after the electrical endurance test. The initial cycles form these cracks, which are covered by Ag–C as the temperature rises. Crack initiation at the brazing interface, shown in Fig. 10, is the same as in Fig. 7. The main difference is crack propagation. Mechanical endurance tests cause the crack to propagate into the copper base due to repetitive impact load.

4.4 Mechanical Changes on Copper Base

A hardness or microhardness test is the only mechanical test conducted on selected samples during failure analysis. It is hard to get UTM or impact testing samples.

4.5 Movable Contact Tip

Before brazing, the movable contact tips' ETP copper is 100 HV. Figure 11a shows that resistance brazing lowers hardness to 60–75 HV (C_1—hardness on a copper base related to distance from the brazed interface). The brazing electrode's direct heating zone lowers hardness near and far from the brazing interface. Copper's mechanical properties change after electrical tests, as shown in the hardness profiles (C_2—short-circuit test), (C_3—mechanical endurance test), and (C_4—hardness profile after electrical endurance test). This gives us a rough estimate of copper base cold work. Since only three shots were taken, the temperature rise during the short-circuit test did not have enough time to soften the copper. The hardness graph after the mechanical endurance test is slightly higher than the untested samples. Copper uniformly softens after the electrical endurance test. In the mechanical endurance test, cold work is more prevalent than in other tests since an impact load repetition of 20,000 increases the hardness range from 60–75 to 70–88 HV after 20,000 repetitions. However, the short-circuit test does not change the hardness, and the electrical endurance test softens it.

4.5.1 Stationary Contact Tip

Pre-brazing, stationary contact tip ETP copper is 100 HV, and annealing in furnace brazing reduces hardness to 55 HV. The brazing interface softens due to filler material melting. Figure 11b shows electrical, mechanical, and short-circuit test results. After short-circuiting, little hardness change is observed, but the region near the next brazing interface deforms and becomes harder. Mechanical endurance test hardness is lower than electrical due to sufficient contact tip material, reducing impact load. In electrical tests, complete material loss occurs, leading to greater deformation. Hardness profiles consider copper grain size below the brazed interface, and unetched micrographs do not show grain boundaries/grains.

5 Conclusion and Recommendation

Micrographs shows the material loss in movable contact after three tests. Highest loss occurred in short-circuit and least in mechanical endurance. Microhardness graphs reveal cold working in copper base: least in short-circuit, most in electrical endurance. Severe conditions in electrical and short-circuit tests include burns, wear, tear, and surface degradation. Comparison is based on 28,000 operations, 1000A AC (55 Hz), and 500 V. By predicting failure analysis results for MCCB contact tips, real service conditions may vary due to factors like short circuits, temperature fluctuations, wear, or contact tip burning.

By studying the failed samples and correlating them with the result data, one can determine the most probable cause of failure. This information can then be used to take appropriate actions and prevent similar incidents in future. While it may not be possible to completely eliminate changes that occur after tests, it is possible to prolong the life span of electrical contacts by implementing proper measures and precautions.

(a) Modify alloying elements for enhanced wear resistance.
(b) Employ short peening to extend life span.
(c) Utilize waffle-shaped design to prevent sliding.
(d) Incorporate aluminum/copper strips with a braze shim for additional support.
(e) Adopt ultrasonic brazing techniques to address insufficient brazing.
(f) Explore grain boundaries through etching for conductivity changes with deformation in contact tip applications.

References

1. Ravindranath B, Chander M (1977) Power system protection and switchgear. New Age International
2. Nag MK, Kumar P (2023) Optimization of cost and performance of the material used in the mechanism unit of air circuit breaker (ACB) based on various analysis approach. In: Proceedings of the institution of mechanical engineers (Part B: Journal of Engineering Manufacture), p 09544054231168682
3. Slade PG (2017) Electrical contacts: principles and applications. CRC Press
4. Nag MK (2023) Lifetime estimation and surface degradation of MCCB contact tips due to multiple electrical interruptions. In: Proceedings of the institution of mechanical engineers (Part C: Journal of Mechanical Engineering Science)
5. Biyik S, Aydin M (2016) Investigation of the effect of different current loads on the arc-erosion performance of electrical contacts. Acta Physica Polonica A 129(4)
6. Nag MK, Kumar P, Paswan MK (2021) Environmental impacts from the system of solar energy. In: Recent advances in power systems: select proceedings of EPREC 2020. Springer, Singapore, pp 453–465
7. Mützel T, Kempf B (2014) Silver tungsten carbide contacts fork circuit breaker applications. In: 2014 IEEE 60th Holm conference on electrical contacts (Holm). IEEE, pp 1–7
8. Kesim MT, Yu H, Sun Y, Aindow M, Alpay SP (2018) Corrosion, oxidation, erosion and performance of Ag/W-based circuit breaker contacts: a review. Corros Sci 135:12–34
9. Wojcik-Grzybek D, Frydman K, Borkowski P (2013) The influence of the microstructure on the switching properties of Ag C, Ag-WC-C and Ag-WC contact materials. Arch Metall Mater 58(4):1059–1065
10. Nag MK, Kumar P (2023) Synthesis and characterization of high-performance blended alkali-activated geopolymer (FA/GBFS) from industrial wastes. Iran J Sci Technol Trans Civil Eng 1–21
11. Zhang PL, Yao S, Ding M, Lu FG, Lou SN (2010) Microstructural analysis in the vacuum brazing of copper to copper using a phosphor–copper brazing filler metal. Int J Mater Res 101(11):1436–1440
12. Yu H, Sun Y, Alpay SP, Aindow M (2015) Microstructure effects in braze joints formed between Ag/W electrical contacts and Sn-coated Cu using Cu–Ag–P filler metal. J Mater Sci 50:324–333
13. https://www.georgevandervoort.com/metallography/specific/20001319-microstructureofcopperandcopper-alloys.html
14. Nag MK (2024) Experimental studies on micro alloyed steel used in oil and gas industries. J. Mech. Sci. Technol. 38(5): 2465–2473
15. Nag MK, Shrivastava A, Solanki K (2022) Active cooling system incorporated in a portable laptop cooling pad. Int J Therm Energy Appl 8(2):1–8
16. Nag MK, Kumar P (2023) Fabrication and characterization of woven and comingled nonwoven sheet polypropylene hybrid composite by recycling and alkali-treated jute waste fibers. In: Proceedings of the institution of mechanical engineers (Part C: Journal of Mechanical Engineering Science)

17. Nag MK, Kumar P (2023) Fabrication and characterization of laminated natural fibers and SS303 wire mesh reinforced epoxy-based hybrid composite. In: Proceedings of the institution of mechanical engineers (Part L: Journal of Materials: Design and Applications)
18. Ray N, Kempf B, Mützel T, Froyen L, Vanmeensel K, Vleugels J (2015) Effect of WC particle size and Ag volume fraction on electrical contact resistance and thermal conductivity of Ag–WC contact materials. Mater Des 85:412–422
19. Nag MK (2023) Significance of concurrent engineering methods used in automotive industries. Int. J. Sci. Res. Eng. Manag. (IJSREM) 7(3):1–4

Optimizing Waste Heat Recovery in the Post-combustion Chamber of a Sponge Iron Plant

N. Kamal, P. Kekarjawelkar, S. Sar, B. Deo, P. Nanda, P. Malakar, S. K. Sahoo, and R. Manchada

1 Introduction

Post-combustion chamber (PCC) of a sponge iron rotary kiln plant at TSLPL, Joda, is associated with a rotary kiln with a constant RPM [1]. Coal, iron ore, and dolomite are used as raw materials. The iron ore is reduced to sponge iron inside the rotary kiln. The coal enters from both ends of the rotary kiln; one end is known as the feed side, and the other is the injection side. The iron ore, dolomite, and coal (nearly 50% of the total coal required) enter from the feed end of the rotary kiln, and coal (balance), in some cases raw dolomite also, enters from the injection end. The PCC is attached to the feed side of the rotary kiln. The gases generated inside the kiln flow from the injection side to the feed side and enter the PCC at a temperature of around 1023–1223 K. It is essential to combust the CO gas and any other carried-over coal with the exit waste gases to recover energy. Carried-over coal fines (in exit gas from the kiln) are mostly less than 3 mm in size.

The air blowers attached to the PCC inject optimum air to combust carbon fines (C) and CO. The combustion of CO and C results in the release of energy which increases the temperature of the gases inside the PCC. The thermocouple named PCC T-11 measures the temperature of these gases far away from the exit of PCC. The location where this temperature is measured is known as the 'elbow duct.' From here on, the gases move toward the boiler where the heat in waste gases is used for steam generation purposes, to be then used in a steam turbine to generate electricity. An increase in the amount of CO and unburnt coal entering PCC, in general, increases the temperature of output waste gas. The upper limit of the PCC outlet temperature

N. Kamal (✉) · P. Kekarjawelkar · S. Sar · B. Deo
School of Minerals, Metallurgical and Materials Engineering, Indian Institute of Technology Bhubaneswar, Bhubaneswar, Odisha, India
e-mail: 20mm01017@iitbbs.ac.in

P. Nanda · P. Malakar · S. K. Sahoo · R. Manchada
Tata Steel Long Products Limited (TSLPL), Joda, Odisha, India

R. Sahu et al. (eds.), *Advancements in Materials Processing Technology, Volume 1*, Springer Proceedings in Materials 48, https://doi.org/10.1007/978-981-97-4958-4_18

is maintained below 1050 °C due to the material property of boiler tubes and elbow duct. When the temperature increases beyond a set value of outlet temperature, water is injected into the PCC by nozzles which are regulated by control valves tuned by PID controllers for sudden cooling. The PCC outlet temperature is thus affected by CO at the kiln inlet, the mass of fines (less than 3 mm size particles) in feed coal, air injected in the PCC, PCC inlet gas temperature, and water injected into the PCC. Owing to fluctuations in the affecting parameters, the PCC outlet temperature is not constant and it varies in the window of 920–1020 °C. If the output temperature goes out of control (i.e., greater than 1050 °C) for a period of 20 s or more, then the emergency cap on the top portion of PCC opens up and prevents any unwanted effects by letting the gas pass into the atmosphere.

The present work focuses on the development of optimal control programs through energy balance, mass transfer, and heat transfer calculations for optimal energy generation. Optimal values of air and water to be added in PCC are found by using the genetic algorithm. These calculations will be described along with the predicted values of air and water to be added to PCC.

2 Method of Analysis

2.1 Energy Balance

The principles of energy balance are described in [2, 3]. The procedure of calculations is as follows: In the PCC system, energy is fundamentally associated with the temperature of the components (gases, and solids in the flue gas) and the energy associated with the chemical reactions like combustion of C and CO. The base temperature for performing required calculations is considered to be room temperature (i.e., 25 °C).

As the components entering from the kiln to PCC are at high temperature, these are bought down from the kiln inlet temperature to reference temperature, and enthalpy changes corresponding to this temperature change are calculated for all components. As all the components in the PCC attain the same final temperature, the combustion reactions of C and CO are carried out and the enthalpy changes associated with these are also calculated. After these calculations, the sum of the enthalpies in the above two processes is utilized to raise the temperature of the system components from room temperature to the final elevated temperature. Now this elevated temperature is maintained in an optimum range (window) by adding a calculated amount of water into the system, which is also estimated by energy balance. The amount of CO and unburnt fixed carbon (FC) in coal entering PCC to be combusted is first calculated using the concepts of volume balance and mass balance described in the next sections.

2.2 Volume Balance: (Estimation of CO)

For the estimation of CO (m^3 per hour), the reading of a multi-gas analyzer is used. Firstly, the volume of CO entering the PCC is some percentage (from the multi-gas analyzer) of the flue gas entering the PCC ($V\text{in}$). Hence, calculation of $V\text{in}$ is done first in order to calculate the amount of CO combusted in PCC.

Considering the PCC system, Vout represents the volume of the flue gases exiting PCC. PAA represents the amount of air added into the system and contributes to the combustion of CO and unburnt carbon in the coal entered in PCC. On combustion of this CO and carbon in unburnt coal, CO2 is produced. This leads to the consumption of CO entering the PCC. For the combustion reaction to take place, the oxygen present in PCC air is also consumed in the system. In any case, the water guns present in the PCC will generally be pumping some minimum water into the system with a 5–10% opening of the valve so that the thermal damage of the material at the nozzle of the pipe is reduced (the hot gases are moving around). PWA represents this quantity. If the guns are taken out of PCC, then this quantity can be assigned with the value of zero. All the above-mentioned parameters are in the unit of Normal m^3 per hour (Nm^3/hr).

$$V\text{out} = V\text{in} + \text{PAA} + \text{CO}_2\text{ produced} + \text{PWA} - \text{CO consumed} - \text{O}_2\text{ consumed} \tag{1}$$

2.3 Mass Balance: (Estimation of Input FC to PCC)

Mass balance is done to find out the amount of fixed carbon (FC) being combusted in the PCC. In order to calculate this, the material balance of FC of the entire plant is considered. The input FC to the plant is calculated from coal entering from both the injection and feed ends [4]. The FC amount present in the coal is input FC to the plant. The effect of moisture in input coal is also to be considered. The moisture in coal directly affects the flame temperature and the heat transfer mechanisms. This input FC is consumed at various stages of operation along the length of the kiln. The stages of consumption include the reduction of iron ore, increasing the temperature of the contents in the kiln (coal, dolomite, iron ore, gases, and moisture), and heat losses from the kiln. Input FC is also used up in the form of dust at electrostatic precipitator (ESP), wet scrapper, back spillage, de-dust, CD char, and FC losses in sponge iron produced (some carbon enters the sponge iron which is approximately around 0.2%).

From the calculations performed, the amount of FC input into the plant, its consumption, and wastage in the form of dust, heat losses, and in sponge iron are estimated. The difference between total input FC and the consumption and wastage of FCs gives the value of the amount of FC which is being combusted in PCC.

$$\text{PCC FC under combustion} = \text{Total input FC} - (\text{Consumed FC} + \text{Wasted FC}) \quad (2)$$

2.4 *Optimization by Using Simple Genetic Algorithm*

The simple genetic algorithm (SGA) is used as a tool to optimize the process parameters [5]. The choice of the correct combination of parameters in the genetic algorithm (mutation, population, crossover) reduces the possibility of local minimum trapping in solving the optimization problems and also helps in faster convergence to optimal solutions.

To enhance the accuracy of calculating the FC reaching the PCC within a specified window, an optimization approach is employed by introducing a reference value. This optimization process involves defining a bandwidth range for the variations in FC input, and consumption. Additionally, the reference values are determined based on the specific type of coal being utilized, which are obtained through data analysis of various coal types used in the sponge iron plant. The selected bandwidth for this analysis is set at 15%, while the reference value is set at 20% of the input feed side FC.

The calculated PCC FC under combustion obtained by using the mass balance analysis is discussed above, and the reference value should be comparable. In other words, the modulus difference between them should be minimal. The objective function of GA is as follows:

$$Y = \min|\text{PCC FC under combustion} - (0.2 * \text{Input FC1})| \quad (3)$$

By considering this, objective function GA returns a solution set of consumption, wastage, and input of FCs such that PCC FC under combustion is close enough to reference value.

3 Estimation of Air/Water Required to Be Added

The calculations of the amount of PCC FC under combustion and the amount of CO entering PCC are performed based on the above-mentioned methods. Combustion of this C and CO by the air entering the PCC system increases the temperature of the components present in the system and takes them to an elevated temperature. Following this, the components have to be bought down to the optimum temperature range, by the addition of water into the PCC. For this, the optimum temperature range has to be given as input to the model. Using this range, the model gives the amount of water to be added as the output by performing the energy balance calculations.

4 Results and Discussions

A 'dynamic graphical user interface (GUI)' has been developed to showcase the water quantity required for adding to the PCC at various time intervals. To illustrate this, Figs. 1, 2, 3 and 4 display the necessary inputs for the model. Additionally, the model requires supplementary inputs in the form of a text or Excel file, such as the composition and inlet temperature of the gases entering the PCC. The flow rate of the gases exiting the PCC is also provided as an input to enable volume balance calculations within the model. The resulting output of the model, which provides insights into the time-dependent water amount added, is presented in Fig. 5.

All input values for feed rates are expressed in units of metric tonnes per hour. The abbreviations MC, VM, and Ash correspond to the moisture content, volatile matter,

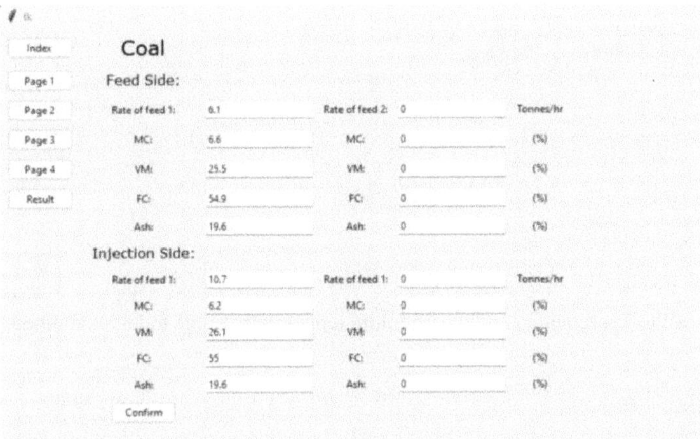

Fig. 1 Quantities and properties of coal input to the kiln from the feed side and injection ends

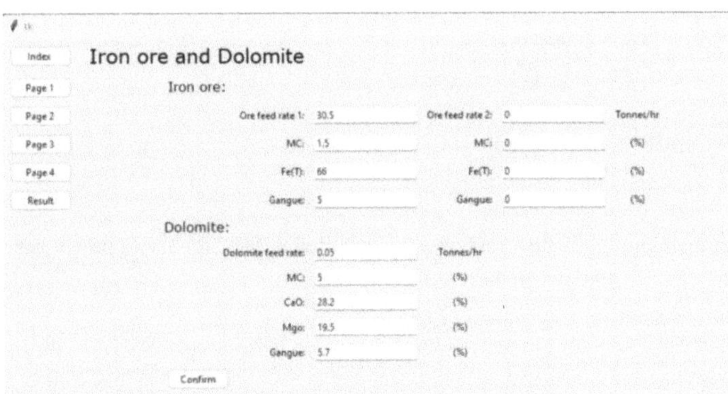

Fig. 2 Quantities of iron ore, dolomite, and their properties that are added to the kiln

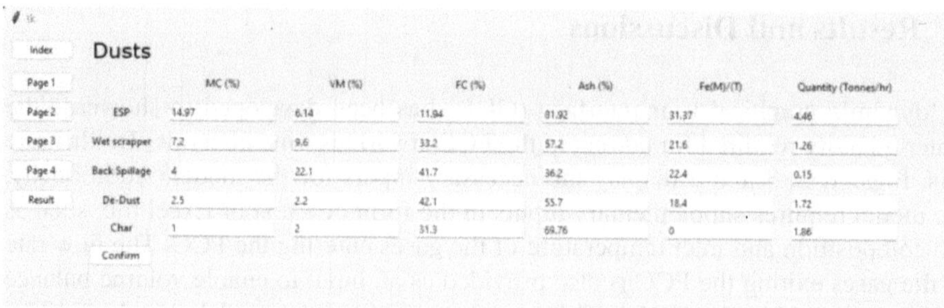

Fig. 3 Quantities and properties of dusts, wastages, and spillages of carbon which are used for performing mass balance for the entire plant

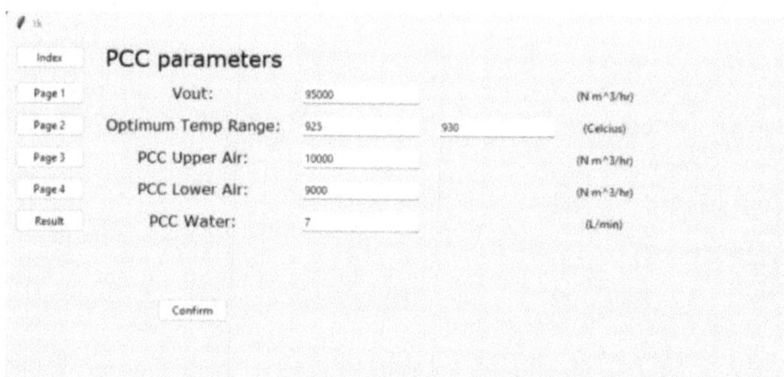

Fig. 4 Input of PCC parameters and the optimum temperature range to be maintained

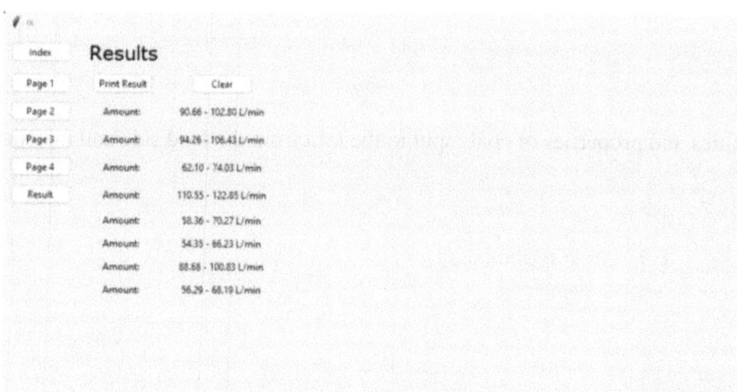

Fig. 5 Output mentioning the flow rate of water required in PCC for stable power generation

and ash content, respectively, of the material. FC stands for fixed carbon content of the coal. Fe(T) stands for the total iron content present in the input ore. As a final output, the model displays the flow rate (liter per minute) of water needed to be added to PCC for stable power generation. The model continues to dynamically display the output at short intervals as we provide the inputs required dynamically but at short intervals.

5 Conclusions

The present work focuses on optimizing waste heat recovery in the post-combustion chamber (PCC) of a coal-fired sponge iron plant. The study employs energy balance, mass transfer, and heat transfer calculations, along with a genetic algorithm for optimization. By considering the inputs and outputs, the model provides a good estimate of air and water quantities to maintain the optimum temperature range of gases fed to the boiler and hence the steam used in the turbine for controlled power generation.

Enhancements to the present work can be made by the integration of a more comprehensive energy and mass balance model for the rotary kiln by incorporating the variable moisture in raw materials, variable humidity, and temperature of input air. This will provide a more precise estimation of the volume of carbon monoxide entering the PCC from the kiln. The thickness of the deposits on PCC walls, and accretions inside the rotary kiln, influences the heat transfer process, thereby indirectly impacting the entire procedure. Additionally, the existing model primarily focuses on thermodynamics. The inclusion of chemical kinetics and dynamic heat transfer and fluid flow phenomena pertaining to the various reactions occurring within the system would further improve the accuracy of the calculations. In spite of the possibilities of further improvements, as mentioned, the findings of the present work are already a good tool to minimize fluctuations in power generation and the overall energy efficiency of the plant.

Acknowledgements This work is a result of a collaborative R&D project between Tata Steel Long Products Limited (TSLPL), Joda, India, and IIT Bhubaneswar, Odisha, India.

References

1. Deo B, Choudhary P, Bose S (2021) Control of post combustion to stabilize power generation in a sponge iron plant. In: 7th virtual international conference on science, technology, and management in energy 2021. Mathematical Institute of the Serbian Acad- emy of Sciences and Arts, Belgrade, Serbia, pp 195–201
2. Morris AE, Geiger G, Fine HA (2012) Handbook on material and energy balance calculations in material processing, 3rd edn. Wiley

3. Cameron I, Sukhram M, Lefebvre K, Davenport W (2020) Introduction to the blast furnace enthalpy balance, blast furnace ironmaking. Elsevier, pp 59–63
4. Cameron I, Sukhram M, Lefebvre K, Davenport W (2020) Introduction to the blast furnace mass balance, blast furnace ironmaking. Elsevier, pp 47–57
5. Popov A (2005) Genetic algorithms for optimization. User manual, Hamburg 2013

A Methodical Approach on the Fundamentals of High Entropy Materials

Premlata Rao, Krishna Kumar, and Upender Pandel

1 Introduction

For ages, the progress of human civilization has always been dependent on natural materials. The beginning of human life was with the Stone Age where stone, clay, wood, and skins were used until the Bronze Age at about 3000 BC, where the copper came into the picture and people become acquainted with the concept of alloying for making copper hard [1]. Alloying is defined as the incorporation of minor elements into the materials, to change their properties [2]. The concept of intermetallic also came from classical times, where mirrors of copper-tin alloys had been used. In the Iron Age at about 1200 BC, iron and steel were used for war purposes due to their increased strength and hardness. Wrought iron was discovered at about 1000 BC, and cast iron/steel was found one millennium thereafter. Steel was unintentionally found with iron-carbon alloy, because of the nonexistence of elemental carbon till that [1].

As the industrial and technological growth expanded in the middle of the eighteenth century, the rate of metal evolution also enlarged (Fig. 1). The researcher's eye has always been stuck on the scanning of new metallic materials having higher productivity since each passing year. For this reason, they started to investigate the chemical composition of alloys. Up to 1990, 30 alloys with one major element have been produced with distinct compositions and processing [3].

P. Rao (✉) · K. Kumar · U. Pandel
Department of Metallurgical and Materials Engineering, Malaviya National Institute of Technology, Jaipur 302017, India
e-mail: premlata10791@gmail.com

R. Sahu et al. (eds.), *Advancements in Materials Processing Technology, Volume 1*, Springer Proceedings in Materials 48, https://doi.org/10.1007/978-981-97-4958-4_19

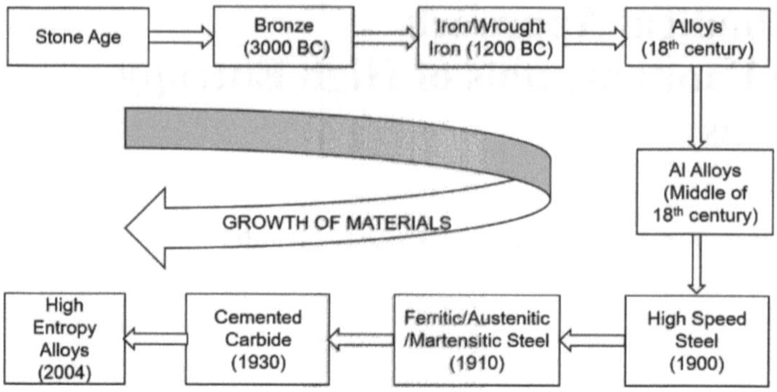

Fig. 1 Progress of materials [1, 3]

1.1 Beginning of High Entropy Alloys

All the above-discussed alloys are based on one major element. Alloys with one major element are beneficial for civil growth, but for the degree of freedom point, it still limits the progress of microstructure, properties, and applications [4, 5]. That is why material science and engineering has continued its conception toward this direction until the multicomponent system has been discovered. In the middle of eighteenth century, a German scientist and metallurgist named Franz Karl Achard talked about the multicomponent alloys with five to seven principal elements. He was the forerunner for the concept of high entropy alloys [6].

In 1981, Professor Cantor with his undergraduate student named Alain Vincent lightens up this new concept by preparing lots of alloy systems with equal atomic composition. Among all the alloys system, only one system $Fe_{20}Cr_{20}Ni_{20}Mn_{20}Co_{20}$ showed a single-phase FCC structure. Highlights of this research were as follows [7]:

(I) This FCC phase with dendritic structure is majorly observed in alloys produced by adding Cu, Nb, Ta, Mo, W, V, Ti elements into the basic alloy, i.e., $Fe_{20}Cr_{20}Mn_{20}Ni_{20}Co_{20}$.
(II) Lower stability of Cu and Ge in the FCC phase with dendritic structure is due to higher electronegativity. Because of this, these elements are dropped into the inter-dendritic region.
(III) Alloy of 20 elements each with 5 at.% and alloy of 16 elements each with 6.25 at.% are found to be multiphase, crystalline, and brittle both in casting and melting state.
(IV) However, alloys with single-phase FCC structure majorly have transition metal elements, which are Co, Cr, Ni, Fe, Mn.

On the parallel side, another research on multicomponent alloys was going on since 1995 by Yeh [8]. Yeh gave his theory that as the entropy of mixing increases, the number of phases in an alloy system decreases while the properties of alloys improved. For the effectiveness of this theory, he started working on it and made the following conclusions:

(I) In the as-cast condition, the dendritic structure was observed.
(II) In cast and annealed conditions, hardness values of all alloys were observed from 590 to 890 HV in which composition and fabrication condition were considered. It was also observed that an increased number of elements increases hardness but with the only exception that alloy with nine elements showed a little bit low hardness value. This can be resolved with the addition of a small quantity of boron. At the end of 2013, professor Yeh along with his co-workers directed 79 MS theses and 10 Ph.D. theses. The main area of focus was high solution hardening due to large lattice distortion and stronger bonding [9] which led to important counsel about the high entropy effect, lattice distortion effect, and sluggish diffusion effect. It was also concluded that due to the presence of passive elements and high entropy (low free energy), these alloys show better corrosion resistance by immersing in 0.01 and 1 M HCl, HNO_3, H_2SO_4, and HF acidic solutions for 24 h [10]. Ranganathan was also working parallelly to explore multicomponent alloys. He disclosed his first article in Current Science in November 2003, titled 'Alloyed pleasures—multi-metallic cocktails' [11] based on conveyance and exchanged views with Professor Yeh. High entropy alloys has become the topic of interest for many more researchers since the open articles published by the above four generators in this area.

2 Definition—High Entropy Alloys

2.1 Composition Based

According to Yeh et al. [8], high entropy alloy is composed of at least five major metal elements, each having an atomic percentage of 5–35%, with the addition of a small atomic percentage of elemental additives (<5%) to improve the properties. This definition is only about composition, and no information about entropy and solid solubility was provided.

2.2 Entropy Based

The name high entropy is all about configurational entropy (entropy value at equilibrium), which depends upon the location of the elements in a crystal. Entropy is greatly influenced by the atomic arrangement. Thermodynamically, entropy is the evaluation of the presence of microscopic states in the system [12]. Basis of entropy alloys is categorized into Table 1.

Table 1 Entropy values of different entropic alloys [12]

S. No.	Alloy	Entropy value
1	Zero entropy alloys	$S_{mix} = 0$
2	Low entropy alloys	$S_{mix} < 0.69R$
3	Medium entropy alloys	$0.69R < S_{mix} < 1.61R$
4	High entropy alloys	$S_{mix} > 1.61R$

Here, R is the gas constant ($R = 1.38 \times 10^{13}$ J/K). According to Boltzmann's arithmetical theory [13], the entropy of a system is given thermodynamically as Eq. 1:

$$\begin{aligned}\Delta \text{Sconf.} &= -nR\left(\frac{1}{n}\ln\frac{1}{n} + \frac{1}{n}\ln\frac{1}{n} + \cdots + \frac{1}{n}\ln\frac{1}{n}\right)\\ &= -R\ln\frac{1}{n}\\ &= R\ln n \end{aligned} \tag{1}$$

Here, n is the number of principal elements present in the system. This equation ensures that a higher value of entropy can be achieved at about equal composition of every element present. With further investigation, a new category of high entropy alloys with multiphase non-equal-atomic solid solution came into the picture [14]. Some points to remember about the first category of HEAs are as follows:

(I) Cantor et al. [7] predicted that temperature dependency of alloys is greatly influenced by nano-twins as these increase the instability time for necking at higher strain rates, also due to high work hardenability which leads to high yield strength.
(II) Senkov et al. [14] predicted the good resistance to high-temperature softening due to increased plasticity, resulting in consistency in yield strength at above 600 °C. The second category of HEAs has at least four principal elements with non-equal composition forming multiphase. These were invented to fulfill the limitations of previously invented HEAs. Some points to remember about second category of HEAs are as follows [14].
(III) Due to the instability of the high-temperature phase, interphase hardening takes place.
(IV) Due to mechanical instability of room temperature phases, transformation-induced hardening takes place due to hindrance of trans-grain and inter-grain slip for increment in strength value, while strain hardening improves ductility.

Other than composition and phases, some other features like kinetics, microstructure, and properties are also important. These features can be explained by the core effects of HEAs [11].

2.3 High Entropy Effect

This is the most significant and important concept for HEAs. It states that a large number of principal elements led to randomness in the system, also ensuring the formation of a solid solution over intermetallic compounds due to higher entropy value. According to Gibbs free energy rule (Eq. 2) [15], high entropy minimizes free energy of the system and promotes solid solution phases particularly at high temperature. More theories explain the high entropy effects deeply [16]. Here, we understand this effect with an XRD plot of binary to septenary alloys shown in Fig. 2.

$$\Delta G \downarrow = \Delta H - T\Delta S \uparrow \tag{2}$$

It is evident from the figure that phases are stable by adding one extra element to the prior one. In the case of alloy designated as (1), there are four major peaks; alloy (2) also has more than two major peaks. Similar behavior is shown by alloys (3) and (4). In the case of alloy (5), intensity of minor peaks started decreasing. Alloy (6) has two clear major peaks. Hence, it is proved that a large number of principle elements promotes a stable solid solution due to high mixing entropy. This effect strongly opposes thermodynamic due to the high possibility of compound formation at the center of the phase diagram. It also vibrates the Gibbs phase rule (Eq. 3) [15] which is as follows:

$$P = C - F + 1 \tag{3}$$

Here, C is the component number, F is the degree of freedom, and P is present phases in an alloy system at equilibrium. According to this rule for a five-component system, number of phases at equilibrium must be six at fixed temperature and pressure variables. But this is not true due to solid solution formation by HEAs [15].

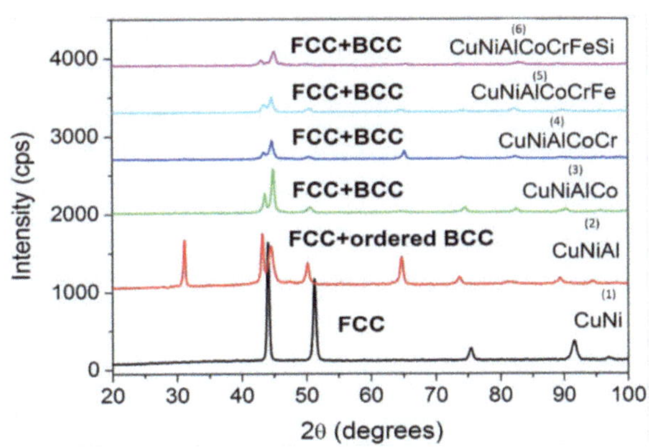

Fig. 2 XRD plot of multicomponent alloy [17]

2.4 Sluggish Diffusion Effect

This is also known as the slow diffusion effect. HEAs show this effect due to the presence of a greater number of principle elements with different atomic vibrations. Every atom in a system tries to gain stability by occupying an atomic site with the lowest free energy. This results in variations of potential energy. These variations affect the diffusion process and also the activation energy required for diffusion. Yeh et al. [8] studied this effect in HEAs and gave a comparison of diffusion coefficients of different materials.

$$\text{HEAs} < \text{Stainless steel} < \text{Pure metal}$$

Hence, HEAs have the smallest diffusion coefficient. The reason is that atoms with high variation in potential energy have greater difficulties to diffuse. Therefore, atoms get trapped by lower energy sites and increase the energy barrier for diffusion and activation. This behavior is clearly explained in Fig. 3. From the above figure, it can be seen that the mean difference is zero in the case of pure metal; as the number of principle elements increases, variation in mean difference starts due to the trapping of atoms. This difference is higher for HEAs because of having a higher number of principle elements. This whole phenomenon is responsible for slow diffusion. The impact of slow diffusion is a reduction in the formation and grain growth of nano-precipitates [18] with a decrement in strength at high temperatures.

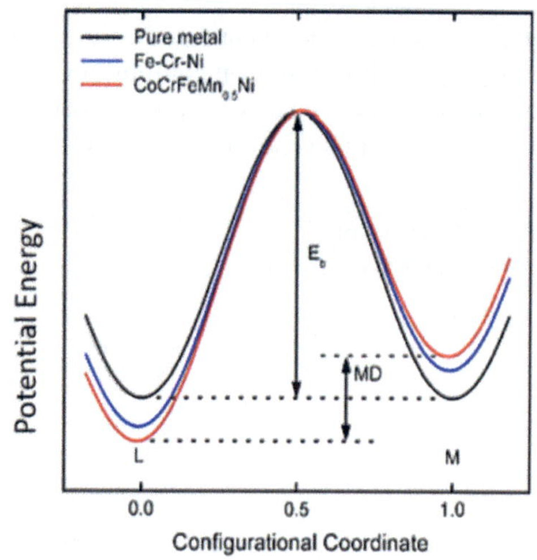

Fig. 3 Schematic plot between potential energy versus configurational coordinate [10]

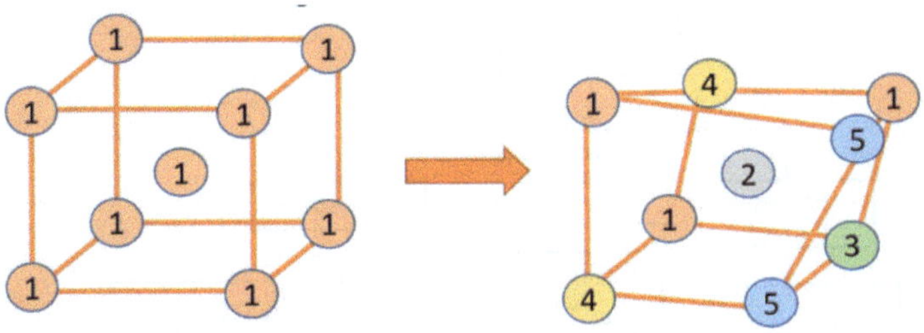

Fig. 4 Lattice distortion effect in HEAs [19]

2.5 *Lattice Distortion Effect*

In the case of conventional alloys, there was a concept of solute and solvent, in that solvent atom trapped into lattice sites whereas solute atom trapped into interstitial sites during mixing. But in the case of HEAs, all the atoms have an equal probability to occupy minimum energy sites. Due to this, severe lattice distortion take place which results in decreased crystallinity, X-ray scattering, and peak intensity [19]. This greatly impacts on structure and properties of HEAs. Lattice distortion phenomenon of HEAs is shown in Fig. 4.

2.6 *The Cocktail Effect*

The inventor of this term 'cocktail' was Ranganathan [11]. This effect does not come in the category of the above-discussed effects. Rather, it is collectively a major effect that influences properties of HEAs, and it is known to all that properties of HEAs depend on the composition of elements and phases formed.

Although the above-discussed effects are not applicable for a heterogeneous solution having high entropy, also diffusion is not slower in the case of heat treatment of HEAs. No strict restriction is given about lattice distortion and cocktail effect [11]. Therefore, a lot of work to be studied in this direction to date.

3 Classes of High Entropy Materials

In this work, high entropy alloys are divided into three main categories—based on phases present in an alloy system, electronic configuration of alloy system, applications of a particular alloy system (Fig. 5). Subclasses of HEAs are briefly discussed below.

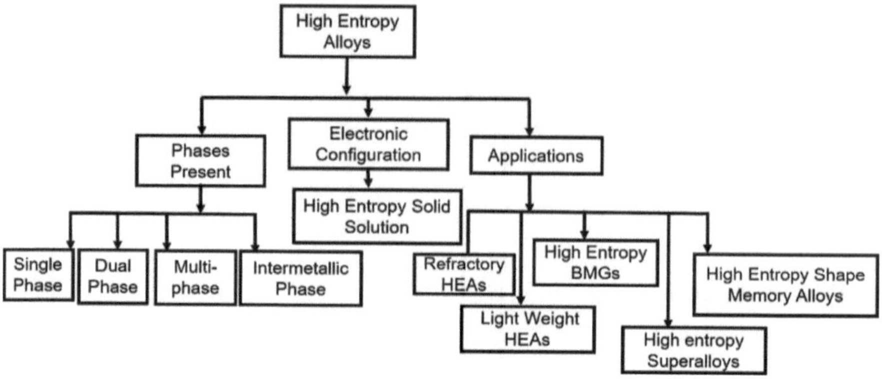

Fig. 5 Classification of high entropy alloys [20]

3.1 *Single-Phase High Entropy Alloys*

To design high entropy alloy, two factors are most important, i.e., thermodynamic and kinetics [21]. Only alloys with minimum free energy can achieve stability under a given situation. This free energy value is dependent on the enthalpy and entropy of the alloy system. The range of these values for single-phase HEAs [20] is given below:

$$-15 < \Delta H\text{mix} < 5\,\text{kJ/mol}$$

$$\Delta\text{Sconf.} > 1.61R \text{ (for } n = 5)$$

Table 2 summarizes the single-phase high entropy alloys with good tensile properties due to nano-twins formation by deformation that results in lower stacking fault energy.

Table 2 List of single-phase high entropy alloys [14, 20–24]

Single-phase high entropy alloys	Phase present	Enthalpy (KJ/mol)	Entropy (k_B per atom)	Strength (MPa)	Ductility (%)	Hardness (HVN)
CuNiFeCrMo	FCC	4.64	1.61	–	–	–
TaNbHfZrTi	BCC	2.72	1.61	1015	50	335
GdTbDyTmLu	HCP	0	1.61	–	–	–
$Al_{0.5}CoCrCuFeNiTi_{0.2}$	FCC	– 4.15	1.86	–	–	567
$Al_{0.3}CoCrFeNi$	FCC	– 3.27	1.54	–	–	325–497
$Al_{0.5}CrCuFeNi_2$	FCC	– 2.51	1.52	–	–	419

Table 3 Strength/ductility plot for dual-phase high entropy alloys [8, 25, 26]

TRIP dual-phase HEA	Phase present	Ω parameter
AlCoCrCuFeNi (A1)	BCC + FCC	14.13
AlCoCrCuFeNiTi (A2)	BCC + FCC	7.31
AlCoCrCuFeNiV (A3)	BCC + FCC	7.00

Table 4 Strength/ductility plot for multiphase high entropy alloys [16, 22, 27–29]

Multiphase HEAs	Phase present	Strength (MPa)	Ductility (%)
$Al_{0.5}CoCrCu_{0.5}FeNiTi_{0.5}$	Multiphase	2389	17
$Al_{0.75}CoCrCu_{0.25}FeNiTi_{0.5}$	Multiphase	2697	12
CoCrCuFeMnNiTiV	Multiphase	1312	< 1
$Al_{11.1}(CoCrCuFeMnNiTiV)_{88.9}$	Multiphase	2431	< 1
$Al_{0.7}Co_{0.3}CrFeNi$	Multiphase	2635	8

3.2 *Dual-Phase High Entropy Alloys*

Dual-phase HEAs were designed to overcome the limitation of single-phase HEAs, i.e., strength and ductility combination by doing small changes in composition. This helps in the creation of strong inter-atomic interactions, resulting in transformation-induced plasticity that promotes second-phase generation. 'Ω' parameter plays an important role in designing these alloys. Ω parameter was given by Zhang et al. [12] and used to determine energy value concerning enthalpy and entropy (Table 3). The optimum combination of strength/ductility is due to the plasticity effect during phase transition.

3.3 *Multiphase High Entropy Alloys*

Multiphase HEAs are the result of local atomic stress/strain atmosphere. Due to this stress/strain behavior, improvement in strength and ductility value has been observed for these alloy systems. Table 4 indicates the mixed-phase region with high strength and a small amount of deduction in ductility value due to composite formation while the FCC region has low strength and high ductility due to solid solution formation.

3.4 *Intermetallic Phase High Entropy Alloys*

In place of simply ordered phases, complex intermetallic phases (laves and sigma) are also present in HEAs having a range of compositions with low configurational

Table 5 Strength/ductility list for intermetallic phase high entropy alloys [30, 31]

Intermetallic phase HEAs	Phase present	Strength (MPa)	Ductility (%)
$AlCo_{0.1}CoCrFeNi$	BCC + intermetallic	2550	11
$AlCo_{0.2}CoCrFeNi$	BCC + intermetallic	2386	9
$AlCo_{0.3}CoCrFeNi$	BCC + intermetallic	2178	8
$AlCo_{0.4}CoCrFeNi$	BCC + intermetallic	2375	7
$AlCo_{0.5}CoCrFeNi$	BCC + intermetallic	2250	6
$AlC_{1.5}CoCrFeNi$	BCC + intermetallic	2083	6

Table 6 List of high entropy alloys solid solution [32]

HEA solid solution	Phase present	Enthalpy (ΔH_{mix})	Hardness (GPa)
CoCrFeNi	FCC	− 3.75	4.45
CoFeMnNi	FCC	− 4.00	5.25
CoCrMnNi	FCC	− 5.50	5.42

entropy [30, 31]. It is evident from Table 5 that the intermetallic phase in HEAs is responsible for better mechanical properties, especially strength and ductility. This is mainly due to precipitation hardening. This is the reason for the structural suitability of these alloys.

3.5 High Entropy Alloy Solid Solution

Atoms having similar electronic configuration and radius fall under this category. This includes 3-D transition metal, block D elements from group 3 to 11 in the periodic table. The transition metals have partially filled d-orbit, which is responsible for delocalization of electrons, resulting in strong metallic bonding. Due to this, these alloys have high hardness values (Table 6).

3.6 Refractory High Entropy Alloys

High entropy alloys are used for structural applications at high temperatures. It was observed from previous literature that BCC is the major phase in these alloys due to its high hardness value (Table 7).

High resistance to thermal softening and high hardness is due to slow diffusion in refractory alloys at temperatures up to 1600 °C. These characteristics make these alloys suitable for high-temperature applications.

Table 7 List of refractory high entropy alloys [14, 33]

Refractory HEAs	Phase present	Temperature (°C)	Hardness (GPa)
$Nb_{25}Mo_{25}Ta_{25}W_{25}$	BCC	600–1600	4.46
WNbMoTa	BCC	600–1600	4.45
WNbMoTaV	BCC	600–1600	5.25
$V_{20}Nb_{20}Mo_{20}Ta_{20}W_{20}$	BCC	600–1600	5.42

Table 8 List of lightweight and high entropy superalloys [22, 24, 34–43]

Lightweight HEAs	Phase present	Density (gm/cm^3)	Hardness (HVN)
$Al_{20}Li_{20}Mg_{10}Sc_{20}Ti_{30}$	FCC	2.67	599
$Mg_{50}(MnAlZnCu)_{50}$	HCP	2.20	173
$AlFeCuCrMg_{0.5}$	BCC + σ	5.7	853
AlFeCuCrMg	BCC + σ	5.3	740
NbVAlTi	BCC	5.59	440
NbVZrTiAl	BCC + Laves	5.76	540
$NbVZrTiAl_{1.5}$	BCC + Laves	5.50	620

3.7 Lightweight High Entropy Alloys

High entropy alloys possess high strength, good resistance to thermal softening, and good structural thermal stability. Despite this, the weight factor is also important in the case of HEAs for various potential applications. Lightweight HEAs include lightweight elements like Al, Mg, Ti, Si, Sc, V, Li, and Ca as shown in Table 8. Other than lightweight elements are added for property improvement.

3.8 High Entropy Bulk Metallic Glass

Bulk metallic glasses have multiple components, unlike high entropy alloys. The only difference is the configurational entropy. High entropy favors glass-forming ability. This is the reason why some HEAs with more negative enthalpy and large radius difference (Table 9) use in place of bulk metallic glass [40, 44]. Formation criterion for BMGs is as follows:

$$\delta \geq 6$$

$$\Delta H\text{mix} \leq -25\ \text{kJ/mol}$$

Table 9 List of high entropy bulk metallic glasses [40, 44, 45]

High entropy BMGs	Phase present	Enthalpy (KJ/mol)	Atomic size difference (δ)
$Ti_{20}Zr_{20}Cu_{20}Ni_{20}Be_{20}$	Amorphous	− 30.24	12.53
ZrHfTiCuNi	Amorphous	− 27.36	10.34
ErTbDyNiAl	Amorphous	− 37.60	13.74
PdPtCuNiP	Amorphous	− 23.68	9.29
ZrHfTiCuCo	Amorphous	− 23.52	10.24

4 Compositional Design of High Entropy Materials

Designing of the alloy is dependent on elements and their composition. After the selection of elements, microstructure prediction is important to ensure the formation of a solid solution, intermetallic, and amorphous structure. And then, the experimental step is to be carried out. Besides these factors, some other parameters like cost, processing ability, thermodynamic and kinetic theory are also considerable [46]. Figure 6 represents the designing concept of HEAs.

4.1 Thermodynamic Theory

The cooling process involves nucleation and growth of grains in the supercooled liquid. Nuclei are stable in supercooled liquid only with a definite value of the radius, which is known as the critical radius (r_c) (Eq. 4) [47], given as:

$$r_c = 2\sigma/\Delta G_{l-s} \tag{4}$$

Here, ΔG_{l-s} is the free energy difference between liquid and solid, and σ is the free energy of the surface for solid and liquid. So, the critical value of energy required for the formation of the nucleus (Eq. 5) [47] is given as:

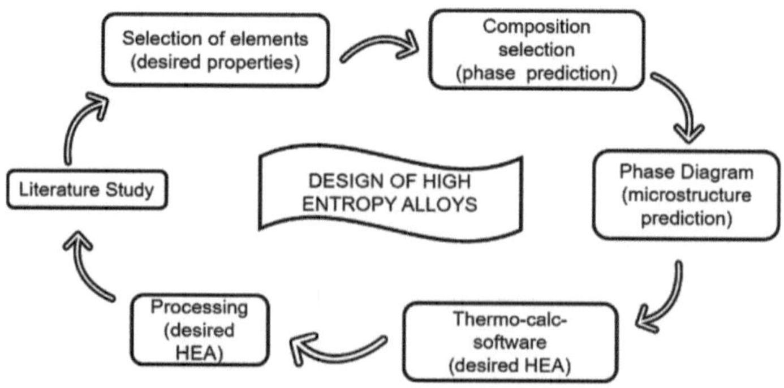

Fig. 6 Design strategy of high entropy alloys [46]

$$E_c = (16\pi\sigma^3)/(3\Delta G_{l-s}^2) \tag{5}$$

We all know that crystallization takes place at a minimum value of free energy. This value of critical energy depends on the free energy difference (given by Thompson (Eq. 6) [47]) which is given as:

$$\Delta G_{l-s} = \Delta S_f^* \times \Delta T \times (2T/(T_m + T)) \tag{6}$$

Here, ΔT is the degree of supercooling, and S_f is the entropy of melting which equals to entropy difference between solid and liquid at melting point. It is evident that ΔG_{l-s} increases when liquid is supercooled gradually. In this situation, crystallization takes place in place of an amorphous structure. It is known to all that high entropy alloys form simple structured solid solution during solidification. This is not possible thermodynamically due to the possibility of the formation of intermetallic compounds and other complex ordered phases in equilibrium. These compounds and phases make the HEA structure brittle and complex.

4.2 Dynamic Theory

Entropy which is a temperature function is mainly the deciding parameter for crystalline and amorphous structure formation. Entropy is divided into two categories: vibrational entropy and configurational entropy [48]. Vibrational entropy plays a major role in liquids only, and the reason is that with the decrement in temperature, atoms come near the equilibrium position which results in fall in vibrational entropy. While due to the large inter-atomic distance, configurational entropy is not so dominating. Here, high entropy meaning is an increment in the possible number of configurations for atoms during melting [47].

Entropy $\uparrow$ Viscosity $\downarrow$ Diffusivity $\uparrow$ better nucleation and growth

The magnitude of entropy greatly affects the microstructure. Poor nucleation and growth in high entropy alloys are due to slow diffusion which leads to a disordered amorphous phase. It would be greatly recommended for an alloy system to have appropriate entropy with a suitable composition.

4.3 Forming Criteria for the Stable Solid Solution of High Entropy Alloys

Due to the dual effects of thermodynamic and kinetic, not all HEAs are capable to form a single solid solution. Properties of HEAs broadly depend on phases present.

Table 10 List of forming criteria of high entropy alloys [47, 48]

S. No.	Factor	Formula
1	Average atomic size difference (δ)	$\delta = 100\sqrt{\sum_{i=1}^{n} c_i\left(1 - \frac{r_i}{\bar{r}}\right)^2}; \bar{r} = \sum_{i=1}^{n} c_i r_i$ $\delta < 6.5\%$
2	Enthalpy (ΔH)	$\Delta H_{\text{mix}} = \sum_{i=1, i \neq j}^{n} c_i c_j \Omega_{ij}; \Omega_{ij} = 4\Delta H_{\text{mix}}^{AB}$ $-11.6 < \Delta H < 3.2$ kJ/mol
3	Ω parameter	$\Omega = \frac{T_m \Delta S_{\text{mix}}}{\lvert\Delta H_{\text{mix}}\rvert}$ $\Omega < 1$; intermetallic compound $\Omega = 1$; about to form solid solution $\Omega > 1$; solid solution form
4	VEC (valance electron concentration)	$\text{VEC} = \sum_{i=1}^{n} c_i(\text{VEC})_i$ VEC $\geq$ 8.6; FCC stable VEC < 6.87; BCC solid solution stable
5	Ø parameter	$Ø = \frac{\Delta G_{ss}}{-\lvert\Delta G_{\max}\rvert} \geq 1; Ø = \frac{S_C - S_H}{\lvert S_E\rvert}; S_H = \frac{\lvert H_a\rvert}{T_m}$ $Ø > 20$; single-phase disordered solid solution $Ø < 20$; multiphase structure

The BCC phase solid solution has high strength and low ductility while vice versa for FCC. Other than the solid solution phase, HEAs are capable enough to form an amorphous phase with high glass-forming ability. It is known to all that amorphous alloys have some unique characteristics like high strength, high elasticity, high catalytic properties, good magnetic properties, and thermo-plasticity [47, 48]. Table 10 includes the forming criteria of high entropy alloys which are given below.

5 Conclusion

High entropy alloys as an emerging class of materials have unique characteristics. This article reviewed the history of multicomponent alloys, definition, and classification of high entropy alloys, core effects of high entropy alloys, compositional design of high entropy materials (including thermodynamic and kinetic theory, forming criteria of high entropy alloys), preparation routes of high entropy alloys, the corrosion behavior of high entropy alloys, and applications of high entropy alloys. The conclusion is as follows:

(I) With the industrial and technological growth, the rate of metal evolution also enlarged. The eye of the researchers has always been stuck on the scanning of new metallic materials since each passing year. For this reason, they started to investigate the chemical composition of alloys. In this series, the concept of a multicomponent system came into the picture.

(II) High entropy alloys have at least five metallic elements with an equal or near-equal atomic percentage of composition in between 5 and 35%. The high entropy alloy system must have the configurational entropy greater than $1.61R$. Classification is majorly based on phases, configuration, and applications.

(III) Core effects of high entropy alloys summarized the effect of higher mixing entropy on the stability of alloy system, slow diffusion kinetics of high entropy alloys, the effect of lattice distortion on properties of high entropy alloys, and significance of individual elements characteristics on the overall performance of alloy system.

(IV) The compositional design of high entropy alloys has briefly summarized the thermodynamics and kinetics of high entropy alloys formation, and also this section contains designing parameters of high entropy alloys properly.

References

1. Hummel RE (1999) Understanding materials science. IEEE Electr Insul Mag 15(3):52
2. Mitomo M, Uenosono S (1991) Gas pressure sintering of β-silicon nitride. J Mat Sci 26(14):3940–3944
3. Ozesmi M, Patiroglu TE, Hillerdal G, Ozesmi C (1985) Peritoneal mesothelioma and malignant lymphoma in mice caused by fibrous zeolite. Br J Ind Med 42(11):746–749
4. Dorin T, Vahid A, Lamb J (2018) Aluminium lithium alloys. Funda Al Met Rec Adv 387–438 (February 2019)
5. Fairhart V (1973) Tool steels. Met Sci Heat Treat 15(7):557–559
6. Alvi SA (2020) Refractory high-entropy alloys and films for high temperature applications 1–64
7. Cantor B, Chang ITH, Knight P, Vincent AJB (2004) Microstructural development in equiatomic multicomponent alloys. Mater Sci Eng A 375–377(1–2) (SPEC, ISS, 213–218)
8. Yeh JW et al (2004) Nanostructured high-entropy alloys with multiple principal elements: novel alloy design concepts and outcomes. Adv Eng Mater 6(5):299–303+274
9. Wu JM, Lin SJ, Yeh JW, Chen SK, Huang YS, Chen HC (2006) Adhesive wear behavior of AlxCoCrCuFeNi high-entropy alloys as a function of aluminum content. Wear 261(5–6):513–519
10. Tsai KC, Hsiao BC, Lai JW, Chen CH, Lin ML, Weng YT. Pseudo dynamic experimental responses of a full scale CFT/BRB composite frame
11. Ranganathan S (2003) Alloyed pleasures: multimetallic cocktails. Curr Sci 85(10):1404–1406
12. Zhang Y, Zhou YJ, Lin JP, Chen GL, Liaw PK (2008) Solid-solution phase formation rules for multi-component alloys. Adv Eng Mater 10(6):534–538
13. Kombaiah B, Jin K, Bei H, Edmondson PD, Zhang Y (2018) Phase stability of single phase Al0.12CrNiFeCo high entropy alloy upon irradiation. Mater Des 160:1208–1216
14. Senkov ON, Wilks GB, Scott JM, Miracle DB (2011) Mechanical properties of Nb25Mo25Ta 25W25 and V20Nb20Mo 20Ta20W20 refractory high entropy alloys. Intermetallics 19(5):698–706
15. Van Poolen LJ (1990) Extended phase rule for non-reactive, multiphase, multicomponent chemical systems. Fluid Phase Equilib 58(1–2):133–145
16. Yeh JW (2006) Recent progress in high-entropy alloys. Ann Chim Sci Des Mater 31(6):633–648

17. Zhang W, Liaw PK, Zhang Y (2018) Science and technology in high-entropy alloys. Sci China Mat 61(1):2–22
18. Cantor B (2014) Multicomponent and high entropy alloys. Entropy 16(9):4749–4768
19. Wilson TW (2008) TRACE: Tennessee research and creative exchange processing, structure , and properties of amorphous aluminum alloys
20. Gao MC, Liaw PK, Yeh JW, Zhang Y (2016) High-entropy alloys: fundamentals and applications. High Entropy Alloy Fundam Appl 1–516
21. Li C, Li JC, Zhao M, Jiang Q (2009) Effect of alloying elements on microstructure and properties of multiprincipal elements high-entropy alloys. J Alloys Compd 475(1–2):752–757
22. Chen MR, Lin SJ, Yeh JW, Chen SK, Huang YS, Tu CP (2006) Microstructure and properties of Al0.5CoCrCuFeNiTix (x = 0–2.0) high-entropy alloys. Mater Trans 47(5):1395–1401
23. Kao YF, Chen TJ, Chen SK, Yeh JW (2009) Microstructure and mechanical property of as-cast,—homogenized, and -deformed AlxCoCrFeNi ($0 \leq x \leq 2$) high-entropy alloys. J Alloys Compd 488(1):57–64
24. Ng C et al (2014) Phase stability and tensile properties of Co-free Al 0.5CrCuFeNi2 high-entropy alloys. J Alloys Compd 584:530–537
25. Hsu US, Hung UD, Yeh JW, Chen SK, Huang YS, Yang CC (2007) Alloying behavior of iron, gold and silver in AlCoCrCuNi-based equimolar high-entropy alloys. Mater Sci Eng A 460–461:403–408
26. Tong CJ et al (2005) Mechanical performance of the AlxCoCrCuFeNi high-entropy alloy system with multiprincipal elements. Metall Mater Trans A Phys Metall Mater Sci 36(5):1263–1271
27. Number S (2012) J Phys Sci Appl 2:2
28. Zhang Y, Qiao JW, Liaw PK (2016) A brief review of high entropy alloys and serration behavior and flow units. J Iron Steel Res Int 23(1):2–6
29. Zhou YJ, Zhang Y, Wang YL, Chen GL (2007) Solid solution alloys of AlCoCrFeNi Tix with excellent room-temperature mechanical properties. Appl Phys Lett 90(18)
30. Tang Z et al (2013) Aluminum alloying effects on lattice types, microstructures, and mechanical behavior of high-entropy alloys systems. Jom 65(12):1848–1858
31. Senkov ON, Wilks GB, Miracle DB, Chuang CP, Liaw PK (2010) Refractory high-entropy alloys. Intermetallics 18(9):1758–1765
32. Youssef KM, Zaddach AJ, Niu C, Irving DL, Koch CC (2014) A novel low-density, high-hardness, high-entropy alloy with close-packed single-phase nanocrystalline structures. Mater Res Lett 3(2):95–99
33. Li R, Gao J, Fa K (2010) Study to microstructure and mechanical properties of Mg containing high entropy alloys. Mater Sci Forum 650:265–271
34. Tsai MH, Yeh JW (2014) High-entropy alloys: a critical review. Mater Res Lett 2(3):107–123
35. Science S (2018) d M cri pt. 1–71 [Online]. https://doi.org/10.1088/1361-6579/ac92bd
36. Manuscript A (2019) Ac d M us. Eur J Phys 0–16
37. Sanchez JM, Vicario I, Albizuri J, Guraya T, Garcia JC (2019) Phase prediction, microstructure and high hardness of novel light-weight high entropy alloys. J Mater Res Technol 8(1):795–803
38. Ma L, Wang L, Zhang T, Inoue A (2002) Bulk glass formation of Ti-Zr-Hf-Cu-M (M=Fe Co, Ni) alloys. Mater Trans 43(2):277–280
39. Takeuchi A et al (2011) Pd20Pt20Cu20Ni20P20 high-entropy alloy as a bulk metallic glass in the centimeter. Intermetallics 19(10):1546–1554
40. Chuang MH, Tsai MH, Wang WR, Lin SJ, Yeh JW (2011) Microstructure and wear behavior of AlxCo 1.5CrFeNi1.5Tiy high-entropy alloys. Acta Mater 59(16):6308–6317
41. Yang M, Liu XJ, Ruan HH, Wu Y, Wang H, Lu ZP (2016) High thermal stability and sluggish crystallization kinetics of high-entropy bulk metallic glasses. J Appl Phys 119(24):0–7
42. Xia L, Wu Q, Zhou K, Han B, Wang Z (2022) Effect of initial grain structure on the concurrent precipitation and recrystallization of Co37Cr20Ni37Ti3Al3 high-entropy alloy. Mater Sci Eng A 854:143828
43. Wang C et al (2022) Development of the γ' phase strengthened high-temperature high-entropy alloys with excellent mechanical properties. Mater Des 221:110940

44. Zhang ZJ et al (2015) Nanoscale origins of the damage tolerance of the high-entropy alloy CrMnFeCoNi. Nat Commun 6:2–7
45. Thompson CV, Spaepen F (1979) On the approximation of the free energy change on crystallization. Acta Metall 27(12):1855–1859
46. Troparevsky MC, Morris JR, Kent PRC, Lupini AR, Stocks GM (2015) Criteria for predicting the formation of single-phase high-entropy alloys. Phys Rev X 5(1):1–6
47. Jiang L, Powers M, Cui Y, Derby BK, Misra A (2021) Microstructure and mechanical properties of nanoscale Cu/(Ta50Nb25Mo25) multilayers. Mater Sci Eng A 799:140200
48. Zhang F, Zhang C, Chen SL, Zhu J, Cao WS, Kattner UR (2014) An understanding of high entropy alloys from phase diagram calculations. Calphad Comput Coupling Phase Diagrams Thermochem 45:1–10

Emissions and Performance Characteristics of Diesel Engines Fueled with Diesel, Biodiesel, and Oxygenated Compound Blends: A Review

Pratik V. Swami, S. C. Kamate, S. N. Topannavar, and Belagur K. Venkanna

1 Introduction

Because of their excellent drivability and mileage, diesel engines are the most common. Diesel vehicles are particularly well suited to applications requiring high torque and strength and improved fuel economy [1]. Some of the emissions emitted by diesel engines include unburned hydrocarbon (HC), nitrogen oxide (NOx), carbon monoxide (CO), carbon dioxide (CO_2), and particulate matter (PM) [2].

Biodiesel is a non-conventional energy source that is gaining popularity around the world as a diesel additive or a direct substitute for diesel engines. As compared to other biofuels, it decreases reliance on sources of nonrenewable energy and also makes that one easier to offset the negative environmental effects of fuel combustion [3]. The use of this biodiesel will substantially reduce the energy [4]. Higher viscosity, nitrogen oxide (NOx) emissions, cloud and pour point, lower energy content, lower engine power, engine compatibility, and high price are some of the disadvantages of biodiesel [5].

Our reliance on fossil fuels is growing as a result of increased energy demand brought on by population and economic growth. Excessive amounts of greenhouse gases and particulate particles are produced while burning fossil fuels [6]. Researchers, governments, and scientists are working to create new, renewable energy sources that can either completely replace fossil fuels or dramatically lessen our need on them. High cetane number, lower pour point, low aromatics, low sulfur content, and greater lubricating efficiency are all characteristics of biodiesel. Biodiesel can be produced by catalyzing the transesterification of triglycerides with

P. V. Swami (✉) · S. C. Kamate · S. N. Topannavar
Department of Mechanical Engineering, HIT, Nidasoshi, VTU Belagavi, Karnataka, India
e-mail: pratik4826@gmail.com

B. K. Venkanna
Department of Mechanical Engineering, BEC, Bagalkot, Karnataka, India

R. Sahu et al. (eds.), *Advancements in Materials Processing Technology, Volume 1*,
Springer Proceedings in Materials 48, https://doi.org/10.1007/978-981-97-4958-4_20

alcohols. The production of biodiesel can utilize both edible and non-edible oils as well as third-generation sources like algae [5, 7, 8].

2 Materials and Methods

Biodiesel is a sustainable and better burning option than conventional diesel fuel that is derived from various feedstocks [1]. The properties of biodiesel feedstocks can vary depending on the source material used. Here are some common biodiesel feedstocks and their properties:

2.1 *Vegetable Oils and Animal Fats*

The most widely utilized vegetable oils for the manufacturing of biodiesel include sunflower oil, soybean oil, rapeseed oil, and palm oil. They are fatty acid-based triglycerides that may be transformed into biodiesel by a procedure known as transesterification. Vegetable oils typically have high energy content and good lubricity, which means they can provide adequate lubrication to engines [7, 9–11].

Animal fats, such as tallow or lard, can also be used as feedstocks for biodiesel production. Like vegetable oils, animal fats are triglycerides that undergo transesterification to convert them into biodiesel. Animal fats generally have similar properties to vegetable oils in terms of energy content and lubricity [12, 13].

2.2 *Waste Cooking Oil and Algae*

It is also known as used cooking oil or yellow grease and is recycled from restaurants, homes, and other food establishments. It is a known feedstock for the production of biodiesel due to its availability and potential to reduce waste [7].

Algae are considered a favorable feedstock for biodiesel due to their high oil content and potential for large-scale cultivation. Algae-based biodiesel, often referred to as "algae oil," can be produced from various algae strains [14].

Table 1a, b shows properties of biodiesel with feedstock.

Table 1 **a** Properties of biodiesel with feedstock [10, 15–22], **b** properties of biodiesel with feedstock [12, 13, 23–28]

oil	Density	Pour point	Heating value	CN	Kinematic viscosity	Cloud point	Flash point	References
	(kg/l)	(°C)	(MJ/kg)		(mm^2/s)	(^0C)	(^0C)	
a								
Rice bran oil methyl ester	0.877	–	39.6	56.3	5.6	–	153	[10, 16]
Canola oil methyl ester	0.88 to 0.9	– 5 to – 1	37.8 to 44.93	53.6 to 56.1	2.56 to 4.4	–	170	[17]
Karanja oil methyl ester	0.876 to 0.89	– 3 to – 1	36.12 to 40.96	57.6 to 60.9	5.52 to 9.6	–	187 to 230	[15]
Corn oil methyl ester	–	– 5 to – 2	44.92 to 45.06	58.37 to 59.0	2.45 to 2.56	–	–	[15]
Beef tallow methyl ester	0.870	–	39.99	61	4.7	–	169	[15]
Pongamia oil methyl ester	0.866 to 0.89	–	36.05 to 39.14	55.4	5.3	–	147	[18]
waste oil methyl ester	0.855 to 0.8903	–	40.5 to 41.24	44.03 to 52	4.57 to 4.918	–	120<	[29]
Mahua oil ethyl ester	–	2	41.6	55	6.2	5	164	[20]
Jojoba oil methyl ester	0.866 to 0.903	–	39.88 to 47.38	52.5 to 63.5	15.3 to 19.2	–	61 to 104	[21]
Jatropha oil methyl ester	0.849 to 0.880	– 13.1 to 2	38.467 to 41.63	51 to 63.27	2.35 to 5.48	– 2.9	172 to 175	[15]
Mahua oil methyl ester	0.865 to 0.916	1 to 6	36.8 to 39.91	51 to 62	3.98 to 5.8	3 to 5	127 to 208	[19]
Palm oil methyl ester	0.872	–	38.29 to 40.48	56.5 to 61.7	4.605	–	184.8	[15, 22]
Soybean oil methyl ester	0.840 to 0.881	–	33.5 to 39.83	45 to 57	4.1 to 4.5	4	130 to 164.9	[22]
b								

(continued)

Table 1 (continued)

oil	Density	Pour point	Heating value	CN	Kinematic viscosity	Cloud point	Flash point	References
	(kg/l)	(°C)	(MJ/kg)		(mm^2/s)	(^{0}C)	(^{0}C)	
Coconut oil methyl ester	0.856	3	39.92	56.7	4.02	4	145.5	[23, 26]
Camelina oil methyl ester	0.885 to 0.888	–	36.75	–	4.53 to 5.35	3	202	[30]
Rapeseed oil methyl ester	0.876 to 0.890	1	39.92	47.2 to 51.10	4.92	3	163	[27, 28]
mutton tallow methyl ester	0.856	− 5	–	59	5.98	− 4	155	[12, 24]
Tire pyrolysis oil methyl ester	0.907	–	45.823	–	2.73	–	>100	[25]
chicken tallow methyl ester	0.687	− 6	–	61	6.25	− 5	161	[12, 24]
Swine lard methyl ester	0.895	–	40.1	–	4.4528	–	134	[13]

3 Properties of Oxygenated Compounds

Biodiesel, an environmentally friendly and sustainable fuel, is produced using edible and non-edible vegetable oils, tallow, algal, and microbial oils. The production method, such as transesterification, depends on geographic location, regional conditions, and feedstock availability. Transesterification, known for its simplicity and effectiveness, is the most widely used technique. Biodiesel is essentially a mixture of long-chain monoalkyl esters derived from unsaturated fats in vegetable and animal fats through alcohol-induced transesterification [10, 14, 31–34]. Table 2 shows properties of oxygenated compounds.

Table 2 Properties of oxygenated compounds [35–40]

Compound	Formula	Oxygen content (wt %)	Density (kg/m^3)	Viscosity (mm^2/s)	Flash point (°C)	Cetane number	References
Methanol	CH_3OH	50.0	796	1.01	12	5	[35]
Ethanol	C_2H_5OH	34.8	789	1.8	17	5–8	[35]
n-butanol	C_4H_9OH	21.3	795.8	3.6	35	25	[35, 36]
Methyl tertiary butyl ether	$CH_3OC(CH_3)_3$	18.2	740	0.47	− 33	–	[35]
Ethyl tertiary butyl ether	$CH_3CH_2OC(CH_3)_3$	15.7	736	0.55	− 19	8	[35]
Methyl butanoate	$C_5H_{10}O_2$	31.3	898	2.26	12	–	[35]
Dimethyl ether	C_2H_6O	35	1970	1.84	− 41	> 55	[35]
Dimethyl carbonate	$C_3H_6O_3$	53.3	1070	0.63	17	35	[35, 37]
(Diglyme)	$C_6H_{14}O_3$	35.8	937	–	57	112	[35]
Glycerine triacetate	$C_9H_{14}O_6$	53.3	1160	–	138	–	[35]
Acetone	C_3H_6O	27.5	791	–	− 17.2	–	[35]
Diethyl succinate	$C_8H_{14}O_4$	36.8	791	1.86	90.56	–	[35]
Diethyl ether	$(C_2H_5)_2O$	21.62	712	0.22	− 40	125	[36]
Hexanol	$C_6H_{14}O$	15.68	819	4.59	59	23.3	[38]
Pentanol	$C_5H_{12}O$	18.18	814.8	2.89	49	20	[39]
Anisole	C_7H_8O	14.81	995	0.922	43	–	[40]

4 Results and Discussion

4.1 Effect of Oxygenated Additives on Performance and Emission Characteristics

The study by Imtenan et al. explores the effects of adding n-butanol and diethyl ether to Jatropha biodiesel–diesel blends. The additives decrease density and viscosity, with 10% blends showing better results than 20% (J20). Diethyl ether outperforms n-butanol in terms of brake-specific fuel consumption (BSFC). Additives reduce nitrogen oxides and carbon monoxide emissions and improve smoke opacity compared to J20. The study concludes that 10% blends of n-butanol and diethyl ether enhance efficiency and emission characteristics [36].

Devarajan experimented with dimethyl carbonate (DMC) and biodiesel blend combustion, emission, and efficiency. The most significant result of this research was that no obstruction was found during or after the experiment. The addition of DMC lowers peak pressure while increasing the net heat release rate. Pure biodiesel emits 8.7% less HC and 7.2% less CO than ethanol, while DMC decreases HC emissions by 2.95% and CO emissions by 3.7%, in contrast to biodiesel (pure). The supplement of DMC takes down the emission of NO_x by 4.5% in contrast to pure biodiesel. Moreover, the addition of DMC reduces smoke opacity. When DMC is combined with pure biodiesel, the BTE rises and the BSFC falls. As DMC was applied to biodiesel, it improved combustion. DMC is also a cetane number enhancer [37].

Ramesh et al. (2019) focused on hexanol's role as an additive in Calophyllum inophyllum biodiesel (CIME) which was conducted using a diesel engine. The research revealed that a blend containing 40% hexanol, 10% biodiesel, and 40% diesel (D40B10H40) exhibited higher brake thermal efficiency (BTE) similar to biodiesel, regardless of the engine load. This improvement was attributed to enhanced fuel atomization and increased oxygen content within the mixture, which closely approached diesel fuel's heat release characteristics. Blends with higher alcohol content, especially the 40% hexanol blend, demonstrated significantly reduced smoke, hydrocarbon (HC) emissions, and lower carbon monoxide (CO) emissions. Notably, NO_x emissions from the 40% hexanol blend were on par with pure diesel. These findings emphasize hexanol as a valuable oxygenated additive with several advantages [38].

Nathagopal et al. test the effect of Calophyllum inophyllum biodiesel when mixed with additives including 1-pentanol and 1-butanol. The results of the experiments show that adding more alcohol to the CIME reduces BTE compared to CIME. The HC emission of alcohol/CIME blends is higher because of the great latent heat of vaporization and the quenching effect in the cylinder's linear mixing region. Moreover, when it contrasts with pure CIME diesel, emissions of NO_x are lower with CIME-alcohol blends. CIME blends with higher alcohol produce more emissions of CO as well as smoke than CIME fuel because of higher alcohol's weak autoignition property. Apart from NO_x emissions, in comparison with pure biodiesel, adding 1-pentanol and 1-butanol to CIME resulted in lower emissions and performance. As a result, higher alcohol concentrations can be mixed with lower concentrations to boost emission and efficiency characteristics [39].

Londhe and his team studied methyl acetate and anisole as fuel additives. Biodiesel-methyl acetate (BM) and biodiesel-anisole (BA) blends showed pressure peaks similar to biodiesel, with improved but slightly slower peak heat release, indicating enhanced combustion. These blends had shorter ignition delays, higher brake-specific fuel consumption (BSFC), lower HC and CO emissions, and slightly varied NO_x emissions. They also exhibited increased exhaust opacity compared to biodiesel. While diesel fuels and blends yielded similar engine load results, the study concluded that despite variations in emissions and fuel consumption, the combustion process resembled that of base oils, with potential significance for certain engine loads [40].

Nabi et al. studied that the blends of diesel–biodiesel have an effect on diglyme as an additive that was investigated. The efficiency, emission, and combustion, parameters of tree bends of coconut oil biodiesel, diglyme, and diesel were viewed. While matched to blends of diesel fuel, the thermal performance and mean effective pressure is lower. The highest rise in the emission of NO*x* is shown in pure biodiesel. However, in contrast to diesel, blends display higher NO*x* emissions. However, the increases in emission of NO*x* for (D70B30), (D70B20DGM10), (D70B10DGM20), and (B100) were 2.47%, 7.8%, 16%, and 16.7%, respectively. The findings show that a lower percentage of diglyme (10 and 20%) has the ability to reduce emissions and may be used in existing engines [41].

Imtenan et al. studied the impact of diethyl ether (DEE), *n*-butanol, and ethanol additives on diesel–biodiesel blend efficiency and emissions. Their findings revealed that these additives enhance viscosity, density, disintegration, and overall performance. DEE exhibited the most significant gains due to its superior calorific value, while *n*-butanol showed similar improvements, and ethanol had reduced impact due to its lower calorific value. The additives notably affected CO and NO emissions, reducing NO due to lower calorific value and higher latent heat of evaporation and lowering CO due to higher oxygen content [42].

Alptekin studied the influence of biodiesel and additives like ethanol and solketal on emission, injection, and combustion characteristics that was investigated. The solketal biodiesel blend possesses an advanced flash point when matched to a blend of ethanol biodiesel, according to test results. The addition of ethanol and solketal to biodiesel has improved BSFC. When contrast to diesel fuel, biodiesel is used up in the engine, and then, it diminishes emissions of HC and CO while increasing NO*x* as well as CO_2 emissions. When solketal and ethanol are used as engine additives, emissions of HC, CO_2, and CO are reduced. All of the findings indicated that the solketal could be used as an ethanol-like additive in a common rail direct injection, i.e., CRDI diesel engine since solketal such as ethanol is readily available [43].

Zheng et al. conducted tests on biodiesel fuel blends (DMF, ethanol, *n*-butanol) to analyze their impact on emissions and combustion in a single-cylinder CI engine. Different load levels and EGR rates were explored. Results indicate lower thermal efficiency at low loads but higher efficiency as load increases compared to diesel. Smoke emissions were lower for biodiesel blends, while NO*x*, HC, and CO varied, with notable reductions in soot emissions for Bu50 and DMF50 [44].

5 Conclusion

As predicted, additive characteristics such as flash point, density, viscosity, cetane number, and oxygen concentration are greatly affected by the kind of biodiesel utilized and its degree of unsaturation.

- Biodiesel contains higher oxygen, viscosity, density, and cetane number than diesel fuel, but has an inferior heating value and compressibility. As biodiesel is blended with oxygenated additives, it resembles diesel fuel in properties. The incorporation of additives diminished the viscosity as well as density of the fuel mix, resulting in better fuel disintegration.
- As linked to (pure) biodiesel, adding oxygenated compounds to a diesel–biodiesel blend increases combustion efficiency and decreases emissions, according to most reports. Most oxygenated additives blend appears to reduce emissions and increase combustion and performance parameters based on engine operating circumstances. If sufficient fuel blending proportions and relevant engine operating conditions are set, it appears that most oxygenated additives blend reduce emissions and increase combustion and performance parameters.
- As compared to the other oxygenated compounds, the blend of diesel–biodiesel with a high oxygen content of oxygenated additives has the advantage of improved thermal performance, lower specific fuel consumption as well as lower emissions, for example NO_X, HC, and CO. The dimethyl carbonate, triacetin, and methanol show better results as compared to the other compounds.

References

1. MattiMaricq M (2007) Chemical characterization of particulate emissions from diesel engines: a review. J Aerosol Sci 38:1079–1118
2. NaniGuarieiro LL, de Almeida Guerreiro ET, dos Santos Amparo KK, Manera VB, Regis ACD, Santos AG, Ferreira VP, Leao DJ, Torres EA, de Andrade JB (2014) Assessment of the use of oxygenated fuels on emissions and performance of a diesel engine. Microchem J 117:94–99
3. Gharehghani A, Mirsalim M, Hosseini R (2017) Effects of waste fish oil biodiesel on diesel engine combustion characteristics and emission. Renew Energy 101:930–936
4. Ashok B, Karuppa Raj RT, Nanthagopal K, Krishnan R, Subbarao R (2017) Lemon peel oil—a novel renewable alternative energy source for diesel engine. Energy Conv Manag 139:110–121
5. Kumar MV, Babu AV, Ravi Kumar P (2018) The impacts on combustion, performance, and emissions of biodiesel by using additives in direct injection diesel engine. Alexandria Eng J 57:509–516
6. Sahin Z, Durgun O, Aksu ON (2015) Experimental investigation of n-butanol/diesel fuel blends and n-butanol fumigation evaluation of engine performance, exhaust emissions, heat release and flammability analysis. Energy Convers Manage 103:778–789
7. Suresh M, Jawahar CP, Richard A (2018) A review on biodiesel production, combustion, performance, and emission characteristics of non-edible oils in variable compression ratio diesel engine using biodiesel and its blends. Renew Sustain Energy Rev 92:38–49
8. Misra RD, Murthy MS (2011) Performance, emission, and combustion evaluation of soapnut oil–diesel blends in a compression ignition engine. Fuel 90:2514–2518
9. Chen H, Xie B, Ma J, Chen Y (2018) NO_x emission of biodiesel compared to diesel: higher or lower? Appl Therm Eng 137:584–593
10. Bora BJ, Saha UK (2016) Experimental evaluation of a rice bran biodiesel e biogas run dual fuel diesel engine at varying compression ratios. Renew Energy 87:782–790
11. Devan PK, Mahalakshmi NV (2009) Performance, emission, and combustion characteristics of poon oil and its diesel blends in a DI diesel engine. Fuel 88:861–867

12. Bhatti HN, Hanif MA, Qasim M, Ata-ur-Rehman (2008) Biodiesel production from waste tallow. Fuel 87:2961–2966
13. Mikulski M, Duda K, Wierzbicki S (2016) Performance and emissions of a CRDI diesel engine fuelled with swine lard methyl esters–diesel mixture. Fuel 164:206–219
14. Ma Y, Gao Z, Wang Q, Liu Y (2018) Biodiesels from microbial oils: opportunity and challenges. Biores Technol 263:631–641
15. Patel RL, Sankhavara CD (2017) Biodiesel production from Karanja oil and its use in diesel engine: a review. Renew Sustain Energy Rev 71:464–474
16. Musthafa MM, Sivapirakasam SP, Udayakumar M (2011) Comparative studies on fly ash coated low heat rejection diesel engine on performance and emission characteristics fueled by rice bran and pongamia methyl ester and their blend with diesel. Energy 36:2343–2351
17. Bari S (2014) Performance, combustion and emission tests of a metro-bus running on biodiesel-ULSD blended (B20) fuel. Appl Energy 124:35–43
18. Bora BJ, Saha UK (2015) Comparative assessment of a biogas run dual fuel diesel engine with rice bran oil methyl ester, pongamia oil methyl ester and palm oil methyl ester as pilot fuels. Renew Energy 81:490–498
19. Raheman H, Ghadge SV (2007) Performance of compression ignition engine with mahua (Madhucaindica) biodiesel. Fuel 86:2568–2573
20. Puhan S, Nagarajan G, Vedaraman N, Ramabramhmam BV (2007) Mahua oil (*MadhucaIndica Oil*) derivatives as a renewable fuel for diesel engine systems in india: a performance and emissions comparative study. Int J Green Energy 4:89–104
21. Saleh HE, Selim MYE (2017) Improving the performance and emission characteristics of a diesel engine fueled by jojoba methyl ester-diesel-ethanol ternary blends. Fuel 207:690–701
22. Mangus M, Kiani F, Mattson J, Tabakh D, Petka J, Depcik C, Peltier E, Stagg-Williams S (2015) Investigating the compression ignition combustion of multiple biodiesel/ULSD (ultra-low sulfur diesel) blends via common-rail injection. Energy 89:932–945
23. Machacon HTC, Matsumoto Y, Ohkawara C, Shiga S, Karasawa T, Nakamura H (2001) The effect of coconut oil and diesel fuel blends on diesel engine performance and exhaust emissions. JSAE Rev 22:349–355
24. Buendía-Tamariz MN, Trejo-Calzada R, Abiola A, Pedroza-Sandoval A, Jacobo-Salcedo R, Reveles-Hernández M (2015) Characterization of biodiesel produced from chicken fat and pennycress oil using different concentrations of basic catalysts. J Agr Environ Sci 4:127–133
25. Hurdogan E, Ozalp C, Kara O, Ozcanli M. Experimental investigation on performance and emission characteristics of waste tire pyrolysis oil diesel blends in a diesel engine. Int J Hydrogen Energy 4:23373–23378
26. How HG, Masjuki HH, Kalam MA, Teoh YH (2018) Influence of injection timing and split injection strategies on performance, emissions, and combustion characteristics of diesel engine fueled with biodiesel blended fuels. Fuel 213:106–114
27. Ileri E, Atmanli A, Yilmaz N (2016) Comparative analyses of n-butanol-rapeseed oil-diesel blend with biodiesel, diesel and biodiesel-diesel fuels in a turbocharged direct injection diesel engine. J Energy Inst 89:586–593
28. Swami PV, Shaikh SM, Yadav RH (2017) Effect of rapeseed oil methyl ester-butanol blend at various EGR and VCR on emission characteristics of diesel engine. IJIRST 4:2349–6010
29. Akar MA, Kekilli E, Bas O, Yildizhan S, Serin H, Ozcanli M (2018) Hydrogen enriched waste oil biodiesel usage in compression ignition engine. IJHE 43:18046–18052
30. EnginOzcelik A, Aydogan H, Acaroglu M (2015) Determining the performance, emission, and combustion properties of camelina biodiesel blends. Energy Conv Manag 96:47–57
31. Herbinet O, Pitz WJ, Westbrook CK (2008) Detailed chemical kinetic oxidation mechanism for a biodiesel surrogate. HAL Arch Ouvertes 154:507–528
32. Singh SP, Singh D (2010) Biodiesel production through the use of different sources and characterization of oils and their esters as the substitute of diesel: a review. Renew Sustain Energy Rev 14:200–216
33. Herbinet O, Pitz WJ, Westbrook CK (2010) Detailed chemical kinetic mechanism for the oxidation of biodiesel fuels blend surrogate. Combust Flame 157:893–908

34. Raheman H, Phadatare AG (2004) Diesel engine emissions and performance from blends of karanja methyl ester and diesel. Biomass Bioenerg 27:393–397
35. Khalife E, Tabatabai M, Demirbas A, Aghbashlo M (2017) Impacts of additives on performance and emission characteristics of diesel engine during steady state operation. Prog Energy Combust Sci 59:32–78
36. Imtenan S, Masjuki HH, Varman M, Rizwanul Fattah IM, Sajjad H, Arbab MI (2015) Effect of n-butanol and diethyl ether as oxygenated additives on combustion–emission-performance characteristics of a multiple cylinder diesel engine fuelled with diesel–jatropha biodiesel blend. Energy Conv Manag 95:84–94
37. Devarajan Y (2019) Experimental evaluation of combustion, emission and performance of research diesel engine fuelled di-methyl-carbonate and biodiesel blends. Atmos Pollut Res 10:795–801
38. Ramesh A, Ashok B, Nanthagopal K, Pathy MR, Tambare A, Mali P, Phuke P, Patil S, Subbarao R (2019) Influence of hexanol as additive with Calophyllum Inophyllum biodiesel for CI engine applications. Fuel 249:472–485
39. Nanthagopal K, Ashok B, Saravanan B, Patel D, Sudarshan B, Aaditya Ramasamy R (2018) An assessment on the effects of 1-pentanol and 1-butanol as additives with CalophyllumInophyllum biodiesel. Energy Conv Manag 158:70–80
40. Londhe H, Luo G, Park S, Kelley SS, Fang T (2019) Testing of anisole and methyl acetate as additives to diesel and biodiesel fuels in a compression ignition engine. Fuel 246:79–92
41. Nabi MN, Rasul MG, Brown RJ (2019) Influence of diglyme addition to diesel-biodiesel blends on notable reductions of particulate matter and number emissions. Fuel 253:811–822
42. Imtenan S, Masjuki HH, Varman M, Arbab MI, Sajjad H, Rizwanul Fattah IM, Abedin MJ, Hasib ASMd (2014) Emission and performance improvement analysis of biodiesel-diesel blends with additives. Procedia Eng 90:472–477
43. Alptekin E (2017) Emission, injection and combustion characteristics of biodiesel and oxygenated fuel blends in common rail diesel engine. Energy 119:44–52
44. Zheng Z, Wang X, Zhong X, Hu B, Liu H, Yao M (2016) Experimental study on the combustion and emission fueling biodiesel/n-butanol, biodiesel/ethanol and biodiesel/2,5-dimethylfuran on a diesel engine. Energy 115:539–549

Parametric Study on the Steady Shear Rheology of Fumed Nanosilica Colloidal Dispersions

Prince Kumar Singh and Neelanchali Asija Bhalla

1 Introduction

Shear thickening fluid belongs to a class of non-Newtonian fluids, which transforms to a high viscosity thickened state when subjected to external forces [1]. There are different theories which are used to explain the shear thickening phenomenon like order–disorder transition theory, hydro-cluster theory, jamming theory, and friction contact theory [2]. For instance, Hoffman [3] conducted a thorough investigation into the shear thickening phenomenon and made an important observation. He reported that when a fluid exhibits shear thickening behavior, there is a structural transformation where the previously ordered flow of particles to a disordered state. Based on this observation, Hoffman proposed that shear thickening occurs due to an "order–disorder" mechanism. Later, it was proved that the "order–disorder" transition is not required for shear thickening to occur [4]. Bossis [5] introduced a mechanism called the "hydro-cluster" model, which effectively explains the phenomenon of continuous shear thickening. According to this model, when subjected to shear forces, suspended particles come together to form sizable clusters due to the dominance of short-range hydrodynamic lubrication forces over repulsive forces generated by neighboring particles. While the "hydro-cluster" mechanism is a relevant factor in the discontinuous shear thickening behavior, it alone is insufficient to fully explain the phenomenon. The transmission of stresses in higher magnitudes plays a crucial role in triggering the discontinuous shear thickening behavior [6]. Seto et al. [4] proposed the frictional contact model to explain the shear thickening mechanism

P. K. Singh · N. A. Bhalla (✉)
Mechanical Engineering Department, Bennett University, Greater Noida, Uttar Pradesh 201310, India
e-mail: neelanchali.bhalla@bennett.edu.in

N. A. Bhalla
Center of Excellence for Nano Sensors and Nano Medicine, Bennett University, Greater Noida, Uttar Pradesh 201310, India

R. Sahu et al. (eds.), *Advancements in Materials Processing Technology, Volume 1*,
Springer Proceedings in Materials 48, https://doi.org/10.1007/978-981-97-4958-4_21

which elucidates that the viscosity experiences a sudden surge due to the particle frictional contact, which is primarily responsible for the discontinuous shear thickening behavior. Recently, Waitukaitis [7] proposed a solidification mechanism that is initiated by an impact event. When there is an impact, it causes a localized front of jamming to propagate in the direction of the impact. As the jamming zone approaches the solid bottom surface, it sends a high normal stress through the jammed material, causing a force to be communicated back to the object that made the collision. This process ultimately leads to solidification and stabilization of the impacted material. It is important to note that shear thickening can be influenced by multiple mechanisms acting simultaneously, and the relative contribution of each mechanism may vary depending on the specific system and conditions. Additionally, the mechanisms proposed for shear thickening are still the subject of ongoing research, and the field is evolving rapidly as new insights and experimental techniques emerge. The rheological properties of STF are influenced by several factors, like particle concentration, size, shape, surface properties, solvent properties, applied shear rate/stress, temperature, time, and pressure [8–11]. These factors can be broadly classified into two groups: static and dynamic. Static factors remain constant after the synthesis of the STF, while dynamic factors change during the operational life of the STF. Static factors that influence the rheological properties of STFs include particle concentration, size, shape, surface properties, and solvent properties. On the other hand, applied shear rate/stress, temperature, time, and pressure are considered as dynamic factors. These dynamic factors can alter the rheological properties of the STF over time, making them important to consider during the design and operation of STF-based systems. By understanding how static and dynamic factors influence the rheological properties of STFs, researchers and engineers can optimize the performance of STFs for a variety of applications.

The primary objective of the present research was to investigate how temperature, particle size, and solvent quality affect the rheological properties of shear thickening fluids (STFs). We can select the most effective STFs with desirable attributes by identifying the best particle size and solvent combination. Since STFs are utilized in a variety of geographical regions and applications, it is critical to assess their efficacy at varied temperatures to ensure their dependability in a variety of circumstances.

2 Materials and Method

Silica nanoparticles, polyethylene glycol (PEG 200 and PEG 400), and ethyl alcohol were used in this investigation. Four different grades of fumed silica were taken for this study, namely A130, A150, A200, and A300 [10], respectively, where the numbers represent the specific surface area, which were obtained from Evonik Degussa. PEG solvents with average molecular weights of 400 g/mole, 200 g/mole were used to dissolve the silica nanoparticles. Ethyl alcohol was used to adequately disperse the fumed silica. Qualikems Fine Chemicals supplied the PEG, while Sigma-Aldrich supplied the ethanol. Each material was used in the form it was supplied

Table 1 Compositions of different STF samples

Sample No.	Solute	Solute % by wt	Solvent
1	Fumed Silica(A200)	20	PEG200
2	Fumed Silica(A200)	20	PEG400
3	Fumed Silica(A300)	20	PEG400
4	Fumed Silica(A300)	20	PEG200
5	Fumed Silica(A150)	20	PEG200
6	Fumed Silica(A150)	20	PEG400
7	Fumed Silica(A130)	20	PEG200
8	Fumed Silica(A130)	20	PEG400

by the above-mentioned suppliers. Table 1 shows the constituents of various STF samples in detail.

The ultrasonic homogenization technique was used to prepare the test samples. PEG and silica nanoparticles in known weight percentages were blended with surplus amount of ethanol. Following mixing, the mixture was treated for three hours with a ultrasonic horn at 20 kHz frequency, power density of 1200 W/cm^2, and amplitude of 40%. The silica weight percentage remained constant at 20% wt. Thereafter, the dispersion was placed in an oven for two hours at 80 °C to evaporate ethanol, after which the produced samples were ready for further testing.

3 Experiment and Results

3.1 Steady-State Rheological Testing

All the synthesized STF samples were tested on Anton Paar rheometer (MCR 301) equipped with a PP20 measuring system (20-mm plate), and a Peltier temperature control device to vary the test temperature precisely. The steady shear testing was accomplished by ramping the shear rate linearly from 0.1 to 1000 s^{-1} [12], at the desired temperature, and plotting viscosity versus shear rate graph on a log–log scale.

3.2 Effect of Surface Area

In this study, colloidal dispersions consisting of PEG 200 with different specific surface areas of fumed silica were examined to explore its effect on the steady shear rheology of fumed silica nanodispersions. Figure 1 depicts the experimental results obtained from rheological testing.

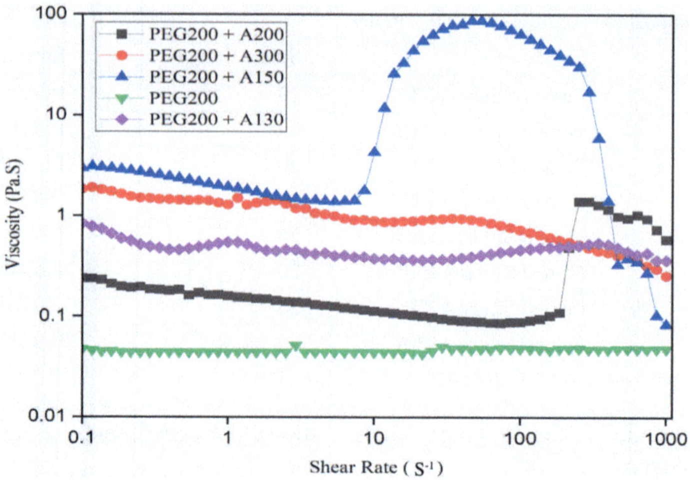

Fig. 1 Viscosity versus shear rate plots of different STFs with fumed silica samples of different specific surface area

From Fig. 1, it is observed that as the SSA increases from A130 to A300, the critical shear rate tends to move upwards. Notably, the A150 fumed silica dispersion was found to exhibit highest post-shear thickening viscosity, followed by A200 dispersion. A130 and A300 fumed silica dispersions displayed minimal shear thickening behavior. PEG 200 dispersion medium did not exhibit any shear thickening behavior as it is a Newtonian fluid.

3.3 Effect of Dispersion Medium

Shear thickening behavior of a fluid is significantly affected by the viscosity of dispersion medium. The interactions between particles in a suspension play a vital part in shear thickening, and these interactions can be affected by various properties of the dispersion medium, including viscosity and intermolecular forces. Consequently, the response of shear thickening is influenced by the strength of these interparticle interactions. Figure 2 depicts the effect of PEG 200 and PEG 400 dispersion mediums on the steady shear rheology of A150 and A200 fumed silica dispersions.

It is observed that the STF prepared from PEG 400 as the dispersion medium exhibited stronger shear thickening behavior in terms of higher post-shear thickening viscosity, as compared to STF synthesized from PEG 200 dispersion medium. This trend is common for both A150- and A200-based shear thickening dispersions.

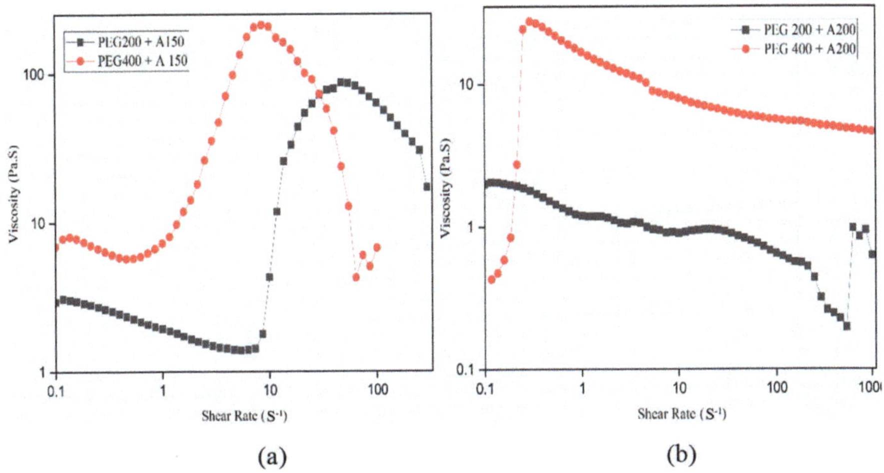

Fig. 2 Rheogram of A150 (**a**) and A200 (**b**) in PEG 200 and PEG 400 dispersion mediums

3.4 Effect of Temperature Variation

The experimental results clearly indicate a distinct behavior of the shear thickening fluid (STF) samples with respect to temperature. Steady-state shear tests were performed at 10 °C, 25 °C, and 45 °C, respectively, and the observations are reflected in Fig. 3. Figure 3a and b depict the rheological behavior of A150 and A200 fumed silica dispersions in PEG200, respectively. It was observed that with the rise in temperature, the severity in shear thickening decreased, resulting in lower post-shear thickening viscosity. Also, the onset of shear thickening was delayed to higher shear rates with the increase in temperature, causing the increase in critical shear rate value.

4 Conclusions

The shear thickening dispersions synthesized in this study were exposed to steady shear rheological testing at different temperatures. The below-mentioned inferences can be obtained from the experimental results:

- With the increase in specific surface area, the onset of shear thickening was delayed to high shear rates, with the decrease in the severity of shear thickening. This was evident from the rheograms of A150 and A200 fumed silica dispersions.
- The shear thickening enhanced with the increase in viscosity as well as the molecular weight of the dispersion medium. This was evident from the experimental results where STF using PEG 200 solvent exhibited lower post-shear thickening viscosity as compared to STF using PEG 400 as the solvent.

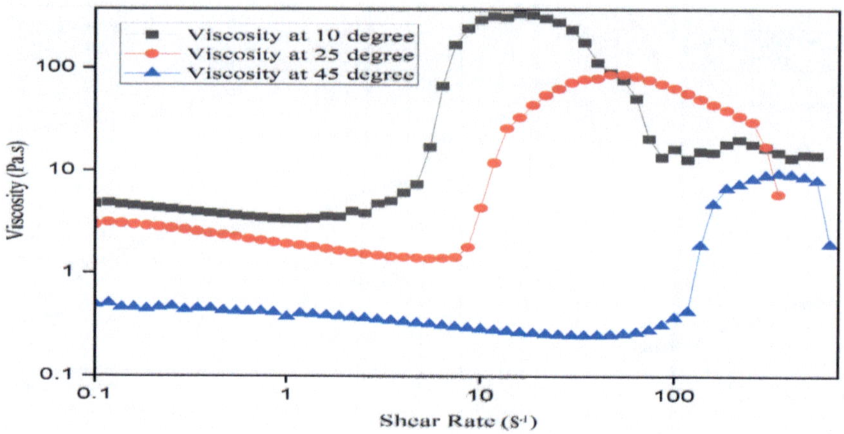

(a)

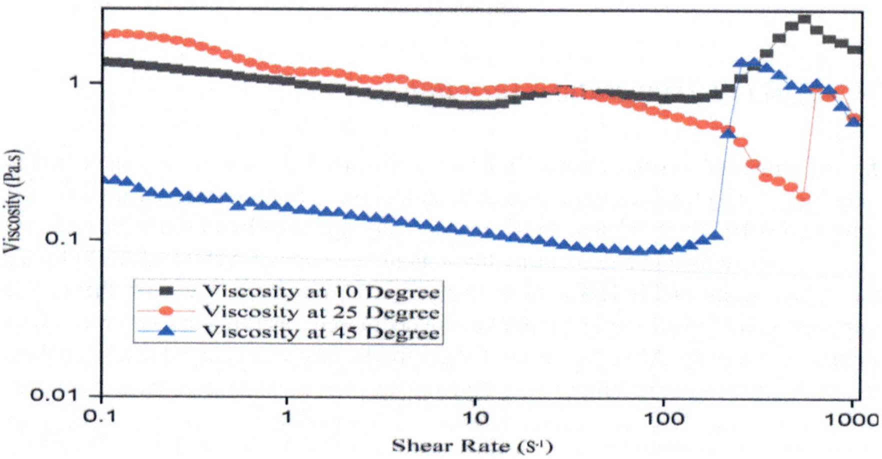

(b)

Fig. 3 Rheogram depicting the effect of temperature variation on the viscosity of **a** A150 and **b** A200 fumed silica dispersion in PEG 200

- The shear thickening decreased as well as the onset of shear thickening behavior was delayed to higher shear rates, with the increase in temperature. This was evident from the rheograms obtained from the testing of A150 and A200 fumed silica dispersions at different temperatures.

Acknowledgements There is no direct funding for this research work. Authors would like to express my sincere gratitude to Dr. Sumit Singh of Anton Paar India Pvt. Ltd. and Dr. Piyush Gaur of the University of Petroleum and Energy Studies for their invaluable contributions and support throughout the course of this research project.

References

1. Barnes HA (1989) Shear-thickening ("dilatancy") in suspensions of nonaggregating solid particles dispersed in Newtonian liquids. J Rheol (N Y N Y) 33:329–366. https://doi.org/10.1122/1.550017
2. Asija N, Chouhan H, Bhatnagar N. Synthesis of shear thickening fluids for liquid armour applications. ISME J Manuf Sci 6:9–17
3. Hoffman RL (1974) Discontinuous and dilatant viscosity behavior in concentrated suspensions. II. Theory and experimental tests. J Colloid Interface Sci 46:491–506. https://doi.org/10.1016/0021-9797(74)90059-9
4. Seto R, Mari R, Morris JF, Denn MM (2013) Discontinuous shear thickening of frictional hard-sphere suspensions. Phys Rev Lett 111. https://doi.org/10.1103/PhysRevLett.111.218301
5. Bossis G (1985) The rheology of concentrated suspensions of spheres in simple shear flow by numerical simulation. J Fluid Mech 155:105–129. https://doi.org/10.1017/S0022112085001732
6. Melrose JR, Ball RC (2004) "Contact networks" in continuously shear thickening colloids. J Rheol (N Y N Y) 48:961–978. https://doi.org/10.1122/1.1784784
7. Waitukaitis SR (2013) Impact-activated solidification of cornstarch and water suspensions
8. Tian T, Li W, Ding J et al (2013) Study of the temperature effect of shear thickening fluid. In: 2013 IEEE/ASME international conference on advanced intelligent mechatronics: mechatronics for human wellbeing, AIM 2013, pp 833–837. https://doi.org/10.1109/AIM.2013.6584197
9. Wei M, Lin K, Sun L (2022) Shear thickening fluids and their applications. Mater Des 216:110570. https://doi.org/10.1016/j.matdes.2022.110570
10. Asija N, Chouhan H, Gebremeskel SA, Bhatnagar N (2017) Influence of particle size on the low and high strain rate behavior of dense colloidal dispersions of nanosilica. J Nanoparticle Res 19. https://doi.org/10.1007/s11051-016-3723-4
11. Hasanzadeh M, Mottaghitalab V (2014) The role of shear-thickening fluids (STFs) in ballistic and stab-resistance improvement of flexible armor. J Mater Eng Perform 23:1182–1196. https://doi.org/10.1007/s11665-014-0870-6
12. Bhalla NA, Chouhan H, Ahmed A et al (2019) Study on the low and high strain rate behavior of fumed nano silica dispersions. Procedia Struct Integr 14:564–570. https://doi.org/10.1016/j.prostr.2019.05.069

References

Hydration and Strength Properties of Alkali-Activated High-Volume Fly Ash Concrete: Incorporation of Water Glass and Sodium Hypochlorite Activators

Pritha Das and Anup Kumar Mondal

1 Introduction and Literature Review

Among the several alternative binders that can replace cement concrete, geopolymer concrete has recently become a highly popular and intriguing binder. Geopolymer is simply a composition of completely replaced by-product components that are activated by an alkaline solution at high temperatures before curing to strengthen properties. According to statistics, OPC generated 8% of CO_2 gas in the environment at the time, which is a serious issue for our ecosystem [1]. Sustainable geopolymer or alkali-activated material, on the other hand, is composed of aluminum silicate material such as fly ash and slag that interacts with alkali solution to generate a binder-like substance [2, 3]. Previous research on geopolymer activation has identified sodium hydroxide (NaOH) and sodium silicate solution (water glass) as well-known alkali activators [4, 5]. As a result, the current research effort seeks to break the tradition of used alkali solution in all prior literature history, as well as the effects of sodium silicate and new alkali liquid on the behavior and strength of paste. There is a limited research history since fly ash-based alkali-activated materials or geopolymer is new and exciting topics to develop as an alternative technology to regular concrete. Some reviews are highlighted as a result of this inclusion.

In early 2000–2002, much and potential experimental work was performed based on characteristics of geopolymer mortar. Author Gorhan G by his experimental work stated that the enhanced strength of geopolymer mortars is observed at temperature > 60 °C than 80 °C for the same molar concentration and also pointed out that at ambient temperature (about 27 ± 2 °C) it provides very poor strength due to slow polymerization [6, 7]. On the other side found that during mixing the fresh geopolymer mortar quickly tends to become very stiff and dry, and showed high viscosity and cohesive nature. Even though it was earlier concluded that Vicat's

P. Das (✉) · A. K. Mondal
Department of Civil Engineering, Techno India University, Kolkata, West Bengal, India
e-mail: prithadas2899@gmail.com

R. Sahu et al. (eds.), *Advancements in Materials Processing Technology, Volume 1*,
Springer Proceedings in Materials 48, https://doi.org/10.1007/978-981-97-4958-4_22

needle apparatus is not applicable to use to measure the set time, in 2003 authors stated that the information available for the quantitative measurement of the setting time of geopolymer material is utilizing the Vicat needle and also in his paper proposed that geopolymer mortars do not set even during initial period setting [8]. In the FA compound, lower level of obtained reactive silica and alumina helps to set off the binding reaction and also stated that a long time of curing at a high temperature can lead to a decrease the compressive strength causing diminished microstructure and increased porosity due to loss of moisture [9, 10].

According to author Chanh et al. [11], compressive strength is an important attribute for all concrete and is affected by curing time and temperature. The compressive strength increases as the curing time and temperature increase. The compressive strength of concrete can be attained in the range of 40–50 N/mm^2 with curing temperatures ranging from 60 to 90 °C and curing times ranging from 24 to 72 h [11]. Furthermore, author Komljenovi'c et al. [12] observed that the amount of fine fly ash particles (less than 43 m) determined the compressive strength of geopolymers. As the fineness of the fly ash grew, so did its compressive strength. As a result, the kind and concentration of activators were important factors in the alkali activation reaction. With a sodium silicate solution as an activator ($n = 1.5$; 10% Na_2O), the best compressive strength was attained. Because it contains dissolved and partially polymerized silicon, sodium silicate is the ideal alkaline activator because it reacts quickly, permeates into the reaction products, and significantly enhances mortar properties [12]. Also noticed that the increased water/binder ratio and decreased aggregate/binder ratio lead to an increased flow of alkali-activated mortars, as mortar flow decreased with aggregate/binder ratios > 2 [13]. They also discovered that at the same mix proportion, slag-based geopolymer mortar had higher compressive strength than fly ash-based geopolymer mortar [14].

Recent research work of year 2015 stated that the molarity of alkali activator influences compressive strength greatly. They took two samples with alkali liquid to fly ash ratios of 0.35 and 0.4 and varied the molarity of sodium hydroxide 8 M, 10 M, 12 M, and 14 M and found that compressive strength increases with sodium hydroxide concentration but the ratio of fly ash to solution has no effect on strength [1]. Author Inti et al. [15] in their research work concluded that the curing method has a significant impact on the strength value of geopolymer concrete, and that while hot air over dying samples produces higher compressive strength values than sunlight curing samples, sunlight curing samples are preferable for practical in-situ work [15, 16]. Also, in fly ash-based geopolymer concrete, ground granulated blast furnace slag (GGBFS) or rice husk ash (RHA) can be partially replaced, because experiment results show that 50% partial replacement of fly ash by GGBFS yields satisfactory compressive strength values, similar to partial replacement by rice husk ash [17].

Authors Rao and Satheesh Kumar [18] investigated whether oven-curing may reduce the time required for GGBS to set, conducting trials on geopolymer concrete built with 8 M, 12 M, and 16 M for 1440 min at room temperature and 60 °C in oven-curing [18]. Many additives have lately been added into geopolymer concrete, with some establishing usefully their safe qualities in geopolymer concrete and many others undergoing additional research. As a result, from 2020 to 2023, several researchers attempted to establish various viable additions such as rick husk ash silicate, steel fiber, polypropylene steel fiber, and metakaolin. Some of these additives have been successfully mixed with geopolymer concrete to increase its mechanical qualities [15, 19]. A number of research have been carried out in order to produce ambient cured fly ash-based geopolymer concrete by including calcium supplemented components, for use in both in-situ and precast construction. The presence of free calcium ions in the solution extends the breakdown of fly ash and enhances the geopolymer strength due to the increased heat of hydration.

2 Materials and Methodology

2.1 Materials

For aluminosilicate precursors, low-calcium class F fly ash with a fineness modulus of 2.06 used. Additionally, OPC is used as a binder substance, and at 40%, 50%, 60%, and 70%, fly ash is substituted. Tables 1, 2, 3, 4, 5, 6, 7 and 8 summarize the characteristics of all materials used in this study.

Table 1 Chemical composition of class F fly ash

SiO_2	Al_2O_3	Fe_2O_3	TiO_2	CaO	MgO	SO_3	Na_2O	LOI*
51.09	28.34	12.5	1.51	1.79	0.84	1.09	0.42	2.42

*Loss in ignition
Note Data collected from Global lab, Kolkata 2023

Table 2 Chemical composition of OPC (43 grade)

Specific gravity	3.15
Loss of ignition, %	2.9
SiO_2, %	19.14
Al_2O_3, %	4.75
CaO, %	64.1
MgO, %	1.65
SO_3, %	3.35
Na_2O, %	0.25
K_2O, %	0.8
Fe_2O_3, %	2.78

Note Data collected from Global Lab, Kolkata 2023

Table 3 Physical properties of rive sand (zone III)

Properties	Values
Fineness modulus	2.16
Specific gravity	2.55
Bulk density	1.45 gm/cc

Note Data obtained from Techno India University laboratory examination

Table 4 Particle size distribution of fine aggregates

Sieve size	wt retained (gm)	% of wt retained	%wt passing	Cumulative %wt retained
4.75 mm	10	2	98	2
2.36 mm	10	2	98	4
1.18 mm	10	2	98	6
600 microns	20	4	96	10
212 microns	430	86	14	96
150 microns	10	2	98	98
				Σ216

Note Data obtained from Techno India University laboratory examination

Table 5 Physical properties of coarse aggregate (20 mm nominal size)

Properties	Values
Fineness modulus	2.635
Specific gravity	2.96
Bulk density	1.52 gm/cc

Note Data obtained from Techno India University laboratory examination

Table 6 Particle size distribution of coarse aggregate

Sieve size (mm)	wt retained (gm)	% of wt retained	%wt passing	Cumulative %wt retained
20	420	21	79	21
12.5	920	46	54	67
10	190	9.5	90.5	76.5
4.75	450	22.5	77.5	99
				Σ263.5

Note Data obtained from Techno India University laboratory examination

Table 7 Physical properties of sodium silicate (Na_2SiO_3)

Specific gravity	1.325
Density	1.4 g/cc
pH	11.56
Na_2O, %	10.4
SiO_2, %	31.2
H_2O, %	50.5

Note Data collected from Omega Lab, Thakurpukur, Kolkata

Table 8 Physical properties of Sodium hypochlorite (NaOCl)

Specific gravity	1.24
Density	1.11 g/cc
pH	12
Molarity	12

Note Data collected from Omega Lab, Thakurpukur, Kolkata

2.2 *Mix Proportion and Preparation of Alkali-Activated High-Volume Fly Ash Concrete Samples*

All materials are mixed in accordance with IS 10262:2019 mix proportioning for M30, M40, M50, and M60 grades with fly ash contents of 40, 50, 60, and 70%, respectively, for high-volume fly ash concrete. The procedure is as follows: first, the job parameters—aggregate characteristics, maximum aggregate size, slump, *w*/cm, admixtures with 2%, 3%, 5%, and 10%—are determined, and then the batch weight is calculated. First, the *w*/*b* ratio is calculated. According to the water/cementitious ratio (*w*/cm) principle, the strength of concrete at a particular age depends on the *w*/cm as the lower the *w*/cm, the higher the concrete strength. However, strength is determined by the *w*/cm, and economy is determined by the proportion of aggregate presence that would still result in a workable mix. The designer's goal should always be to achieve maximum strength with the least amount of cement and acceptable workability. IS code is used to determine the amount of air and water content. Because the mix was designed for mild climatic conditions, the entrapped air space accounted for 0.01 m^3 of the total volume of wet concrete. W/cm is then chosen using an IS code table plotted graph. As the presence of pozzolans can reduce strength, the *w*/cm values for M30, M40, M50, and M60 concrete are 0.43, 0.35, 0.3, and 0.25, respectively. The cement content is calculated by multiplying the amount of water to be added by the *w*/cm. The usage of cementitious material is limited because shrinkage caused by the high volume of finer material and other materials are calculated to correlate. And all casted cubes cured in a water bath at room temperature for 7, 28, and 90 days. The mix proportions are detailed in Table 9.

Six samples were cast for each sample according to the Indian standard code of concrete, followed by the stated mix proportion. To obtain a homogeneous mixture,

Table 9 Mix proportioning of the alkali-activated high-volume fly ash concrete (calculate in accordance with IS 10262:2019) (Das, P. 2023, Unpublished data, Techno India University)

Samples notation	Cement (gm)	Fly ash (gm)	Cementitious materials (C + 75% of FA) (gm)	Alkali solution (gm)	Coarse agg. (gm)	Fine agg. (+25% fly ash)(gm)	W/B ratio	Water (gm)
AM30_ FA40	275.106	183.404	412.66	8.25(2%)	1104.36	539.784	0.43	177.44
AM30_ FA50	229.255	229.255	401.2	12.036(3%)	1092.1	533.79	0.43	172.516
AM30_ FA60	183.404	275.106	389.734	19.5(5%)	1080	527.8	0.43	167.6
AM30_ FA70	137.51	321	378.303	37.83(10%)	1051.78	514.08	0.43	162.67
AM40_ FA40	338	225.324	507	10.14(2%)	1016.72	496.974	0.35	177.45
AM40_ FA50	281.655	281.655	492.895	14.8(3%)	1015.1	490.1	0.35	172.51
AM40_ FA60	225.31	338	478.824	24(5%)	1000	488.76	0.35	167.6
AM40_ FA70	168.99	394.32	464.733	46.5(10%)	950	464.4	0.35	162.7
AM50_ FA40	394.32	262.88	591.48	11.83(2%)	964.13	471.24	0.3	177.444
AM50_ FA50	328.6	328.6	575.05	17.25(3%)	944.56	461.7	0.3	172.52
AM50_ FA60	262.88	394.32	528.62	26.431(5%)	938.81	459	0.3	158.6
AM50_ FA70	197.16	460.04	542.19	54.22(10%)	876.48	428.4	0.3	162.66
AM60_ FA40	473.184	315.456	709.8	14.2(2%)	876.48	428.4	0.25	177.45
AM60_ FA50	394.32	394.32	690.06	20.7(3%)	841.91	411.264	0.25	172.515
AM60_ FA60	315.44	473.2	670.36	33.52(5%)	818.63	400.13	0.25	167.6
AM60_ FA70	236.6	552.05	650.632	65.1(10%)	753.8	368.424	0.25	162.66

10 min of proper hand-mixing was required, followed by mixing with a mixer machine. It should be noted that new pastes may lack air bubbles or lumps. Following that, the paste was put into an iron concrete cube mold and placed in a vibration machine to guarantee that all air bubbles were entirely removed and that the paste in the cube was totally compacted. The cubes needed to rest for 48 h after being

prepared in the molds. The cast samples were cured in a water tub for 7, 28, and 90 days.

Observation during Mixing of Materials. During the preparation of fresh concrete mixes using the mix proportions used in this technique, the corresponding concrete materials such as OPC, fly ash, sand, coarse aggregates, and alkali solutions proved to be slightly insufficient to accomplish the required mold size. In contrast to polymer activation, such as modified sulfo-naphthalene activator, cementitious binder underwent dissolution in the presence of alkali ($Na_2SiO_3 + NaOCl$) activation. The most likely cause of such dissolutions can be attributed to the reaction described below.

$$\overset{\text{(CEMENT)}}{(CaO + SiO_2 + Al_2O_3)} + \overset{\text{(ALKALI ACTIVATOR)}}{(Na_2SiO_3 + NaOCl)} \xrightarrow{\text{(WATER)}} \text{Dissolved sub-stains}$$

The dissolved binder acquired as a product is most likely to blame for the lower resultant volume of concrete mix. Because of the dissolved binder, the matrix becomes denser and fills the interstitial pores. In addition, when Na_2SiO_3 was introduced during the fresh concrete mix preparation, a novel observation was made about the increased heat of hydration that evolved. Alkali activator, unlike water fly ash, operates as a catalyst, rapidly boosting the heat of hydration of the hydraulic binder.

3 Results and Discussions

3.1 *Workability of Fresh Concrete*

Various concrete combinations are created using machine mixing with low water demand. Significant differences in workability were seen for concrete samples ranging from ordinary strength to high strength lean concrete when the alkali activator dosage was increased with increased fly ash concentration. The findings of the experiments reveal that increasing the amount of fly ash substitution above 40% *b*/*w* of cement resulted in decreased workability. When activators were applied, the workability of fresh concrete increased with increasing percentages of OPC replacement with class F fly ash from 40 to 70% cement replacement. Concentrations of activator were 2%, 3%, 5%, and 10%. According to the results of the experiments, 3% activator provided appropriate and satisfactory workability results, whereas 10% activator is preferable for higher slump values, C.F. values, and slump flow values. The workability outcomes are given in Table 10.

Table 10 Results on workability

Samples	*W*/*B* ratio	Alkali content	Slump flow (mm)	Compacting factor (C.F.)	Flow spread (mm.)
AM30_FA40	0.43	(2%)	100	0.89	380
AM30_FA50	0.43	(3%)	120	0.92	400
AM30_FA60	0.43	(5%)	150	0.93	420
AM30_FA70	0.43	(10%)	180	0.95	450
AM40_FA40	0.35	(2%)	110	0.88	320
AM40_FA50	0.35	(3%)	130	0.92	440
AM40_FA60	0.35	(5%)	180	0.94	500
AM40_FA70	0.35	(10%)	220	0.95	520
AM50_FA40	0.3	(2%)	130	0.94	250
AM50_FA50	0.3	(3%)	180	0.95	470
AM50_FA60	0.3	(5%)	210	0.90	490
AM50_FA70	0.3	(10%)	230	0.95	530
AM60_FA40	0.25	(2%)	130	0.82	300
AM60_FA50	0.25	(3%)	160	0.89	430
AM60_FA60	0.25	(5%)	200	0.93	480
AM60_FA70	0.25	(10%)	230	0.95	510

Data were obtained from Techno India University laboratory examination

3.2 Compressive Strength of Concrete After Curing Period

After 7, 28, and 90 days of water bath curing, the graph in Fig. 1 shows the compressive strength values of M30 concrete grades that are significant ALM30 with various percentages of fly ash substitution, i.e., 40, 50, 60, and 70%. It shows that concrete samples do not obtain acceptable strength early in the curing process, but gradually achieve satisfactory strength later on.

Their results suggest that M40, M50, and M60 with fly ash percentages of 40, 50, and 60% exhibit satisfactory strength, i.e., 50–70%, which can be related to literature studies to be a safe strength to use in structural applications. It was also shown that concrete containing 70% fly ash showed declining values from early to late periods, which could be due to low cement content levels participating in the hydration process, as present C_3S in cement contributes the majority of growing strength. Also, there is insufficient mention of a 70% replacement of fly ash in the literary history. Also, increasing the percentage of alkali with increasing the percentage of fly ash implementations in concrete shows great increasing values as increasing the level of external alkali incorporating fills the gap of decreasing the level of cement as external alkali helps to up lite the hydration process faster, which helps to gain compressive strength and consistency at an early stage of curing.

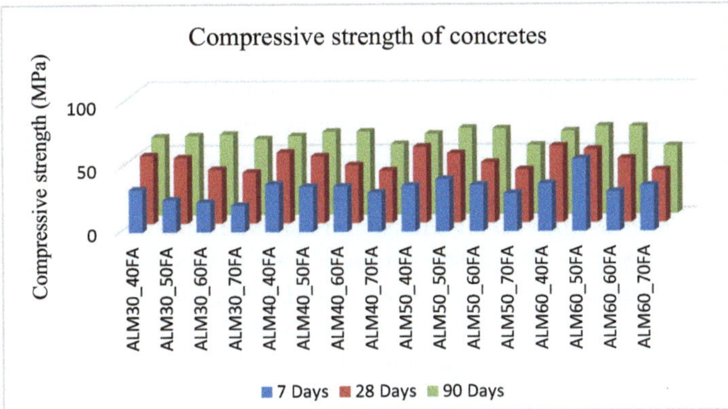

Fig. 1 Compressive strength of all casted samples after 7, 28, and 90 days of curing (Das, P. 2023, Unpublished data graph) (*Note* The column graph was developed using compressive strength's results which is gathered from laboratory examination at Techno India University)

The Impact of Adding Sodium Silicate and Sodium Hypochlorite Solution on the Hydration of Concrete. According to the literature, replacing cement with fly ash/slag alone reduces initial setting time, strength characteristics, and workability [20, 21]; therefore, external silica content is required to increase workability in terms of strength properties and initial setting time. Based on the higher grade of concrete, the alkali content in the concrete mix was taken at 2, 3, 5, and 10% to improve workability, but it was determined that 10% incorporating alkali content in concrete does not improve the strength qualities but improves the slump and workability properties. It was also discovered that 3% and 5% sodium silicate considerably improve the hydration reaction over NaOCl. While prior research has indicated that sodium silicate and sodium hydroxide/potassium hydroxide are superior for hydration and mechanical reactions in concrete [21], this study shows that the newly discovered activator NaOCl is ineffectual when compared to sodium hydroxide.

4 Conclusion

The current study attempts to determine whether or not a newly employed alkali activator such as sodium silicate and sodium hypochlorite solution combination has the capacity to activate the polymerization reaction.

As the replacement increases above 60% high-volume fly ash, both strength generation and strength drop. It is not always possible to increase strength or durability by decreasing *w*/cm. For best results, keep the *W*/cm at 0.3. Due to the slower strength production rate and lower evolution hydration heat of high-volume fly ash concrete, formwork removal can be critical during construction. Some of the strength ratings, however, are unsuitable for concrete. This error could be caused by two things:

first, errors in computation according to the IS code, and second, material quality or quantity issues.

To activate strong qualities in alkali solution, sodium silicate or water glass solution predominately participates in heat of hydration reaction and polymeric reaction. In contrast, increasing the amount of sodium silicate decreases strength values on later days while raising strength values on early days for a faster heat of hydration process. However, using this alkali with sodium silicate and sodium hypochlorite in concrete is less expensive. Because sodium silicate is less expensive than sodium hydroxide on the market, each concrete will cost between Rs. 3000 and Rs. 5000. It has some drawbacks that show up in the data, but overall performance indicates that the new alkali activator can make concrete more economical and environmentally beneficial for the concrete industry.

References

1. Provis JL, van Deventer JS (2014) Alkali activated materials. State of the art report of RILEM TC 224-AAM. Springer/RILEM, Dordrecht
2. Xu H, van Deventer JSJ (2002) Geopolymerisation of multiple minerals. Miner Eng 15(12):1131–1139
3. Davidovits J (1984) Pyramids of Egypt made of man-made stone, myth or fact? In: Symposium on Archaeometry 1984. Smithsonian Institution, Washington, DC
4. Davidovits J (2008) Geopolymer chemistry and applications. Institute Géopolymère, Saint-Quentin, France
5. Yip CK, Lukey GC, Provis JL (2008) Effect of calcium silicate sources on gopolymerization. Cem Concr Res 38:554–564
6. Gorhan G, Kurklu G (2011) The influence of the NaOH solution on the properties of the fly ash based geopolymer mortar cured at different temperature. Compos Part B 58:371–377
7. Barbosa FFV, Mackenzie JDK, Thaumaturgo C (2000) Synthesis and characterisation of materials based on inorganic polymers of alumina and silica: sodium polysialate polymers. Int J Inorg Mater 2000:309–314
8. Hardjito D (2005) Studies on fly ash-based geopolymer concrete, November 2005
9. Fernández-Jiménez A, de la Torre AG, Palomo A et al (2006) Quantitative determination of phases in the alkaline activation of fly ash. Part II: Degree of reaction. Fuel 85:1960–1969
10. Fernández-Jiménez A, Palomo A, Sobrados I et al (2006) The role played by the reactive alumina content in the alkaline activation of fly ashes. Microporous Mesoporous Mat 91:111–119
11. Chanh NV, Trung BD, Tuan DV (2008) Recent research geopolymer concrete. In: The 3rd ACF international conference—ACF/VCA
12. Komljenovic M, Bascarevic Z, Bradic V (2010) Mechanical and microstructural properties of alkali-activated fly ash Geopolymer. J Hazardous Matter 181:35–42
13. Yang KH, Song JK, Lee KS, Ashour AF (2009) Flow and compressive strength of alkali activated mortars. ACI Mater J 106:50–58
14. Yang KH, Song JK (2009) Workability loss and compressive strength development of cementless mortars activated by combination of sodium silicate and sodium hydroxide. J Mater Civ Eng ASCE 21:119–127
15. Inti S, Sharma M, Tandon V Dr (2016) Ground granulated blast furnace slag (GGBS) and rice husk ash (RHA) uses in the production of geopolymer concrete. Geo-Chicago 2016, ASCE J
16. Iftiqar Ahmed A, Siddiraju S Dr (2016) Strength properties on fly ash based geo polymer concrete with admixtures. Int J Civ Eng Technol (IJCIET) 7(3), May–June 2016

17. Panias D, Giannopoulou IP, Perraki T (2007) Effect of synthesis parameters on the mechanical properties of fly ashbased geopolymers. Colloids Surf, A 301(1–3):246–254
18. Rao GM, Satheesh Kumar K, Poloju KK, Srinivasu K (2020) An emphasis of geopolymer concrete with single activator and conventional concrete with recycled aggregate and data analyzing using artificial neural network. In: Proceedings of the 3rd international congress on advances in mechanical sciences, Hyderabad, India, 27–28 August 2020
19. Pimraksa K, Chindaprasirt P, Rungchet A, SagoeCrentsil K, Sato T (2011) Lightweight geopolymer made of highly porous siliceous materials with various Na_2O/Al_2O_3 and SiO_2/Al_2O_3 ratios. Mater Sci Eng, A 528(21):6616–6623
20. Panda RK, Dhal JP, Mishra SC (2012) Effect of sodium silicate on strengthening behaviour of fly ash compacts. Int J Curr Res 4(2):244–246
21. Ishwarya G, Singh B, Deshwal S, Bhattacharyya SK. Effect of sodium carbonate/sodium silicate activator on the rheology, geopolymerization and strength of fly ash/slag geopolymer pastes. https://doi.org/10.1016/j.cemconcomp.2018.12.007

Effect of Multiple Hardening on Bearing Steel to Improve the Fatigue Life

P. Rajesh Kannan, Ayush Jain, and Nitesh Vashishtha

1 Introduction

Automotive industries are demanding higher performance bearings with increased load ratings to cater to the ever-increasing expectations of extended warranties. Bearings are subjected to repeated loads, hence they should possess high strength, wear resistance, and fatigue resistance [1–4]. Apart from the bearing design, material cleanliness, forging ratio, surface roughness, lubrication, rotating speed, and special heat treatment process, bearing rolling contact fatigue life property plays a vital role in the bearing performance [5]. In the material cleanliness, non-metallic inclusions in the material are the key initiators of rolling contact fatigue damage. The nature of inclusion and its shape and size plays a role in the premature failure of the bearing by acting as stress raisers and nucleation sites for fatigue crack initiation. Owing to the advancements in modern steel-making processes, ultra-clean steels are available for bearing applications and it has helped to improve the fatigue life on that score [6, 7]. However, an attempt is made to tailor the conventional heat treatment process for enhancing the bearing life by reducing the prior austenitic grain size (PAGS). Refining the PAGS is one of the effective ways of improving the material properties that can be achieved in several ways. Micro-alloying with alloying element additions to restrict the grain growth during austenitization, and thermo-mechanical treatment, thereby the material is deformed at elevated temperature to form fine grains, and double hardening was adopted widely to refine the PAGS [2–4].

The yield strength of the tempered martensitic steel is explained using different mechanisms, i.e., dislocation strengthening, grain boundary strengthening, solid solution strengthening, and precipitation hardening. Liu et al. stated that the effects of dislocation strengthening and solid solution strengthening are similar and the

P. Rajesh Kannan · A. Jain (✉) · N. Vashishtha
Research and Development, National Engineering Industries Ltd., Jaipur 302006, India
e-mail: Ayush.jain@nbcbearings.in

R. Sahu et al. (eds.), *Advancements in Materials Processing Technology, Volume 1*,
Springer Proceedings in Materials 48, https://doi.org/10.1007/978-981-97-4958-4_23

precipitation hardening and the grain boundary strengthening influence mechanical properties [1]. Due to improving the strength of material without reducing its ductility the grain refinement technique is a widely exercised approach by the grain boundary strengthening method [4, 8]. In the grain refinement technique, the steel samples are subjected to austenitization twice. During the first austenitization and hardening, the steel is attaining a homogeneous microstructure, and the dissolution of the coarse precipitates takes place. In the second austenitization and hardening, the experienced temperature is lesser than the first austenitization and it is resulting the homogeneous and finer grains than the first austenitization. In this process, fine grains lead to increased total grain boundary area and it hinders the dislocation slide and microcrack propagation. Lath width and precipitate size were also reduced in the double hardening treatment [1, 4, 8, 9]. Double hardening treatment attributes to attain a good combination of strength and toughness via grain refinement and the refinement of precipitates [1, 4, 8–12]. Moreover, it reduces the tempering martensite embrittlement and increases the retained austenite (RA) level [13, 14]. Khani et al. stated that grain boundary segregation of the impurity elements such as P, S, Sb, and Sn could lead to temper martensite embrittlement. In double hardening treatment, improvement in toughness is associated with lowering the effective concentration of impurities and its uniform distribution at the prior austenite grain boundaries due to the more grain boundaries [2]. Lee et al. reported that a combination of nitrocarburizing and double hardening treatments resulted in an increase in strength, impact toughness, and high cycle fatigue properties [3].

RA and compressive residual stress plays a vital role in the fatigue behavior of the material. RA is beneficial to material fatigue resistance by blunting the crack propagation tips in the material [11–17]. Compressive residual stresses (CRS) help to increase the fatigue life of the material [16, 18–20].

The current investigation aims to study the influence of the heat treating methods, i.e., single hardening and double hardening on the rolling contact fatigue behavior of the SAE 52100 bearing steel.

2 Materials and Methods

2.1 Material and Heat Treatment Process

SAE 52100 material was used for this present investigation and the chemical composition was analyzed using spark emission spectrometry. The resulting chemical composition of the SAE 52100 is summarized in Table 1. In the conventional heat treatment process, through hardening was done at 860 °C and hardened in hot oil at 110 °C and subsequently tempered at 180 °C in a continuous mesh belt furnace (single hard). However, in the double hardening process, carbonitriding at 930 °C

Table 1 Chemical composition of SAE 52100 material

Elements	C	Mn	Si	Cr	S	P	Ni	Cu	Al	Fe
wt%	1.00	0.37	0.33	1.53	0.005	0.012	0.010	0.019	0.016	Bal

and hardening at 860 °C was performed, and subsequently, the samples were rehardening in the austenitizing temperature 840 °C, i.e., less than the conventional through hardening temperature and hardened in oil at 110 °C.

2.2 *Rolling Contact Fatigue (RCF) Test*

2.2.1 Test Specimens

Rolling contact fatigue test specimens were prepared in the dimensions of 50 mm length and 12.5 mm diameter from the 60 mm spheroidized annealed bar. After hardening treatments, the RCF test specimens (Φ12 mm and 22 mm length) were prepared from the heat-treated rollers. The test specimens were ground and roller surfaces were honed to maintain surface roughness of less than 0.07 μm Ra. Cylindrical RCF test specimens were tested under clean lubrication conditions.

2.2.2 RCF Test Setup

The schematic sketch of the rolling contact fatigue test rig is shown in Fig. 1. Test roller was driven by a driving roller and supported by two balls made of SAE 52100 material. The balls were supported and guided by a guide roll as shown in Fig. 1.

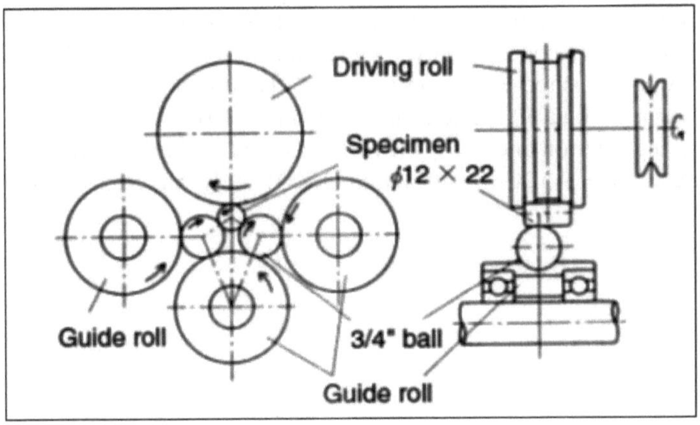

Fig. 1 Schematic sketch of rolling contact fatigue (RCF) test rig [15]

Table 2 RCF test conditions

Specimen size	Cylindrical specimen (Φ12 × L22)
Steel ball size	3/4″ (19.05 mm) Grade 20
Max. Hertzian contact stress, GPa	5.88
Loading roller rotation speed, rpm	4080
Loading cycles on specimen, times/min	46,240
Steel ball rotation speed, rpm	14,566
Lubricating oil	ISO VG 68 (mineral oil)

Thus the rollers were supported by two balls, for every rotation of the roller two contact points were generated. The load was applied and a maximum hertzian stress of 5.88 GPa was maintained during testing and tests were continued till failure. Rollers were lubricated by splashing the oil over the rollers continuously during the entire test duration and test conditions are tabulated in Table 2. Six specimens of each with single and double-hardened conditions were tested and L_{10} life was calculated. The roller specimen rotates at 23,120 rpm and is supported by two balls as shown in Fig. 1, establishing 46,240 contacts per minute. The test rig is equipped with vibration sensors and a provision to stop automatically once flaking occurs during the test. A timer is also provided to record the running time of the roller. After completing the test, the failed rollers were characterized using optical and scanning electron microscopy.

2.3 Metallurgical Characterization

The cut sections of the rollers were prepared for metallographic study as per ASTM E3. The polished samples were etched with 5% Nital solution. After etching, microstructures of the polished samples were observed at different magnifications using an optical microscope. Moreover, Picral etchant was used to evaluate the PAGS. The PAGS was measured as per ASTM E112. Hardness measurement was made across the sample with a 500 gf. load and dwell time of 10 s using a Clemex hardness tester. Scanning electron microscopy (SEM) was used to study the crack propagation and fracture surface. An X-ray diffraction (XRD) study was performed to measure the RA level.

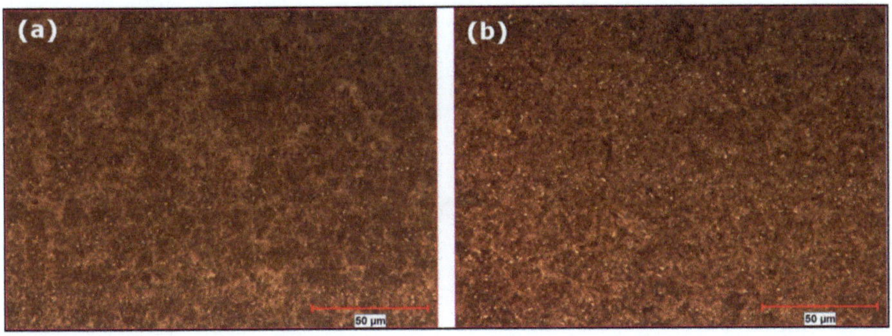

Fig. 2 Microstructure of **a** single quenched and **b** double quenched samples

3 Results and Discussion

3.1 Metallurgical Analysis

The microstructure of the SAE 52100 steel comprises of tempered martensitic structure with uniformly distributed carbides. The microstructure of single and double-hardened specimens is shown in Fig. 2.

During austenitization, these primary carbides are partially dissolved and reprecipitated as finely dispersed carbides in hardened and tempered conditions. The density of uniformly distributed carbides is relatively higher in a single-hardened structure. Subsequent reaustenitization during second hardening would have led to the further dissolution of carbides in the matrix resulting in the observed lower density of carbides. This is corroborated by the fact that a slightly higher amount of RA is observed on the double-hardened specimen (20–25%) when compared with single-hardened specimen (in the range of 12–15%).

There was no significant hardness difference between single- and double-hardened specimen, i.e., hardness values were in the range of 61–63 HRc. The hardness profile of the single-hardened and double-hardened samples is shown in Fig. 3.

3.2 Grain Size Distribution

Microstructures of the single- and double-hardened samples are shown in Fig. 4a and b, respectively.

Moreover, SEM images reveal the PAGS of the single and double-hardened samples and it is shown in Fig. 5. It is clearly evidenced that the distributions of the carbides in the matrix. Number of hardening i.e., single and double hardening influences the grain size, and the grain size of the samples is estimated by the linear intercept method (ASTM E112). The average grain diameters of the

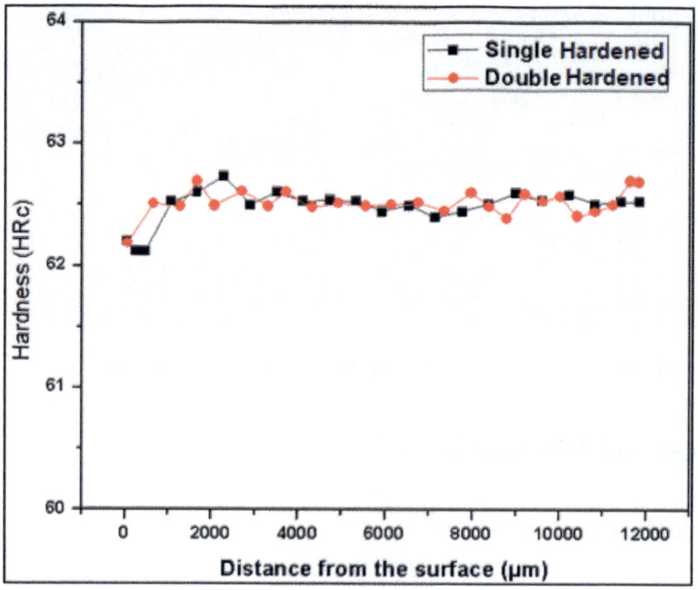

Fig. 3 Hardness profile of single and double quenched samples

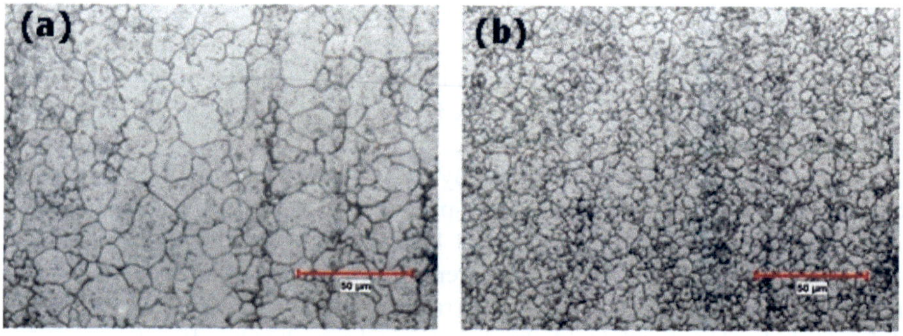

Fig. 4 Prior austenitic grain size **a** single-hardened and **b** double-hardened samples, respectively

samples single-hardened and double-hardened were found to be 16.5 μm and 10 μm, respectively.

The grain size distributions in both samples are shown in Fig. 6. Fine grains hinder dislocations and enhance the strength of the material [15].

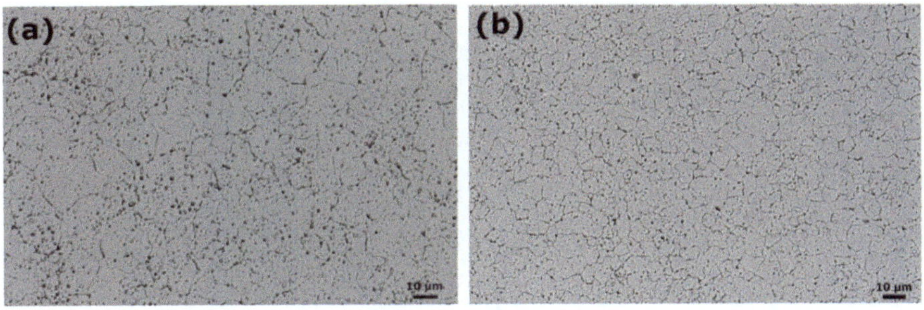

Fig. 5 SEM micrographs **a** single-hardened and **b** double-hardened samples, respectively

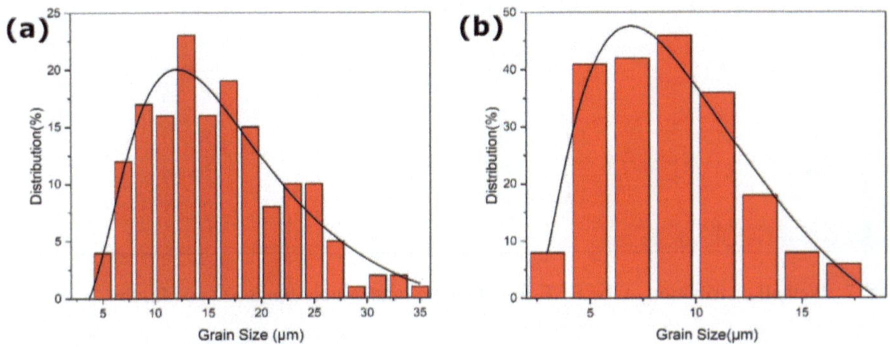

Fig. 6 Grain size distribution of **a** single-hardened and **b** double-hardened samples

3.3 RCF Analysis

The computed RCF L_{10} life of single-hardened and double-hardened specimen is 8206×10^4 cycles and $11{,}212 \times 10^4$ cycles, respectively. Weibull graph (Fig. 7) indicates that the fatigue life of the double-hardened condition samples are increasing 36% higher than the single-hardened samples.

An optical micrograph of the single-hardened failed roller cut section is shown in Fig. 8a and the SEM image (Fig. 8b) is clearly depicting the progression of flaking.

In correlation with RA and fatigue life, the fatigue life is increased with an increase in the amount of RA. Owing to the good plastic deformability of RA, it attributes to reduce the stress generated by the martensitic transformation by rendering the cushion effect. Also, in strain-induced transformation, the fatigue crack propagation is retarded by the RA [21]. After the RCF test, RA is decreased on the rolling contact area of the roller surface. Owing to the repeated contacts during the RCF test, the RA was transformed into stress-induced martensite [21]. It is resulting that, the RA of the failed RCF specimens was about 7.2% and 9.6% in single-hardened and double-hardened specimens, respectively. The stressed zone of the failed roller is shown in Fig. 9a and also evidenced the initiation of the micro-cracks and the elongated

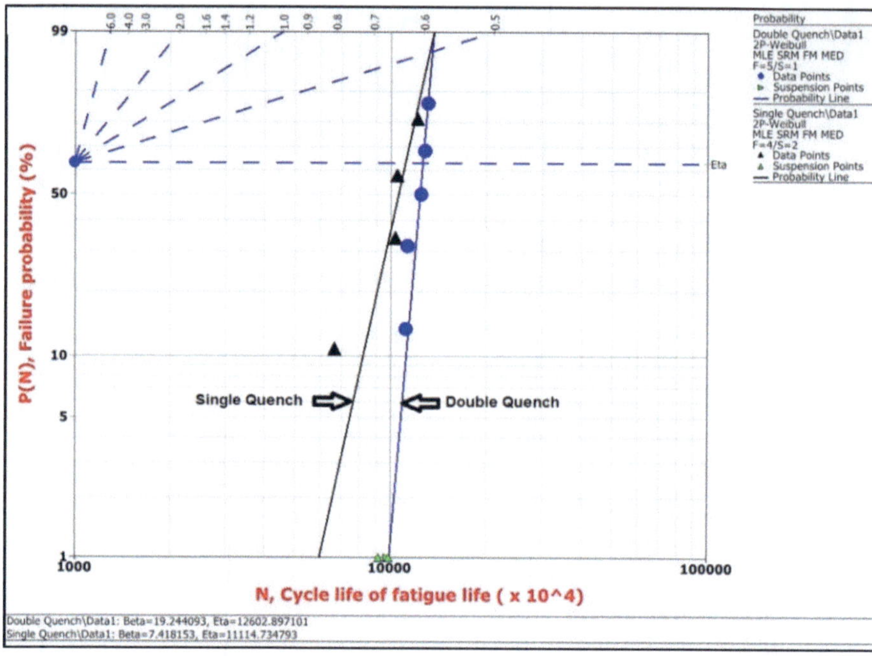

Fig. 7 Weibull distribution of the RCF-life of single- and double-hardened conditions

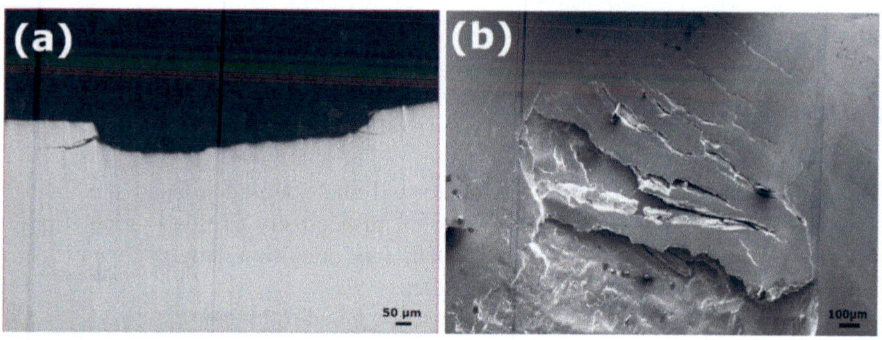

Fig. 8 **a** Cut section of failed single-hardened roller and **b** SEM image depicting the progression of fatigue

grains. As a classical phenomenon of rolling contact fatigue crack initiation beneath the surface in the stressed zone was observed and shown in Fig. 9b.

Both transgranular and intergranular modes of crack propagation are observed in the RCF-tested roller specimens and it is shown in Fig. 10. Presumably, the finer grains hinder the crack propagation and the RA eases the stress concentration around the crack tip. The stress-induced transformational effect of the RA prevents cracks

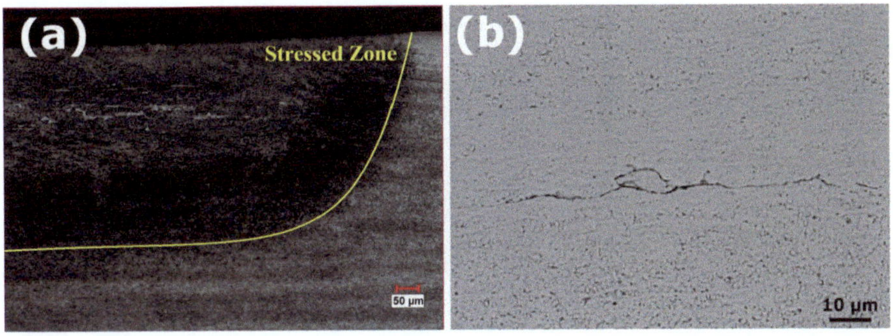

Fig. 9 **a** Cut section of failed roller depicting the stressed zone and **b** SEM image of crack propagation

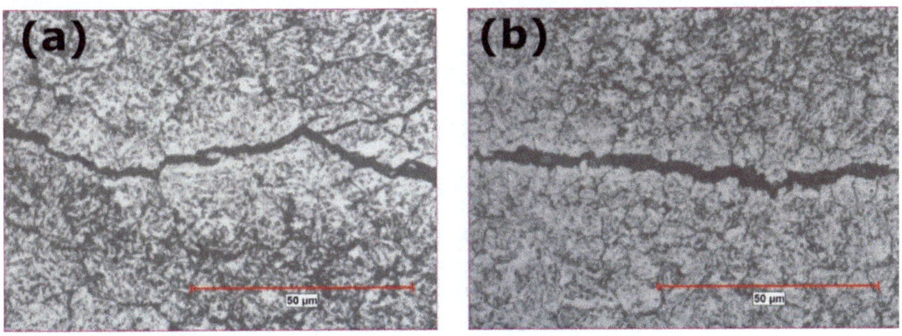

Fig. 10 Crack propagation in **a** single-hardened and **b** double-hardened samples

from fatigue crack propagation [21, 22]. This synergistic effect of fine grain and the RA helps to improve the fatigue life of the double-hardened samples.

4 Conclusion

The effect of hardening methods such as single hardening and double hardening on the rolling contact behavior of the SAE 52,100 steel was studied. The results are leading to the following conclusions:

i. Double hardening attributes the fine grain structure and higher amount of RA than the single-hardened samples.
ii. Double hardening leads to attaining fine grains than the single-hardened condition. The average grain sizes are 16.5 μm and 10 μm for single-hardened and double-hardened, respectively.
iii. Rolling contact fatigue life of the double-hardened samples improved by 35% higher than the single-hardened samples.

Conflict of Interest The authors did not receive any funds, grants, or other support for conducting this study.

References

1. Liu J, Hao Y, Zhou T, Song C, Zhang K (2014) Effect of double hardening and tempering heat treatment on the microstructure and mechanical properties of a novel 5Cr steel processed by electro-slag casting. Mater Sci Eng A 619:212–220. https://doi.org/10.1016/j.msea.2014.09.063
2. Khani Sanij MH, Ghasemi Banadkouki SS, Mashreghi AR, Moshrefifar M (2012) The effect of single and double hardening and tempering heat treatments on the microstructure and mechanical properties of AISI 4140 steel. Mater Des 42:339–346. https://doi.org/10.1016/j.matdes.2012.06.017
3. Lee KO, Hong SK, Kang YK, Yoon HJ, Kang SS (2009) Grain refinement in bearing steels using a double hardening heat-treatment process. Int J Automotive Tech 10:697–702. https://doi.org/10.1007/s12239-009-0082-5
4. Karthikeyan T, Dash MK, Ravikirana, Mythili R, Panneer Selvi S, Moitra A, Saroja S (2017) Effect of prior-austenite grain refinement on microstructure, mechanical properties and thermal embrittlement of 9Cr-1Mo-0.1C steel, J Nucl Mater 494:260–277. https://doi.org/10.1016/j.jnucmat.2017.07.019
5. Ito S, Tsushima N, Muro H. Accelerated rolling contact fatigue test by a cylinder-to-ball rig. ASTM International
6. Cao Z, Shi Z, Feng Y, Sugimoto K-I, Cao W, Weng Y (2019) Effects of double hardening on fatigue properties of high carbon bearing steel with extra-high purity. Int J Fatigue 128:105176. https://doi.org/10.1016/j.ijfatigue.2019.06.036
7. Hashimoto K, Fujimatsu T, Tsunekage N, Hiraoka K, Kida K, Santos EC (2011) Study of rolling contact fatigue of bearing steels in relation to various oxide inclusions. Mater Des 32:1605–1611. https://doi.org/10.1016/j.matdes.2010.08.052
8. Karthikeyan T, Thomas Paul V, Saroja S, Moitra A, Sasikala G, Vijayalakshmi M (2011) Grain refinement to improve impact toughness in 9Cr–1Mo steel through a double austenitization treatment. J Nucl Mater 419:256–262. https://doi.org/10.1016/j.jnucmat.2011.08.010
9. Xiong X, Yang F, Zou X, Suo J (2012) Effect of twice hardening and tempering on the mechanical properties and microstructures of SCRAM steel for fusion application. J Nucl Mater 430:114–118. https://doi.org/10.1016/j.jnucmat.2012.06.047
10. Kamada Y, Ohnishi K, Bessyo K (1991) Effect of double-hard processing on toughness of heavy-gauge ultrahigh strength steel plate. J Mater Sci Lett 10:598–601. https://doi.org/10.1007/BF00724451
11. Yan P, Liu Z, Bao H, Weng Y, Liu W (2014) Effect of normalizing temperature on the strength of 9Cr–3W–3Co martensitic heat resistant steel. Mater Sci Eng A 597:148–156. https://doi.org/10.1016/j.msea.2013.12.068
12. Wang C, Wang M, Shi J, Hui W, Dong H (2008) Effect of microstructural refinement on the toughness of low carbon martensitic steel. Scr Mater 58:492–495. https://doi.org/10.1016/j.scriptamat.2007.10.053
13. Sarikaya M, Steinberg G, Thomas G (1982) Optimization of Fe/Cr/C base structural steels for improved strength and toughness. Metall Trans A 13:2227–2237. https://doi.org/10.1007/BF02648394
14. Narasimha Rao BV, Thomas G (1980) Structure-property relations and the design of Fe-4Cr-C base structural steels for high strength and toughness. Metall Trans A 11:441–457. https://doi.org/10.1007/BF02654568
15. Nishihara T, Nakashima H, Tsushima N, Ito S (1990) Factors that affect rolling contact fatigue life of ceramics and rolling contact fatigue life of ceramic balls and rollers. Presented at the

gas turbine and aeroengine congress and exposition—June 11–14, 1990—Brussels, Belgium. https://doi.org/10.1115/90-GT-377

16. Morris D, Sadeghia F, Singh K, Voothaluru R (2020) Residual stress formation and stability in bearing steels due to fatigue induced retained austenite transformation. Int J Fatigue 136:105610. https://doi.org/10.1016/j.ijfatigue.2020.105610
17. Antolovich SD, Chanani GR (1972) Subcritical crack growth of TRIP steels in air under static loads. Eng Fract Mech 4(4):765–776. https://doi.org/10.1016/0013-7944(72)90014-8
18. Zhu D, Wang F-X, Cai Q-G, Zheng M-X, Cheng Y-Q (1985) Effect of retained austenite on rolling element fatigue and its mechanism. Wear 105(3):223–234. https://doi.org/10.1016/0043-1648(85)90069-9
19. Shen Y, Moghadam SM, Sadeghi F, Paulson K, Trice RW (2015) Effect of retained austenite—compressive residual stresses on rolling contact fatigue life of carburized AISI 8620 steel. Int J Fatigue 75:135–144. https://doi.org/10.1016/j.ijfatigue.2015.02.017
20. Xie ZJ, Ren YQ, Zhou WH, Yang JR, Shang CJ, Misra RDK (2014) Stability of retained austenite in multi-phase microstructure during austempering and its effect on the ductility of a low carbon steel. Mater Sci Eng, A 603:69–75. https://doi.org/10.1016/j.msea.2014.02.059
21. Yajima E, Miyazaki T, Sugiyama T, Terajima H (1974) Effects of retained austenite on the rolling fatigue life of ball bearing steels. Trans JIM 15(3):173–179. https://doi.org/10.2320/matertrans1960.15.173
22. Dong Z, Fu-Xing W, Qi-Gong C, Ming-Xin Z, Yin-Quian C (1985) Effect of retained austenite on rolling element fatigue and its mechanism. Wear 105:223–234. https://doi.org/10.1016/0043-1648(85)90069-9

Multi-objective Optimization of WEDM Process Parameters for TiNi Shape Memory Alloy

Rakesh R. Kolhapure, Duradundi S. Badkar, and Adik M. Takale

1 Introduction

TiNi is a smart material with distinct qualities including corrosion resistance, strength, and form recovery [1]. Residual tension and tool wear present problems for conventional methods [2, 3]. A solution provided by wire electrical discharge machining (WEDM), which correctly cuts complicated forms with little stress. Due to TiNi unique characteristics, the researcher is investigating WEDM for it [4, 5]. Maximum work carried out on WEDM using different wire materials of thin copper, brass, and tungsten [6]. Process parameters in WEDM affect machining performance, so the right selection is crucial. Current research involves studying Ti50Ni50 machining with different process parameters using mixed-mode OA for experimental design. For current research work, reusable molybdenum wire is used as an electrode rather than other wire material for machining of Ti50Ni50 SMA which makes this work distinct from other research work.

R. R. Kolhapure (✉)
Research Scholar, Shivaji University Kolhapur and DKTES TEI, Ichalkaranji, Kolhapur 416004, India
e-mail: rockkolhapure@gmail.com

D. S. Badkar
Fabtech Technical Campus COE & Research, Sangola 413307, India

A. M. Takale
Sharad Institute of Technology, COE, Ichalkaranji, YadravMaharashtra 416121, India

R. Sahu et al. (eds.), *Advancements in Materials Processing Technology, Volume 1*,
Springer Proceedings in Materials 48, https://doi.org/10.1007/978-981-97-4958-4_24

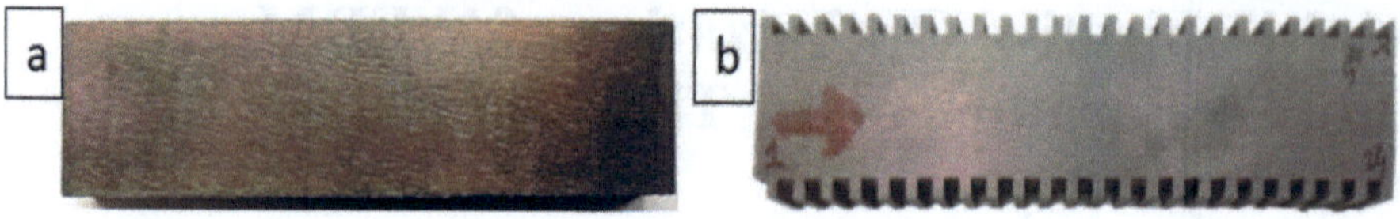

Fig. 1 TiNi workpiece **a** before machining **b** after machining

2 Literature Survey

The effects of WEDM on NiTi smart alloy result show that high-quality surface on NiTi super elastic alloy is heavily dependent on T_{on} and SV. Surface roughness (Ra) increases as T_{on} grows, but it reduces as T_{off} and SV increase [6]. T_{off} was the key factor influencing surface roughness (Ra), whereas SV was the most influential parameter for material removal rate (MRR) [7]. In ANOVA, the current was the most significant factor for both material removal rate (MRR) and surface roughness (SR), accounting for 81.10% of MRR and 81.02% of SR [3]. Higher T_{on} resulted in higher material removal rates (MRR), but poor surface quality, notably for producing a superior surface finish [8]. Higher T_{on} produced poor surface quality, but higher SV produced better surface roughness [9]. Increasing the pulse-on duration, input power, server voltage, and wire tension increases MRR but decreases surface roughness [10]. Higher servo voltage, on the other hand, lowers surface roughness [11]. According to the analysis of variance (ANOVA), the most important component is pulse-on time, which accounts for 52.90% of the effect on MRR and SR [12, 13].

3 Experimental Investigation

3.1 Workpiece Material

The Ti50Ni50 of 100 mm × 20 mm × 10.3 mm plate has been used as a workpiece for current research work as shown in Fig. 1a. For experimental investigation, wire electrical discharge machining (WEDM) is used and 2 mm × 2 mm × 10.3 mm size of punch is prepared from the workpiece as shown in Fig. 1b.

3.2 Experimental Arrangement

A WEDM machine used for the experimentation as shown in Fig. 2. Table 1 presents different machining process parameters with their levels. Throughout the experiment, the following variables remained constant: wire tension, deionized water as a working fluid, wire diameter 0.18 mm, and thickness of workpiece 10.3 mm. Table 2 represents the analyzed and *S*/*N* ratio data with MRR larger and KW, SR, and DD are smaller the better. The signal-to-noise (*S*/*N*) ratio for responses is calculated by using Eqs. (1) and (2) for "larger the better" and "smaller the better," respectively.

$$n = -10 \times \log 10\left[\text{mean of sum square of reciprocal of measured data}\right] \quad (1)$$

$$n = -10 \times \log 10\left[\text{mean of sum of square of measured data}\right] \quad (2)$$

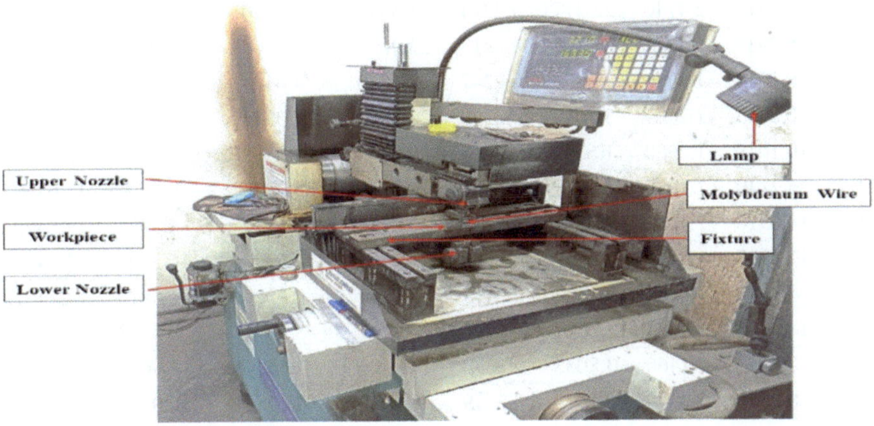

Fig. 2 WEDM experimental setup (Concord United Products Pvt. Ltd.)

Table 1 Process constraints and levels

Sr. No.	Process constraints	Symbol	Level 1	Level 2	Level 3
1	Voltage	V	85	100	–
2	Pulse-on time	T_{on}	40	60	80
3	Pulse-off time	T_{off}	10	14	18
4	Current	A	4	5	6
5	Wire feed	WF	25	40	50

Table 2 Analyzed and *S*/*N* ratio data

Exp. No.	Analyzed data				*S*/*N* ratio (DB)			
	MRR	SR	DD	KW	MRR	SR	DD	KW
1	24.0916	1.6950	1.0833	0.2340	27.6373	− 4.5834	− 0.6952	12.6157
2	20.4741	1.2910	1.5495	0.2240	26.2241	− 2.2185	− 3.8038	12.9950
3	10.1793	1.3783	2.5533	0.2269	20.1543	− 2.7869	− 8.1422	12.8833
4	15.6439	2.1797	2.9817	0.2265	23.8869	− 6.7679	− 9.4892	12.8986
5	11.9966	1.4170	2.9483	0.2241	21.5812	− 3.0274	− 9.3915	12.9912
6	15.0163	1.5333	1.9950	0.2260	23.5313	− 3.7125	− 5.9989	12.9178
7	19.0311	2.4740	2.6317	0.2214	25.5893	− 7.8680	− 8.4047	13.0964
8	15.7939	1.4717	3.5167	0.2213	23.9698	− 3.3564	− 10.9227	13.1004
9	8.1040	1.7070	3.3833	0.2247	18.1740	− 4.6447	− 10.5868	12.9679
10	10.8455	1.2644	3.4017	0.2340	20.7050	− 2.0377	− 10.6339	12.6157
11	10.1048	1.2660	1.5383	0.2416	20.0905	− 2.0487	− 3.7408	12.3381
12	8.2140	1.7980	2.5617	0.2334	18.2911	− 5.0958	− 8.1706	12.6380
13	11.0576	2.4231	2.9983	0.2356	20.8732	− 7.6874	− 9.5375	12.5565
14	11.8805	1.3500	3.2850	0.2358	21.4967	− 2.6067	− 10.3307	12.5491
15	7.1147	1.3704	2.1333	0.2470	17.0431	− 2.7369	− 6.5810	12.1461
16	14.0025	1.4688	3.0017	0.2440	22.9241	− 3.3393	− 9.5473	12.2522
17	7.7598	1.1460	3.8033	0.2390	17.7970	− 1.1837	− 11.6032	12.4320
18	8.8650	2.9440	3.1670	0.2219	18.9535	− 9.3788	− 10.0130	13.0769

4 Results and Discussion

4.1 The Influence of Process Factors on Responses

The study used ANOVA to analyze the data with a 95% confidence level. Variables with a *p*-value below 0.05 were considered significant at this confidence level. According to Table 3, voltage followed by T_{off} and WF has the greatest impact on MRR, while T_{on} and current do not significantly affect MRR. Table 4 presents that the process component that has the extreme impact on SR is current, afterward T_{off}. The SR is unaffected by voltage, T_{on}, and WF. From Table 5, the process component that has the greatest impact on DD is a T_{on}, followed by WF. The DD is unaffected by the voltage, T_{off}, and current. Table 6 presents that the process component that has the greatest impact on KW is voltage. The KW is unaffected by the other parameters.

Table 3 ANOVA for MRR

Parameters	DOF	Adj. SS	Adj. MS	*F*	*P*	%*C*
Voltage	1	141.606	141.606	30.05	0.001	36.55
T_{on}	2	12.963	6.481	1.38	0.307	3.33
T_{off}	2	115.602	57.801	12.27	0.004	29.84
Current	2	4.122	2.061	0.44	0.660	1.06
WF	2	75.412	37.706	8.00	0.012	19.46
Error	8	37.696	4.712	–	–	9.73
Total	17	387.401	–	–	–	100

Table 4 ANOVA for SR

Parameters	DOF	Adj. SS	Adj. MS	*F*	*P*	%*C*
Voltage	1	0.00075	0.00075	0.01	0.935	0.0174
T_{on}	2	0.54017	0.270087	2.56	0.139	12.560
T_{off}	2	1.17091	0.585455	5.54	0.031	27.230
Current	2	1.37572	0.687862	6.51	0.021	31.995
WF	2	0.36680	0.183398	1.74	0.237	8.530
Error	8	0.84539	0.105674	–	–	19.66
Total	17	4.29975	–	–	–	100

Table 5 ANOVA for DD

Parameters	DOF	Adj. SS	Adj. MS	*F*	*P*	%*C*
Voltage	1	0.58588	0.58588	2.14	0.182	5.933
T_{on}	2	3.87801	1.93901	7.08	0.017	39.27
T_{off}	2	0.06143	0.03071	0.11	0.895	0.622
Current	2	0.69331	0.34665	1.27	0.333	7.020
WF	2	2.46606	1.23303	4.50	0.046	24.97
Error	8	2.19023	0.27378	–	–	22.17
Total	17	9.87491	–	–	–	100

Table 6 ANOVA for KW

Parameters	DOF	Adj. SS	Adj. MS	*F*	*P*	%*C*
Voltage	1	0.000594	0.000594	22.01	0.002	52.28
T_{on}	2	0.000055	0.000027	1.01	0.406	4.841
T_{off}	2	0.000021	0.000010	0.38	0.694	1.848
Current	2	0.000229	0.000115	4.24	0.055	20.158
WF	2	0.000022	0.000011	0.41	0.679	1.936
Error	8	0.000216	0.000027	–	–	19.014
Total	17	0.001136	–	–	–	100

4.2 Multi-objective Parametric Optimization

The Taguchi technique uses the signal-to-noise ratio to optimize factors, aiming for higher ratios for closer-to-ideal specifications. However, it can only optimize a single response, making it unsuitable for projects with multiple objectives. For such cases, multi-response optimization is needed, achieved through the GRA procedure [14]. GRA involves normalizing process factors to determine the gray relational coefficient, aiding in identifying optimal solutions. Using Eqs. (3) and (4), the resulting *S/N* ratio numbers are transformed to normalized values where $x_i^*(k)$ is the value after gray relational generation and $\max_i^0(k)$ and $\min x_i^0(k)$ are the maximum and minimum value $x_i^0(k)$.

$$x_i^*(k) = \frac{x_i^0(k) - \min x_i^0(k)}{\max x_i^0(k) - \min x_i^0(k)} \tag{3}$$

$$x_i^*(k) = \frac{\max x_i^0(k) - -x_i^0(k)}{\max x_i^0(k) - \min x_i^0(k)} \tag{4}$$

Order of Deviations ($\Delta_{oi}(k)$) and Gray Relational Coefficient (GRC)

Equation (5) was used to determine it, and Table 7 lists its values. Equation (6) can be used to calculate the gray relationship coefficient. Table 8 presents the values.

$$\Delta_{oi}k = \left|x_0^*(k) - x_i^*(k)\right| \tag{5}$$

$$\gamma_{0,i}(k) = \frac{\Delta_{\min} + \xi.\Delta_{\max}}{\Delta_{0i}(k) + \xi.\Delta_{\max}} \tag{6}$$

where $\Delta_{\min}$ and $\Delta_{\max}$ = least and highest weight, respectively, of $0i\ (k) = \min_i \min_k$ $|x_0{}^*(k) - x_i{}^*(k)|$, $x_0{}^*(k)$ is the ideal normalized *S/N* ratio, $x_i{}^*(k)$ is the normalized comparability sequence, and ξ is distinguishing coefficient factor, taken as 0.5 [15].

Gray Relational Grade (GRG)

Through average gray relational coefficient values for every achievement trait, gray relational score is determined. Equation (7) can be used to determine it, and numbers are given in Table 8.

$$\gamma(x_0, x_i) = \frac{1}{p}\sum_{i=1}^{p} \gamma(x_0(k), x_i(k)) \tag{7}$$

where $\gamma(x_O, x_i)$ is the GRG for the *j*th experiment and p is a variety of success traits.

Estimation of GRG Optimal Factors (η opt)

Equation (8) used to find out the optimal factor.

Table 7 Normalization and deviation data

Exp. No.	Normalization of *S/N* ratio				Deviation sequence			
	MRR	SR	DD	KW	MRR	SR	DD	KW
1	1.0000	0.4148	0.0000	0.5079	0.0000	0.5852	1.0000	0.4921
2	0.8666	0.1263	0.2850	0.1104	0.1334	0.8737	0.7150	0.8896
3	0.2937	0.1956	0.6827	0.2275	0.7063	0.8044	0.3173	0.7725
4	0.6460	0.6814	0.8062	0.2114	0.3540	0.3186	0.1938	0.7886
5	0.4284	0.2250	0.7972	0.1145	0.5716	0.7750	0.2028	0.8855
6	0.6124	0.3086	0.4862	0.1913	0.3876	0.6914	0.5138	0.8087
7	0.8067	0.8156	0.7068	0.0041	0.1933	0.1844	0.2932	0.9959
8	0.6538	0.2651	0.9376	0.0000	0.3462	0.7349	0.0624	1.0000
9	0.1067	0.4223	0.9068	0.1388	0.8933	0.5777	0.0932	0.8612
10	0.3456	0.1042	0.9111	0.5079	0.6544	0.8958	0.0889	0.4921
11	0.2877	0.1055	0.2792	0.7988	0.7123	0.8945	0.7208	0.2012
12	0.1178	0.4774	0.6853	0.4846	0.8822	0.5226	0.3147	0.5154
13	0.3615	0.7936	0.8106	0.5700	0.6385	0.2064	0.1894	0.4300
14	0.4204	0.1736	0.8833	0.5777	0.5796	0.8264	0.1167	0.4223
15	0.0000	0.1895	0.5396	1.0000	1.0000	0.8105	0.4604	0.0000
16	0.5551	0.2630	0.8115	0.8888	0.4449	0.7370	0.1885	0.1112
17	0.0712	0.0000	1.0000	0.7004	0.9288	1.0000	0.0000	0.2996
18	0.1803	1.0000	0.8542	0.0247	0.8197	0.0000	0.1458	0.9753

$$\eta_{\text{opt}} = \underline{T} + \left(\underline{a}_{(1/2/3)} - \underline{T}\right) + \left(\underline{b}_{(1/2/3)} - \underline{T}\right) + \left(\underline{c}_{(1/2/3)} - \underline{T}\right) \tag{8}$$

$\underline{T}$ = Overall response mean; $\underline{a}_{(1/2/3)}, \underline{b}_{(1/2/3)}, \underline{c}_{(1/2/3)}$ = Response averages at the 1st, 2nd and 3rd of parameters *a*, *b* and *c* correspondingly.

$$\eta_{\text{opt}} = 0.6593$$

Table 9 presents GRG response; Table 10 and Fig. 3 give comparison study for WEDM.

Table 8 GRC and GRG

Exp. No.	Gray relational coefficient (GRC)				GRG	Rank
	MRR	SR	DD	KW		
1	1.0000	0.4608	0.3333	0.5040	0.5745	7
2	0.7894	0.3640	0.4115	0.3598	0.4812	15
3	0.4145	0.3833	0.6118	0.3929	0.4506	18
4	0.5855	0.6108	0.7207	0.3880	0.5762	6
5	0.4666	0.3921	0.7115	0.3609	0.4828	14
6	0.5633	0.4197	0.4932	0.3821	0.4646	17
7	0.7212	0.7306	0.6303	0.3343	0.6041	3
8	0.5909	0.4049	0.8891	0.3333	0.5545	9
9	0.3589	0.4640	0.8429	0.3673	0.5083	12
10	0.4331	0.3582	0.8491	0.5040	0.5361	11
11	0.4124	0.3586	0.4096	0.7131	0.4734	16
12	0.3617	0.4889	0.6137	0.4924	0.4892	13
13	0.4392	0.7078	0.7253	0.5376	0.6025	4
14	0.4631	0.3770	0.8108	0.5421	0.5483	10
15	0.3333	0.3815	0.5206	1.0001	0.5589	8
16	0.5292	0.4042	0.7262	0.8181	0.6194	2
17	0.3499	0.3333	1.0000	0.6253	0.5771	5
18	0.3789	1.0000	0.7742	0.3389	0.6230	1

Table 9 Response table for GRG

Levels	Voltage	T_{on}	T_{off}	Current	Wire feed
1	0.5219	0.5008	0.5855	0.5447	0.5480
2	0.5587	0.5389	0.5196	0.5471	0.5466
3	-	0.5811	0.5158	0.5289	0.5262
Max	0.5587	0.5811	0.5855	0.5471	0.5480
Min	0.5219	0.5008	0.5158	0.5289	0.5262
Delta	0.0368	0.0802	0.0697	0.0182	0.0217
Rank	3	1	2	5	4
Total mean value of GRG is 0.5403					

Table 10 Comparison of responses for WEDM on different materials

Authors	Material	Material notation	MRR	SR
[1]	$Ni_{50.6}Ti_{49.4}$	*A*	9.655	2.664
[2]	Ti $Ni_{55.78}$	*B*	0.5262	1.27
[3]	$N_{54.1}Ti$	*C*	5.502	1.489
[4]	$Ni_{50.6}Ti_{49.4}$	*D*	2.537	1.296
[6]	$Ti_{44.13}$ $Ni_{55.74}$	*E*	2.65	2.0188
Current paper	$Ti_{50}Ni_{50}$	*F*	12.128	2.75

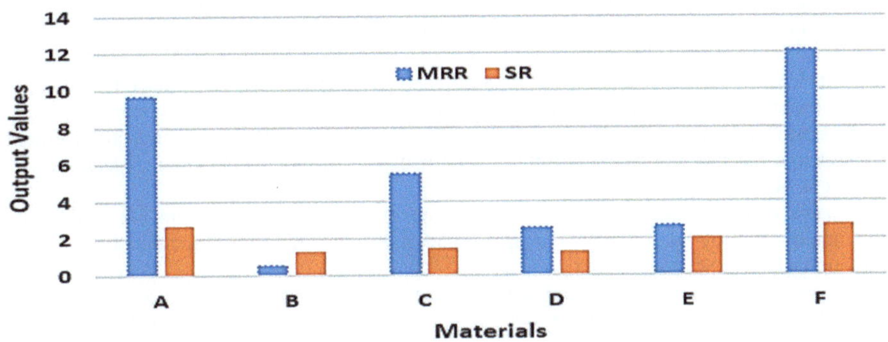

Fig. 3 Different materials versus MRR and SR

5 Conclusion

1. Key process parameters have been identified. Both pulse duration and current have a substantial impact on SR (31.995%) and DD (26.92%). Voltage is quite important for both MRR (63.55%) and KW (52.28%).
2. With the altered A2B3C1D1E1 arrangement, GRA rise initial grade A2B3C3D2E1 from 0.623 to 0.6619, i.e., an increase in grade of 0.04.

References

1. Kesavan J, Velmurugan C, Senthilkumar V, Dinesh S (2020) Optimization of WEDM parameters on surface integrity characteristics of NiTi shape memory alloy. Mater Today Proc 43:183–190
2. Chaudhari R et al (2021) Experimental investigations and optimization of MWCNTs-mixed WEDM process parameters of nitinol shape memory alloy. J Mater Res Technol 15:2152–2169
3. Gupta DK, Dubey AK (2021) Multi process parameters optimization of Wire-EDM on shape memory alloy (Ni54.1Ti) using Taguchi approach. Mater Today Proc 44:1423–1427

4. Takale A, Chougule N (2019) Optimization of process parameters of wire electro discharge machining for Ti49.4Ni50.6 shape memory alloys using the Taguchi technique. Int J Struct Integr 10:548–568
5. Kaynak Y, Karaca HE, Noebe RD, Jawahir IS (2013) Tool-wear analysis in cryogenic machining of NiTi shape memory alloys: a comparison of tool-wear performance with dry and MQL machining. Wear 306:51–63
6. Kulkarni VN, Gaitonde VN, Hadimani V, Aiholi V (2018) Analysis of Wire EDM process parameters in machining of NiTi superelastic alloy. Mater Today Proc 5(9):19303–19312
7. Meshri H, Akar S, Seyedzavvar M, Kilic S (2021) Micro-WEDM of Ni55.8Ti shape memory superalloy: experimental investigation and optimisation. Int J Mech Manuf Syst 14:18–38
8. Manjaiah M, Narendranath S, Basavarajappa S (2015) Wire electro discharge machining performance of TiNiCu shape memory alloy. Silicon:1–9
9. Bisaria H, Shandilya P (2018) Study on effect of machining parameters on performance characteristics of Ni-rich NiTi shape memory alloy during wire electric discharge machining. Mater Today: Proc 5:3316–3324
10. Soni H, Narendranath S, Ramesh M (2018) Effects of wire electro-discharge machining process parameters on the machined surface of Ti50Ni49Co1 shape memory alloy. Silicon:1–7
11. Arikatlaa S, Mannanb K, Krishnaiah A (2017) Parametric optimization in wire electrical discharge machining of titanium alloy using response surface methodology. Mater Today: Proc 4:1434–1441
12. Sonawane S, Kulkarni M (2018) Optimization of machining parameters of WEDM for Nimonic-75 alloy using principal component analysis integrated with Taguchi method. J King Saud Univ Eng Sci 30:250–258
13. Gaikwad V, Jatti V (2018) Optimization of material removal rate during electrical discharge machining of cryo-treated NiTi alloys using Taguchi's method. J King Saud Univ Eng Sci 30:266–272
14. Kolhapure R, Shinde V, Kamble V (2017) Geometrical optimization of strain gauge force transducer using GRA method. Meas J Int Meas Confed 101:111–117
15. John J, Devjani D, Ali S, Abdallah S, Pervaiz S (2023) Optimization of 3D printed polylactic acid structures with different in fill patterns using Taguchi-grey relational analysis. Adv Ind Eng Polym Res 6:62–78

Sustainable Composites Based on Spent Coffee Grounds and Bio PBS: Thermal and Mechanical Analysis

Karri Santhosh Kumar and Inderdeep Singh

1 Introduction

Conventional plastic composites have manifold applications in many sectors, such as automobile, construction, packaging, furniture, household, and biomedical industries. These composites are made of traditional petroleum-based polymers with synthetic fibers and are the primary source of waste generation and consume more energy during fabrication. As a result, both research and industry fraternity emphasized on development of environmentally friendly and renewable materials to reduce the usage of petroleum-based materials [1–3]. The rapid interest in these sustainable materials over petroleum-based materials is due to their superior properties, such as lightweight, high specific strength and stiffness, availability, and biodegradability. Therefore, the ongoing research is focused on sustainable materials to accomplish the scarcity of non-renewable resources. The global biomass production in one year is the same as the production of petroleum, leading to exhaustive research in the field of utilization of biomass [4].

From this perspective, spent coffee ground (SCG) fillers are chosen as a biomass reinforcement in the development of bio-based poly (butylene succinate) (bio PBS) composites. Coffee is the second most consumed beverage in the world after water [5], so it generates plenty of waste. SCG has occupied the majority portion of waste generated by the coffee industry. SCG is derived from the hot water treatment of coffee powder to prepare instant coffee. The annual production of SCG is about 8 billion kg per year [6, 7]. SCGs consist of lignocellulosic constituents such as cellulose (12%), hemicellulose (39%), and lignin (24%) and also contain beneficial organic compounds such as proteins, fatty acids, polyphenols, and carbohydrates that have

K. S. Kumar (✉) · I. Singh
Department of Mechanical and Industrial Engineering, Indian Institute of Technology Roorkee, Roorkee, India
e-mail: karri_sk@me.iitr.ac.in

R. Sahu et al. (eds.), *Advancements in Materials Processing Technology, Volume 1*,
Springer Proceedings in Materials 48, https://doi.org/10.1007/978-981-97-4958-4_25

attracted scientists and researchers for unique bio-resources [8]. These components are renowned for providing exceptional stiffness and flexibility to biocomposites [9].

Bio PBS is one of the most promising biodegradable and renewable polymers. It is in the viscoelastic state at room temperature. It has superior mechanical and thermal properties and good processability with traditional plastic processes such as extrusion and injection molding. Due to these properties, bio PBS can replace petroleum-based polymers like polypropylene (PP) and polyethylene (PE). Bio PBS has potential applications in packaging, medical, furniture, construction, and household appliances [10, 11]. However, the relatively low melt strength, insufficient stiffness, and high cost of bio PBS hamper its applications in the automobile and aerospace industries [7]. To overcome the drawbacks of bio PBS, it is blended with other biodegradable polymers or incorporated with fillers to fabricate biocomposites. Incorporating fillers into polymer has been considered more effective than polymer blends in enhancing the properties [7].

Pagalan et al. [12] reported a study about using SCG to extract activated carbon to treat water contaminated by aniline yellow dye. The results of the morphology of activated carbon showed micro and mesopore structures and various functional groups. Hence, the authors concluded that activated carbon extracted from SCG could be a promising adsorbent. Marto et al. [13] studied the development of new-generation sunscreens from SCG by-products and also measured the biological effects of using the oil fraction of SCG in developed sunscreens. The results showed that SCG oil shows excellent characteristics to enhance sunscreen performance. García-García et al. [6] and Essabir et al. [9] studied the effect of surface treatments and compatibilizers on the mechanical, thermal, and morphological properties of PP/SCG composites. The results showed that flexural modulus slightly increased for developed composites with treated/compatibilized and untreated SCG at 20 wt% filler loading. The thermal stability of PP/SCG composites was improved compared to neat PP, and SCG particles have shown good dispersion and distribution in PP. Seo et al. [8] developed SCG fillers reinforced polyester-based epoxy composites and evaluated thermal, rheology, and mechanical properties. The developed composites showed good dimensional stability. The cured neat epoxy established thermoset behavior while incorporating SCG into the matrix enabled reforming and observed a spontaneous change in the shape.

To date, SCG has been used as a source for cosmetic applications [13], adsorbents [12], antioxidants, and biodiesel production [14]. Since the last decade, it has also been used as a natural filler in polymer composites. However, limited research is available on incorporating SCG fillers into the biopolymers to develop biodegradable composites. The present study aims to develop and characterize mechanical, thermal, and morphological properties, crystallinity, and dynamic mechanical behavior of SCG filler-reinforced bio PBS matrix biocomposites.

2 Materials and Methods

2.1 Materials

Spent coffee grounds (SCG) were procured from Café Coffee Day (CCD), IIT Roorkee, India. SCGs were grounded and sieved into filler size 150–212 μm. Bio PBS (grade: FZ71PB) was procured in pellets form from Tanhim Enterprises Pvt. Ltd., Greater Noida, India, (Density: 1.26 g/cm^3, MFI: 22 g/10 min (190 °C, 2.16 kg)).

2.2 SCG Characterization

The morphology of SCG fillers was examined by SEM analysis using FESEM (Make: ZEISS, Model: Gemini 1). The thermal degradation behavior of fillers was analyzed with the help of a thermogravimetric analyzer (TGA) (EXSTAR 6300). The test was conducted in the nitrogen atmosphere with a maximum temperature of 600 °C and a heating rate of 10 °C/min.

2.3 Fabrication of Biocomposites

Bio PBS pellets were dried and blended with SCG with 30 wt% filler content using a single screw extruder (Make: Sai Extrumech Model: SAI-25) at a temperature and screw speed of 135 °C and 45 rpm, respectively. The blended composite filament was pelletized into pellets. The obtained pellets were employed in injection molding (Make: Electronica Model: ENDURA-60) to develop final specimens following ASTM standards. The optimized process temperature of heaters (from hopper to nozzle) is as follows 155–155–160–155 °C.

2.4 Characterization of Developed Composites

The mechanical properties of fabricated composites were evaluated using a UTM (Instron-5982) with a capacity of 100kN. Tensile (crosshead speed: 2 mm/min, gauge length: 50 mm) and flexural (crosshead speed: 2 mm/min, span length: 60 mm) tests were performed in accordance with ASTM D3039M-14 and ASTM D790-10, respectively. The TGA (Make: EXSTAR 6300) was employed to measure the thermal degradation behavior of biocomposites. The viscoelastic properties of developed composites were evaluated by DMA (Make: NETZSCH, Model: DMA 242 C) analysis with ramping temperature and frequency of 5 °C/min and 1 Hz, respectively. Degree of crystallinity of biocomposites was assessed using a thin film XRD (Make:

Smart Lab Model: Rigaku). The specimen's crystallinity was calculated using Eq. (1) [15].

$$\text{Crystallinity}(\%) = \frac{\text{Crystalline area}}{\text{Total area under the curve}} \times 100 \quad (1)$$

3 Results and Discussion

3.1 Characterization of Spent Coffee Grounds

SEM analysis. Figure 1 depicts the SEM analysis of grounded SCG fillers at different magnifications. At lower magnification, the irregular shape of SCG particles was observed since SCG is a solid residue obtained from the hot water brewing process. At higher magnification, SCG filler shows an intense porous granular structure as some of the constituents of coffee grounds dissolve during the brewing process [16].

Thermal Stability. Figure 2a and b depict TG (mass loss with respect to temperature) and DTG (rate of mass loss) curves, respectively. The TGA curve shows three major stages of mass loss. Stage 1 (up to 130 °C) is attributed to moisture evaporation, and stage 2 and stage 3 mass loss due to cellulose, hemicellulose, and lignin decomposition. The DTG curve shows corresponding temperature peaks for degradation stages mentioned in the TGA curve. The peak around the temperature of 302 °C attributed to the degradation of hemicellulose and cellulose, and a peak around 396 °C determined the degradation of lignin in SCG fillers [10, 16].

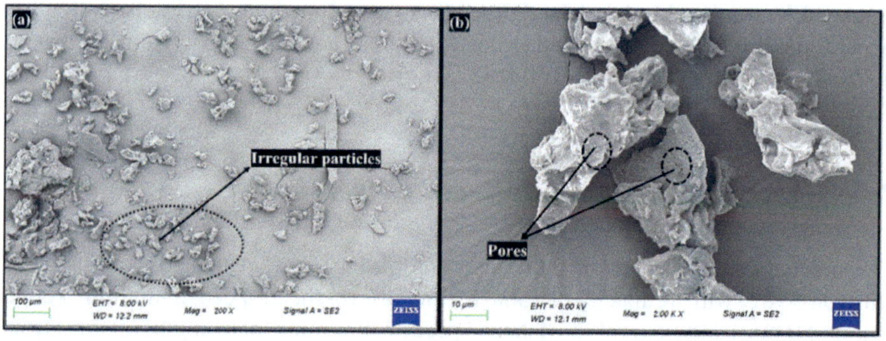

Fig. 1 SEM micrographs of SCG powder. **a** × 200 magnification, **b** × 2000 magnification

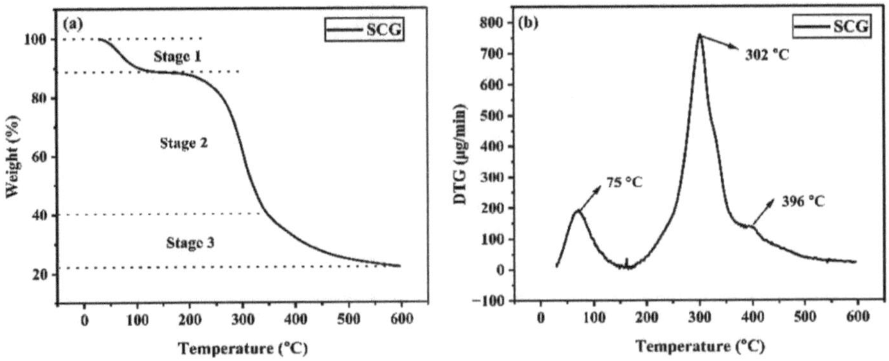

Fig. 2 **a** TGA and **b** DTG curves of SCG

3.2 *Characterization of Biocomposites*

Mechanical Properties. The tensile and flexural properties of developed biocomposites are shown in Fig. 3. The tensile and flexural strength of SCG filler-reinforced biocomposite (30SCG/bio PBS) decreased compared to neat polymer. The decrease in tensile and flexural strength is attributed to the poor interfacial bonding between the SCG fillers and bio PBS and incorporating SCG fillers act as stress concentration sites rather than reinforcement in the biocomposites [4, 6]. On the other hand, the tensile and flexural modulus of the developed biocomposite increased with the addition of SCG fillers because these stiff lignocellulosic fillers restricted the mobility of biopolymeric chains. Hence, modulus of biocomposites increased [17].

Thermal Properties. Figure 4 depicts the thermographs of developed composites. TGA curves illustrate the thermal stability of neat polymer and its composite. The thermal stability of SCG/bio PBS composite is almost like neat bio PBS up to 200 °C. After that stability of the biocomposite decreased because the chemical components (cellulose, hemicellulose, and lignin) of SCG fillers started degrading at elevated

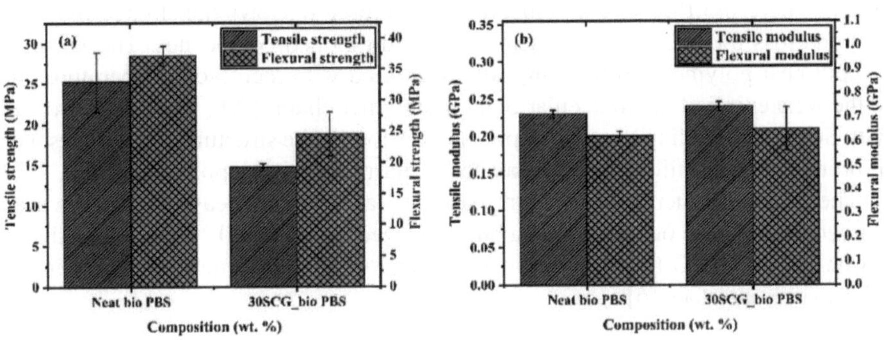

Fig. 3 **a** Strength and **b** modulus of SCG/bio PBS biocomposites

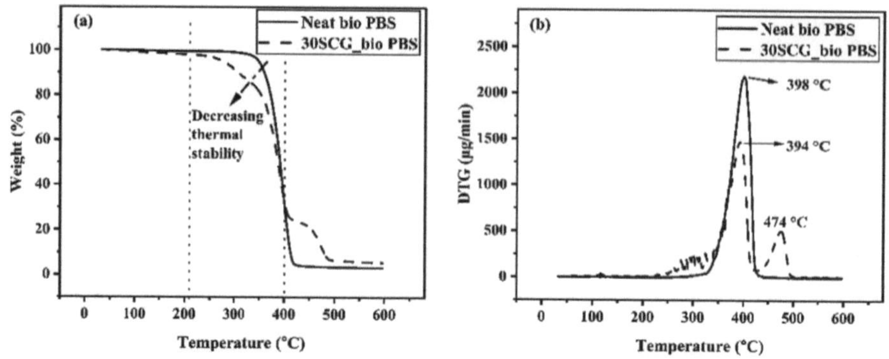

Fig. 4 **a** TGA and **b** DTG thermographs of SCG/bio PBS composites

temperatures [10]. DTG curves showed the temperature peaks where the maximum mass loss rate (thermal degradation) occurred. The neat bio PBS has demonstrated a single degradation step around 398 °C, whereas SCG/bio PBS has shown two degradation steps. The first step is around 394 °C due to the degradation of the polymer, and the second step of degradation (474 °C) is due to the presence of SCG fillers in the biocomposites.

XRD Analysis. Figure 5 shows the XRD patterns of pristine polymer and its biocomposite (30SCG/bio PBS). The pristine bio PBS has three crystalline peaks as 19.58, 21.56, and 22.54°. The crystalline peaks (19.58, 21.72, and 22.58) of the SCG/bio PBS composite are altered by incorporating SCG fillers since a crystalline peak (21.2°) [18] of SCG interferes with the peaks of neat polymer. The degree of crystallinity was evaluated using the peak analyzer in origin pro software. The crystalline percentages of neat bio PBS and SCG/bio PBS were 45.26% and 34.03%, respectively. The degree of crystallinity of biocomposite decreased with the addition of SCG filler since fillers increased the mean distance between molecules in the biopolymer [19].

DMA Behavior. The dynamic mechanical analyzer evaluated the viscoelastic properties of pure bio PBS and its biocomposite. Figure 6 shows the storage modulus (E') and loss modulus (E'') characteristics of developed composites. The E' value of both neat polymer and biocomposite decreased with increased temperature due to the weakening of the molecular bond in polymer chains [20]. Also, E' decreased with SCG filler loading because of pores in biocomposite structure and the presence of proteins in SCG fillers, which may act as plasticizers in biopolymers [21].

The E'' values increased to their peak level and then decreased with increased temperature. Figure 6b shows that a hump between − 25 and 0 °C due to the distribution of solid SCG fillers in the matrix restricted the movement of chains in the filler-matrix interface [6].

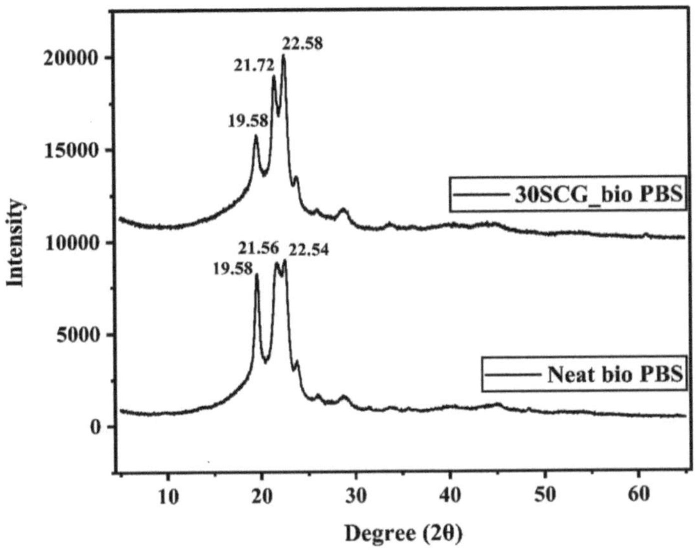

Fig. 5 XRD patterns of SCG/bio PBS composites

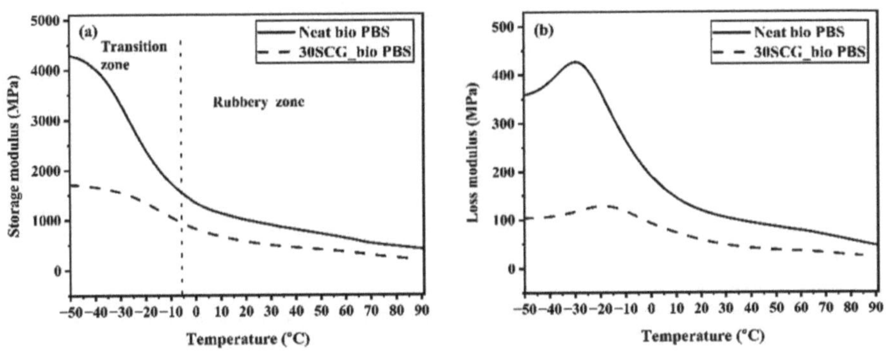

Fig. 6 **a** Storage modulus and **b** loss modulus of biocomposites

4 Conclusion

In summary, a comprehensive analysis of mechanical and thermal properties as well as crystallinity behavior of SCG filler-reinforced bio PBS polymer composites have been reported. The developed biocomposite's tensile and flexural modulus was improved compared to neat polymer. On the contrary, tensile and flexural strength and storage modulus decreased due to poor interfacial bonding between SCG fillers and bio PBS. The addition of SCG fillers reduced the thermal stability and degree of crystallinity of the SCG/bio PBS composite. Positive environmental effects and the cost advantage of SCG fillers balance this moderate decrease in mechanical and thermal properties of developed biocomposites.

References

1. Balaji A, Kannan S, Purushothaman R, Mohanakannan S, Maideen AH, Swaminathan J, Karthikeyan B, Premkumar P (2022) Banana fiber and particle-reinforced epoxy biocomposites: mechanical, water absorption, and thermal properties investigation. Biomass Conv Bioref:2190–6823. https://doi.org/10.1007/s13399-022-02829-y
2. Kumar P, Singh J, Kumari N, Jurail SS, Verma D, Maurya AK (2022) Study of mechanical and thermal behavior of alkali modified groundnut shell powder reinforced ABS composites. Polym Compos 43:4569–4587. https://doi.org/10.1002/pc.26713
3. Gomez-Caturla J, Balart R, Ivorra-Martinez J, Garcia-Garcia D, Dominici F, Puglia D, Torre L (2022) Biopolypropylene-based wood plastic composites reinforced with mango peel flour and compatibilized with an environmentally friendly copolymer from Itaconic acid. ACS Appl Polym Mater 4:4398–4410. https://doi.org/10.1021/acsapm.2c00373
4. Sohn JS, Ryu Y, Yun CS, Zhu K, Cha SW (2019) Extrusion compounding process for the development of eco-friendly SCG/PP composite pellets. Sustainability 11:1720. https://doi.org/10.3390/su11061720
5. Araújo MN, Azevedo AQPL, Hamerski F, Voll FAP, Corazza ML (2019) Enhanced extraction of spent coffee grounds oil using high-pressure CO_2 plus ethanol solvents. Ind Crops Prod 141:111723. https://doi.org/10.1016/j.indcrop.2019.111723
6. García-García D, Carbonell A, Samper MD, García-Sanoguera D, Balart R (2015) Green composites based on polypropylene matrix and hydrophobized spend coffee ground (SCG) powder. Compos Part B: Eng 78:256–265. https://doi.org/10.1016/j.compositesb.2015.03.080
7. Fang Y, Jiang Z, Zhao X, Dong J, Li X, Zhang Q (2022) Spent coffee grounds/poly(butylene succinate) biocomposites with robust mechanical property and heat resistance via reactive compatibilization. Compos Commun 29:101003. https://doi.org/10.1016/j.coco.2021.101003
8. Seo J, Kearney LT, Toomey MD, Keum JK, Naskar AK (2023) Polyester-based epoxy vitrimer integrating spent coffee ground as a natural filler. Compos Part B: Eng 260:110756. https://doi.org/10.1016/j.compositesb.2023.110756
9. Essabir H, Raji M, Laaziz SA, Rodrique D, Bouhfid R, El Kacem Qaiss A (2018) Thermo-mechanical performances of polypropylene biocomposites based on untreated, treated and compatibilized spent coffee grounds. Compos Part B: Eng 149:1–11. https://doi.org/10.1016/J.COMPOSITESB.2018.05.020
10. Gaidukova G, Platnieks O, Aunins A, Barkane A, Ingrao C, Gaidukovs S (2021) Spent coffee waste as a renewable source for the production of sustainable poly(butylene succinate) biocomposites from a circular economy perspective. RSC Adv 11:18580–18589. https://doi.org/10.1039/d1ra03203h
11. Totaro G, Sisti L, Fiorini M, Lancellotti I, Andreola FN, Saccani A (2019) Formulation of green particulate composites from PLA and PBS matrix and wastes deriving from the coffee production. J Polym Environ 27:1488–1496. https://doi.org/10.1007/S10924-019-01447-6/FIGURES/11
12. Pagalan E, Sebron M, Gomez S, Salva SJ, Ampusta R, Macarayo AJ, Joyno C, Ido A, Arazo R (2020) Activated carbon from spent coffee grounds as an adsorbent for treatment of water contaminated by aniline yellow dye. Industr Crops Prod 145:111953. https://doi.org/10.1016/J.INDCROP.2019.111953
13. Marto J, Gouveia LF, Chiari BG, Paiva A, Isaac V, Pinto P, Simões P, Almeida AJ, Riberio HM (2016) The green generation of sunscreens: using coffee industrial sub-products. Industr Crops Prod 80:93–100. https://doi.org/10.1016/J.INDCROP.2015.11.033

14. Tongcumpou C, Usapein P, Tuntiwiwattanapun N (2019) Complete utilization of wet spent coffee grounds waste as a novel feedstock for antioxidant, biodiesel, and bio-char production. Industr Crops Prod 138:111484. https://doi.org/10.1016/J.INDCROP.2019.111484
15. Naik TP, Gairola S, Singh I, Sharma AK (2022) Microwave hybrid heating for moulding of sisal/jute/HDPE composites. J Nat Fibers 19:13524–13538. https://doi.org/10.1080/15440478.2022.2100553
16. Zarrinbakhsh N, Wang T, Rodriguez-uribe A, Misra M, Mohanty AK (2016) Characterization of wastes and coproducts from the coffee industry for composite material production. BioResources 11:7637–7653. http://ncsu.edu/bioresources. ISSN: 1930-2126
17. Sarasini F, Tirillò J, Zuorro A, Maffei G, Lavecchia R, Puglia D, Dominici F, Luzi F, Valente T, Torre L (2018) Recycling coffee silverskin in sustainable composites based on a poly(butylene adipate-co-terephthalate)/poly(3-hydroxybutyrate-co-3-hydroxyvalerate) matrix. Industr Crops Prod 118:311–320. https://doi.org/10.1016/J.INDCROP.2018.03.070
18. Wu CS (2015) Renewable resource-based green composites of surface-treated spent coffee grounds and polylactide: characterisation and biodegradability. Polym Degrad Stab 121:51–59. https://doi.org/10.1016/J.POLYMDEGRADSTAB.2015.08.011
19. Pradhan P, Satapathy A (2022) Physico-mechanical characterization and thermal property evaluation of polyester composites filled with walnut shell powder. Polym Polym Compos 30. https://doi.org/10.1177/09673911221077808
20. Gümüş BE, Yağci Ö, Taşdemir M (2022) High-density polyethylene/artichoke leaf powder polymer composites: dynamic mechanical, morphological and thermal properties. Iranian Polym J (English Edition) 31:787–797. https://doi.org/10.1007/S13726-022-01031-1/FIGURES/9
21. Hejna A, Barczewski M, Kosmela P, Mysiukiewicz O (2023) Comparative analysis of the coffee and cocoa Industry by-products on the performance of polyethylene-based composites. Waste Biomass Valor 1:1–16. https://doi.org/10.1007/S12649-023-02041-7/FIGURES/3

Experimental Investigation on the Effectiveness of Flocculant in Sediment Plumes

Soubhik Mondal, Anooja Sara Mathew, and Vishnu Chandrasekharan Nair

1 Introduction

Over the past decade, there has been an increase in global concern over the long-term availability of secure and adequate supplies of critical raw materials. The rising market prices of metals, together with technological improvements, have driven several countries and industries to prospect for the deep-sea environment for mineral resources, including polymetallic nodules on the ocean floor [1]. Polymetallic nodules are rich in metals and rare earth elements lying on the surface of the seafloor [2]. Due to this, deep-sea mining has gathered considerable attention in recent years. India became the first investor in this situation since the International Seabed Authority (ISA) granted it access to the polymetallic nodules (PMN) site in the Central Indian Ocean Basin (CIOB) for deep-sea mining exploration and technology development [3]. Although there are abundant polymetallic nodules on the benthic floor that could be a significant source of strategic metals such as copper, nickel, cobalt, and manganese, their extraction has negative ecological implications that need to be carefully addressed. The environmental effects of deep-sea mining, particularly the sediment plumes created during substrate cutting and vessel processing, pose a severe threat to the benthic environment. Polymetallic nodule mining can generate 50,000 tons of sediment per day, of which 4000 tons are transported to the surface. These plumes have a longer settling time, which may last from a few to several years, and can lead to ecological damage. The findings from numerical experiments conducted by Rolinski et al. [4] indicated that around 90–95% of fine-grained sediments took 3–14 years to settle depending on the water depth at which they are released and could be carried by ocean currents for hundreds of kilometers, whereas coarser sediments got settled shortly. They observed that the properties of the released material influence the impact of deep-sea mining [4]. To

S. Mondal · A. S. Mathew · V. C. Nair (✉)
Indian Institute of Technology, Kharagpur, West Bengal, India
e-mail: vishnuc@iitkgp.ac.in

R. Sahu et al. (eds.), *Advancements in Materials Processing Technology, Volume 1*,
Springer Proceedings in Materials 48, https://doi.org/10.1007/978-981-97-4958-4_26

tackle this issue, it is critical to develop solutions to reduce these problems. Precise and direction-oriented studies must be conducted to determine the ecological and environmental viability of deep-sea mining projects.

Elevated turbidity levels caused due to the resuspension of sediments during extraction of minerals from the ocean floor can adversely impact marine ecosystems by reducing light penetration, inhibiting photosynthesis, and affecting sediment transport. Flocculation can be one of the mitigation measures to reduce the impact of turbidity caused by deep-sea mining activities in shallow and deep oceanic depths. Sharma et al. [5] described flocculation as the process of clustering of suspended particles with the addition of polyelectrolytes, which can be cationic, anionic, or ampholytic in nature. According to him, this aggregation occurs primarily through two mechanisms: bridging and patching [5]. Ali et al. [6] introduced the effect of flocculation on the deep-sea sediment plume propagation, and the study was conducted using an organic flocculant for lock exchange experiments. The head velocity was determined from the experiments and concluded that the head velocity significantly got decreased in the saline water because of the denser flocs as compared to the freshwater floc's formation (because of the electrostatic repulsion between the organic matter and clay as the polyelectrolyte is less coiled than in saline water) [6]. Addition of flocculants can accelerate the settling velocity of suspended particles by the formation of flocs. According to Costine et al. [7], when compared to identical sizes made with greater molecular weights, the lower molecular weights created denser flocs under mild mixing [7]. The flocculation efficiency of various flocculants of synthetic and natural origin must be determined on a laboratory scale before its introduction into the pelagic environment. Polyacrylamide, a water-soluble polymer, has shown promise in reducing turbidity and sediment settling velocities by improving sediment flocculation and settling behavior. The paper presents the results of flocculation studies conducted to investigate the flocculation efficiency of polyacrylamide in bentonite clay suspensions with synthetic sea water.

Polyacrylamide is a synthetic polymer that possesses unique properties, such as high molecular weight, water solubility, and flocculation capability. Its effectiveness in reducing turbidity and sediment settling velocities stems from its ability to form larger flocs through charge neutralization and bridging mechanisms. Polyacrylamide (PAM) has been widely studied and utilized in various fields due to its remarkable properties as a flocculant and coagulant. One area where PAM has shown significant promise is in the treatment of suspended solids in aqueous environments. When combined with clay, PAM can effectively reduce the settling velocity of particles, leading to improved sedimentation and separation processes. In mining and mineral processing, PAM-clay composites have also shown promising results. A study by Chen et al. [8] examined the application of PAM-clay mixtures in tailings dewatering processes and observed that the addition of PAM to clay-based tailings significantly reduced the settling velocity of solid particles, leading to improved dewatering efficiency and increased water recovery [8]. The minimum and maximum dosage needs to be determined since after a certain limit the anionic PAM may reverse the flocculation process by increasing the turbidity as it has higher density of repulsive force [9].

It is essential to balance economic development with minimal ecological damage to ensure sustainable growth. The presented methodology and detailed results of the study can provide useful insights for addressing the environmental concerns associated with deep-sea mining. The addition of PAM into the bentonite clay suspensions was found to accelerate the settling speed of particles by forming flocs, resulting in low turbidity.

2 Materials and Methods

Oil and Natural Gas Corporation Limited (ONGC) grade bentonite clay sample was used for the study. Polyacrylamide (PAM) was obtained from Envitech Chemical Specialties Pvt. Ltd. Pune, India. Salts for the preparation of synthetic sea water were procured from Sisco Research Laboratories Pvt. Ltd. (SRL), India.

2.1 Characterization of Bentonite Sample

The physical properties such as particle size, bulk density, and moisture content of the bentonite clay sample used in the study were obtained from Oil and Natural Gas Corporation Limited, India.

2.2 Characterization of Polyacrylamide (PAM)

Polyacrylamide (PAM), a polymer molecule formed from the polymerization of acrylamide molecules, was used as the flocculant in this study. It comes under the category of synthetic organic flocculant and can form a soft gel immediately when it meets water.

2.3 Characterization of Synthetic Sea Water

The whole set of experiments was carried out with synthetic sea water to resemble the oceanic environment. The sample was prepared in the laboratory as per ASTM (D 1141—98) standards [10].

2.4 Flocculation Studies

Preparation of Bentonite Suspension

Bentonite suspension was prepared with bentonite clay mixing in synthetic sea water. The concentration of bentonite was kept constant at 1% by weight for the entire set of tests. Each sample was prepared by mixing 2 gm of bentonite in approximately 200 ml of synthetic sea water.

Flocculation by PAM Addition

For flocculation, studies were done with polyacrylamide addition in concentrations varying from 0 to 200 ppm into the bentonite suspension. The effectiveness of PAM as flocculant was analyzed by determining the turbidity values of the different sample solutions. After addition of flocculant into the solution, a uniform stirring was given and was allowed for floc formation.

Turbidity Measurements

In this laboratory experiment, the turbidity of the bentonite suspensions with different dosages of PAM flocculant was measured in Nephelometric Turbidity Units (NTUs) using a turbidity sensor with Arduino. The possibility to monitor continuous turbidity readings is the advantage of turbidity sensors over other turbidimeters. The functioning principle of the sensor relies on the correlation between the level of contaminants in water and the transmission of light through a sample of it. As the water passes through the sensor, the presence of dirt or solid particles affects the amount of light that can pass through. Consequently, as the depth of solid particles increases, the transmitted light decreases accordingly. By quantifying the amount of light that passes through the sample, the sensor provides an indication of the water's turbidity, which is a measure of its clarity or cloudiness caused by suspended particles.

Working Principle of the Turbidity Sensor

The Arduino turbidity sensor from DfRobot utilizes gravity-based technology to measure turbidity levels, providing valuable information about the water's condition. The sensor operates by analyzing the transmittance and scattering rate of light, which is affected by the concentration of total suspended solids (TSS) in the water. By assessing the light's interaction with suspended particles, the sensor can identify and quantify the presence of these particles in the water. As the TSS concentration increases, the turbidity level of the liquid also rises, indicating a higher degree of cloudiness or opacity.

The output signal modes for this liquid sensor include analog and digital. When in digital signal mode, the threshold can be altered. We must set the minimum and maximum values of the turbidity sensor in order to comprehend the behavior of our sensor. For our particular situation, we examined and determined that if the sensor outputs 4.2 V approximately at zero turbidity (clear water), it is only accurate between 2.5 and 4.2 V (3000 to 0 turbidity).

Table 1 Properties of bentonite sample (data given by ONGC limited)

Properties	Numerical value
Particle size, D50 (μm)	3.2
Specific gravity	2.50
Moisture content (%)	13.50
Apparent viscosity (cP)	< 12

Turbidity Sensor Voltage into NTU

The relationship between the turbidity sensor voltage and the turbidity in NTU was established by Dfrobot [11] where we can simply generate our voltage values in terms of NTU values.

The relationship can be represented by the formula given below [11].

$$\text{ntu} = -1120.4 * (\text{Turbidity Sensor Voltage})^2 + 5742.3 * (\text{Turbidity Sensor Voltage}) - 4352.9$$

We have plotted our own graph based on this relationship for the various levels of concentration of PAM (Fig. 3). Here, we have tried to correlate the nature of the graph with the one based on the above equation to check its correctness.

We need to include this formula in the Arduino code to convert the sensor voltage into turbidity in terms of NTU. We have applied this method in our study to get the final turbidity values with time.

3 Results and Discussions

3.1 Physical Properties of Bentonite Clay

The physical properties of bentonite clay given in Table 1 were obtained from Oil and Natural Gas Corporation Limited, India, with which the whole set of the experiments were carried out.

3.2 Properties of Polyacrylamide (PAM)

The physical and chemical properties of anionic PAM DR8400 were studied in the laboratory and is given in Table 2. It is a white granular powder with the following properties.

Table 2 Properties of polyacrylamide (PAM) (results from, DSCEPE, IIT KGP)

Properties	Numerical value
Particle size (85 mesh)	> 50%
Degree of hydrolysis (%)	26.95
Molecular weight (MDa)	17.20
Moisture content (%)	12.53

3.3 Properties of Synthetic Sea Water

The following properties of synthetic sea water were found in the laboratory. The prepared sample had pH 7.8, salinity 35.89 g/kg, and density 1.034 kg/L.

3.4 Flocculation Studies

Bentonite clay is made up of tiny platelets with a large surface area that offers lots of sites for particle attachment. These platelets have a negative charge, which enables them to combine electrostatically with positively charged particles to attract and bind them, resulting in the creation of flocs. A synthetic polymer with outstanding flocculating qualities is polyacrylamide (PAM). Because of its high molecular weight and capacity to create sizable, well-structured flocs, it is commonly used as a flocculant. PAM promotes the aggregation of suspended liquid particles into bigger clusters that settle more readily by binding to them. This enhances the processes of clarification, sedimentation, and filtration.

The ideal dosage and environmental factors for achieving the desired flocculation efficiency are being investigated in the flocculation study using PAM and bentonite clay. To make the process as efficient as possible, variables including pH, temperature, mixing vigor, and settling time can be carefully regulated and tracked. In our case, we have considered only the different PAM concentrations varying from 0 to 200 ppm keeping other parameters fixed.

The application of PAM in bentonite suspensions decreased the turbidity of the solution leading to a positive output for its application in deep-sea mining activities. Bentonite clay and PAM worked well together because of their numerous benefits. It increases the rate of sedimentation, shortens the time needed for settling, and increases the liquid's clarity of the ocean water which is monitored and recorded in our experiment as a positive response of the PAM application.

In this study, each set of experiments was carried out for a period of 30 min, and the data was recorded with the interval of 10 s using the turbidity sensor to get the smooth trend of the changes in turbidity with the time.

Turbidity Sensor Calibration

Before starting the experiment, we need to check whether the output voltage appears near 4.21 or not for a clear solution (distilled water in our case). Once it achieves the voltage, your sensor is ready for the experiment. Otherwise, the sensor should be calibrated by rotating the potentiometer present on the sensor A&D converter circuit.

In case of changing the PAM concentration, after a certain period, the strength of the PAM will not have a noticeable impact to decrease the turbidity further. Turbidity will decrease but that will be a slow process and will tend to get stable. After that time, if we try to increase the PAM concentration, the settling process will not be induced, which is shown in Figs. 1 and 2.

In terms of the strength, every PAM has its own strength limit according to the solution, pH, and temperature, and this strength varies with respect to the concentration levels. It is possible to trace the behavior of how the turbidity of a particular solution changes with respect to the concentration of a particular flocculant. In this

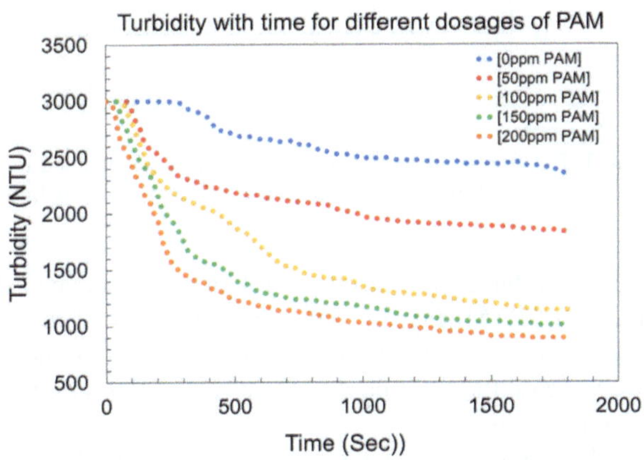

Fig. 1 Time series line plots regarding the changes in turbidity with PAM concentrations varying from 0 to 200 ppm (results from, DSCEPE, IIT KGP)

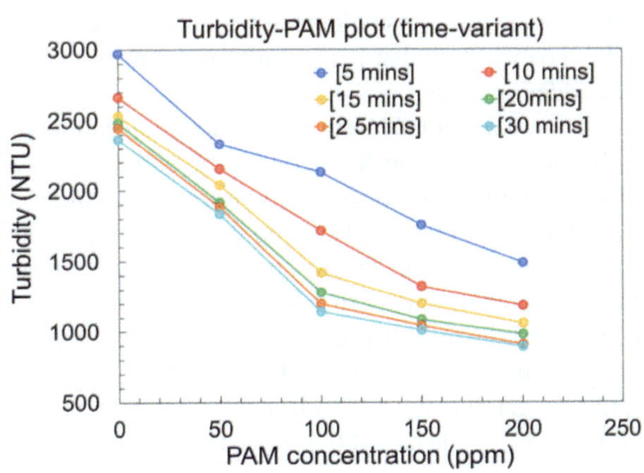

Fig. 2 Relationship between the turbidity and the concentration of PAM at every 5 min of intervals from the beginning (results from, DSCEPE, IIT KGP)

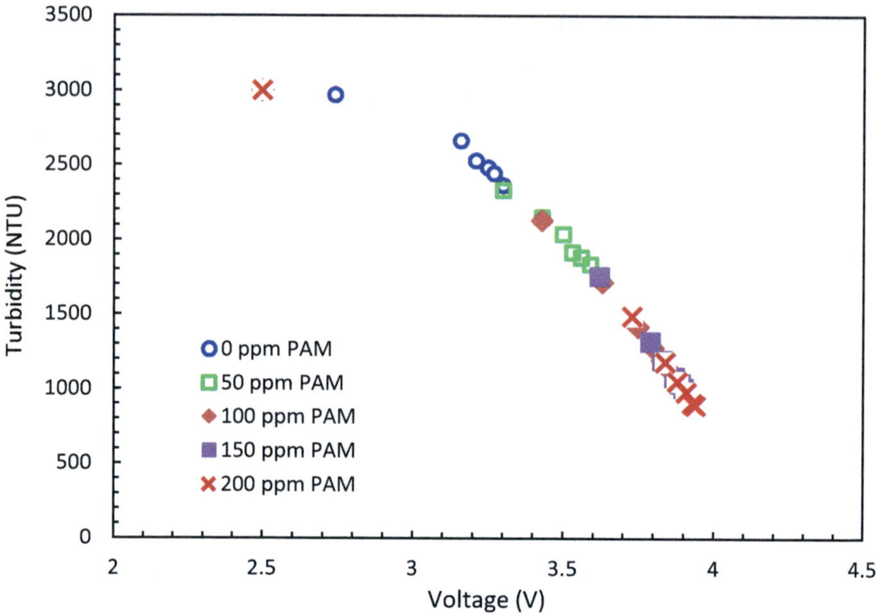

Fig. 3 Relationship between the output sensor voltage and the turbidity for different dosages of PAM at different time intervals from the beginning (results from, DSCEPE, IIT KGP)

study, we have introduced the relationship between the turbidity and the concentration of PAM (Fig. 2) to find the efficiency of the flocculant we are using. Also, Fig. 2 shows a significance regarding the fact that the strength cannot be increased if we increase the flocculant concentration after a particular time which is not clearly visible in Fig. 1.

Here, in our case, we see that after 15 min the strength of the flocculant started to decrease and the changes in turbidity were not so significant. Even after 20 min, the changes were so small since the entire process of particle settlement turned out to be terribly slow that it does not matter how much the concentration of the PAM was added initially.

Since we are considering the output voltage as an input to give rise to the turbidity at the end, the relationship between the output voltage and the turbidity is explained (Fig. 3) where turbidity and the voltage are having a certain considerable range. For the turbidity, we have considered the range 0 to 3000 NTU and 0 to 4.21 V for the output voltage range. In this study, the voltage is approaching 4 V indicating the clear solution at the experiment's end. The graph represents the relationship between voltage and the turbidity at the interval of 5 min from the beginning ($t = 0, 5, 10, 15$, 20 25, 30 min) for each of the dosages of PAM (0, 50, 100, 150, 200 ppm). Since we could not get that much of a clear solution, the voltage has not achieved its maximum limit as well as the turbidity is also decreasing accordingly to signify the application

Fig. 4 Visual representation of the effectiveness of PAM with respect to different concentration levels varying from 0 to 200 ppm (results from, DSCEPE, IIT KGP)

of PAM to reduce the cloudiness. But the higher dosage of PAM (200 ppm) produces better clarity than other dosage levels of PAM.

It is difficult to differentiate the turbidity level by the naked eye, but still we can visualize the effect of the PAM with different concentrations for a particular time (15 min in our case). One can differentiate the solution easily where the PAM dosage is quite low or high, but difficulties will appear if the difference in the PAM concentration is low. Here, we have tried to visualize the effect of PAM application for different dosages after 15 min (Fig. 4).

Though we cannot trace the turbidity level by the naked eye, we are having a trend of reduction of cloudiness in the solution as we are increasing our PAM concentration (Fig. 4). There is a certain limit also up to which we can increase the particle settling velocity indicating the clearer solution.

4 Conclusion

Deep-sea mining operations are gaining significant attention as a potential solution to meet the growing demand for mineral resources. However, these activities can have adverse environmental impacts, particularly related to the suspension of sediment particles and the subsequent increase in turbidity. Adding flocculants in minimal dosages can be opted as one of the mitigation measures to reduce the impact of turbidity caused in shallow and deep oceanic depths during deep-sea mining activities. In this context, we have conducted laboratory-scale experimental study to determine the flocculation efficiency of polyacrylamide (PAM) in bentonite suspensions in synthetic sea water. Keeping other parameters constant, the concentration of PAM varied. The decrease in turbidity with increased PAM concentration was obtained but up to a certain limit beyond which there was no noticeable reduction in turbidity. Further conclusions obtained are as follows:

- The settling velocity of the suspended particles will get induced with higher dosages of the flocculant.

- The particles will settle down according to the dosage, and the solution will have more clarity with time, but after a certain period, the process will become slow and tend to get stable.
- Even if we vary the flocculant dosage, there will be one concentration at which we will see the best performance. In our study, the PAM with 200 ppm of concentration reduced the turbidity in a fast way within 5 min. But the PAM with 100 ppm dosage kept decreasing the turbidity for a longer time. Hence, 100 ppm PAM concentration can be considered to have higher strength than the other concentrations even though the initial settling was not so rapid, but its activity lasted for a longer period.
- Since we have considered all the data for up to 30 min, we could not get a much clearer solution. The process was slow beyond 30 min. As we recorded all the data points with an interval of 10 s, we did not consider further. Otherwise, it would affect the time series plot with unnecessary errors.
- Another thing is the difference in methodology. If we use the turbidimeter, we will only consider the surface level solution to measure the data points. But in our case, we have introduced the turbidity sensor interfaced with Arduino UNO where we need to dip the sensor into the solution at a depth just to make sure that the transmitter and the receiver points are submerged into the solution. The depth is the major concern to get a much clearer solution. It is hard to maintain the height of your sensor every time during the entire experiment in the laboratory. The higher depth will indicate higher turbidity during the process. In the real system also, we must consider the environment inside the water column of the ocean rather than the surface layer only.
- The advantage of using the turbidity sensor interfaced with Arduino over other methods is the ability to get the turbidity values with the interval of a millisecond to an hour. There will be many data points that are significant to generate a smooth trend with fewer errors.
- Future work: The turbidity varies with respect to time and depth. In this study, we have measured the turbidity from a fixed depth of the water column considering only the time as a variable. By setting multiple sensors at varying depths of the water column, it will be able to trace how the settling velocity varies with respect to depth and particularly up to what depth we can expect the major impact of using PAM.

References

1. Hein JR, Koschinsky A (2014) Deep-ocean ferromanganese crusts and nodules. In: Treatise on geochemistry, 2nd ed. Elsevier, 2014, pp 273–291. https://doi.org/10.1016/B978-0-08-095975-7.01111-6
2. Gillard B, Purkiani K, Chatzievangelou D, Vink A, Iversen MH, Thomsen L (2019) Physical and hydrodynamic properties of deep sea mining-generated, abyssal sediment plumes in the

Clarion Clipperton Fracture Zone (eastern-central Pacific). Elem Sci Anthr 7(January). https://doi.org/10.1525/elementa.343
3. Government of India. Ministry of earth science, Environmental Impact State-ment, 2020 Environmental conditions and likely impact in the area selected for nodule col-lection trials at the Indian PMN site in the Central Indian Ocean Basin. https://moes.gov.in/writereaddata/files/India_PMN_EIS_Jan_2020.pdf
4. Rolinski S, Segschneider J, Sündermann J (2001) Long-term propagation of tailings from deep-sea mining under variable conditions by means of numerical simulations. Deep Sea Res Part II Top Stud Oceanogr 48(17–18):3469–3485. https://doi.org/10.1016/S0967-0645(01)00053-4
5. Sharma BR, Dhuldhoya NC, Merchant UC (2006) Flocculants—an ecofriendly approach. J Polym Environ 14(2):195–202. https://doi.org/10.1007/s10924-006-0011-x
6. Ali W, Enthoven D, Kirichek A, Helmons R (2022) Can flocculation reduce the dispersion of deep sea sediment plumes? In: Proceedings of the world dredging conference, WODCON XXIII, Copenhagen, Denmark, WODA, 2022, pp 1–9
7. Costine A, Cox J, Travaglini S, Lubansky A, Fawell P, Misslitz H (2018) Variations in the molecular weight response of anionic polyacrylamides under different flocculation conditions. Chem Eng Sci 176:127–138. https://doi.org/10.1016/j.ces.2017.10.031
8. Chen G, Li C, Ruan Z, Bürger R, Hou H (2023) Research on floc structure and physical properties based on pipeline flocculation. J Water Process Eng 53(November 2022):103627, July 2023. https://doi.org/10.1016/j.jwpe.2023.103627
9. Shamlooh M, Rimeh A, Nasser MS, Al-Ghouti MA, El-Naas MH, Qiblawey H (2020) Enhancement of flocculation and shear resistivity of bentonite suspension using a hybrid system of organic coagulants and anionic polyelectrolytes. Sep Purif Technol 237(Aug 2019):116462, Apr 2020. https://doi.org/10.1016/j.seppur.2019.116462
10. ASTM International (1999) Standard for the preparation of substitute ocean water. D 1141, vol. 98, no. Reapproved, pp 98–100
11. Fahad E. Turbidity sensor with Arduino for water quality monitoring. Turbidity Meter. https://www.electroniclinic.com/turbidity-sensor-with-arduino-for-water-quality-monitoring-turbidity-meter/

Campbell Diagrams of Some Advanced Composite Blades: A Comparative Analysis

Souvik Singh Rathore, Vishesh Ranjan Kar, and Sanjay

1 Introduction

Rotating turbine blades are a critical component of any gas turbine. These blades are subjected to harsh operating conditions such as high temperature, pressure, and vibrations due to the engine's moving components. Blades are often coated to protect them from such harsh conditions. Hence, vibrational analysis of these coated blades becomes crucial for their reliable and safe operation.

Vibration analyses of blades have been carried out in the past by assuming various mathematical models, such as beam, plate, and shell models. The plate and shell models can predict the vibrational characteristics in low aspect ratio blades and when operating at high frequencies [1]. In many research articles published in the past, the blade materials were isotropic and homogeneous. Such blades can be attributed to uncoated blades capable of operating at lenient thermal and environmental conditions. In contrast, composite blades that are lighter and indistinguishable from actually coated blades have been studied to report their vibrational characteristics. Rostami et al. [1] studied the free vibration characteristics of rotating composite plate based on classical plate theory and compared the results evaluated via extended Kantorovich method and extended Galerkin method. In another similar work, Rostami et al. [2] analysed the free vibration characteristics of rotating composite plates assuming first-order-shear-deformation theory (FSDT). Kee and Kim [3] analysed the rotating graphite/epoxy composite blade, assuming it was a cylindrical shell. They reported its vibration characteristics using FSDT kinematics and used the finite element analysis (FEA) approach. Rout and Karmakar [4] showcased the vibration characteristics of carbon nanotube reinforced composite blades

S. S. Rathore (✉)
ARKA JAIN University, Gamharia, Jamshedpur, Jharkhand 832108, India
e-mail: souviksinghrathore@gmail.com

V. R. Kar · Sanjay
National Institute of Technology Jamshedpur, Jamshedpur, Jharkhand 831014, India

R. Sahu et al. (eds.), *Advancements in Materials Processing Technology, Volume 1*,
Springer Proceedings in Materials 48, https://doi.org/10.1007/978-981-97-4958-4_27

by assuming them as rotating twisted cylindrical shells in thermal environments based on FSDT kinematics and FEA.

Many researchers have also analysed functionally graded material (FGM) blades to address the harsh temperature environment encountered by the blades. Li and Zhang [5] developed a novel methodology to capture the vibration characteristics of an FGM rectangular plate based on classical plate theory (CLPT) and von-Karman theory. While Parida and Mohanty [6] analysed the nonlinear vibration characteristics of FGM plate that was based on higher-order-shear-deformation theory, von-Karman geometric nonlinearity, and FEA approach, Ansari et al. [7] adopted FSDT and isogeometric approach to appreciate the vibrational characteristics of FGM blade.

The thermal and environmental barrier-coated blades can be closely approximated as rotating sandwiched structures with metal alloy sandwiched between the ceramic outer layers. A few researchers have analysed FG sandwich blades and reported their vibrational characteristics. Cao et al. [8] analysed the vibrational behaviour of the FG sandwich blades based on FSDT kinematics, von-Karman theory, and Chebyshev–Ritz method. Liu et al. [9] analysed the vibrational behaviour of the FG sandwich blade using FSDT kinematics and Rayleigh–Ritz method. Similarly, Neves et al. [10] carried out vibration analysis of three different types of plates (based on material distribution through the thickness), including FG sandwich plates and involving HSDT and Carrera's Unified Formulation. Chen et al. [11] analysed the vibration characteristics of the FG sandwich blade in the thermal environment based on FSDT and modified Fourier spectral approach. Vibration characteristics of the FG sandwich blades based on FSDT and FEA have also been reported by the author [12, 13].

In the above literature survey, many research articles report the vibration analysis of the plate/shell models but with isotropic material. A few research articles have reported the vibration analysis of the blade model made of advanced composite materials, such as laminated composites and functionally graded materials, but at low temperatures. Sandwich blades with ceramic outer layers closely approximate the TBC-coated blades, but research articles on such blades are rare. Therefore, this article analyses blades with FG sandwich material to relate to the advanced TBC-coated blades used in the state-of-the-art turbines. Three different core materials made of stainless steel (SUS304), titanium alloy (Ti6Al4V), and nickel alloy (IN718) have been compared based on their vibrational frequencies at different blade rotating speeds (Campbell diagrams) for the first time. For this, the blade model for the FG sandwich blade is developed using FSDT kinematics and Voigt's micromechanical model. Then, the equation of motions was derived via Hamilton's principle and further solved to obtain the vibration frequencies using the Block Lanczos algorithm and FEA approach.

2 Mathematical Formulations

The present blade model (Fig. 1) assumes a cantilevered shell with length = a, width = b, and thickness = h (representing a blade with span = a, chord = b, and thickness = h). The blade model is located at an offset R_0 from the Y-axis, representing the blade on a hub rotating about the Y-axis. The blade is twisted about the X-axis by an angle ϕ.

Figure 2 depicts through-the-thickness visualisation of material distribution in the FG sandwich blade. The core is a metal alloy, while the outer skins are FGM. The set of equations governs the volume fraction of the metal and the ceramic constituents (see Eqs. 1 and 2) [10],

$$V_m(z) = \begin{cases} \left(\frac{h+2z}{h-h_m}\right)^l & z \in [z1, z2] \\ 1 & z \in [z2, z3] \\ \left(\frac{h-2z}{h-h_m}\right)^l & z \in [z3, z4] \end{cases} \tag{1}$$

$$V_c(z) = 1 - V_m(z) \tag{2}$$

The material distribution in the outer skins of the FGM follows the power law distribution governed by the parameter l.

The equivalent single layer theory governs the overall property of the material via Voigt's micromechanical model as,

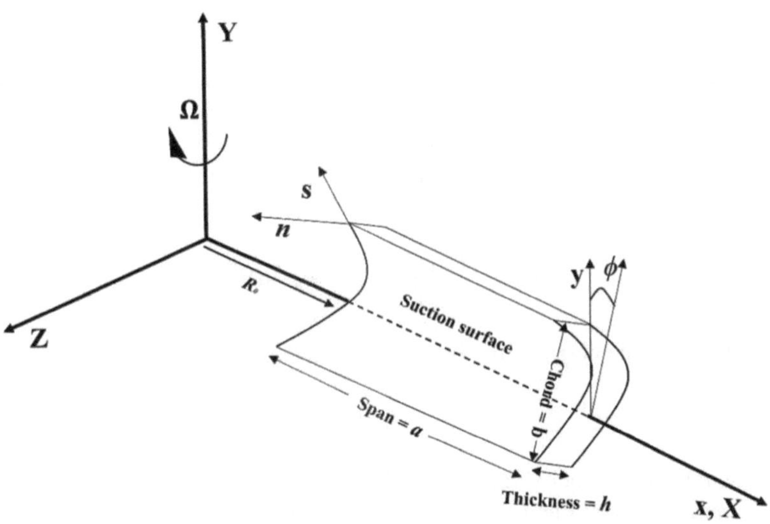

Fig. 1 Model of a pretwisted blade

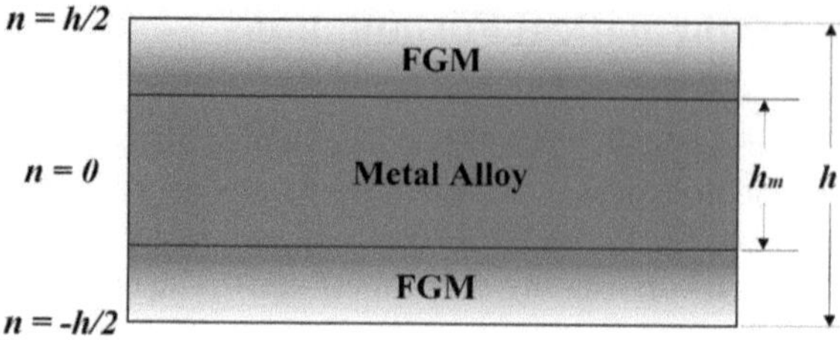

Fig. 2 Distribution of metal alloy and ceramic in a section of FG sandwich blade

$$P(z) = P_m V_m(z) + P_c V_c(z) \tag{3}$$

The displacement field of the blade model is based on the FSDT kinematics as,

$$\left.\begin{aligned} u &= u_0 + z\phi_x \\ v &= v_0 + z\phi_y \\ w &= w_0 \end{aligned}\right\} \tag{4}$$

The strain terms are expressed in terms of the membrane and the curvature terms as,

$$\{\varepsilon\} = \{\varepsilon_{0x}\} + z\{\kappa\} \tag{5}$$

The stress–strain relationship is expressed as Reddy [14],

$$\{\sigma\} = C\{\varepsilon\} \tag{6}$$

where D is the equivalent elastic property matrix.

The strain energy terms due to elastic deformation and centrifugal force are depicted with the help of Eqs. (7) and (8), respectively, as,

$$U_E = \frac{1}{2}\iiint \left(\{\epsilon\}^T\{\sigma\}\right)\mathrm{d}V \tag{7}$$

$$U_R = \frac{1}{2}\iiint \left(\{\epsilon_R\}^T\{\sigma_R\}\right)\mathrm{d}V \tag{8}$$

While the kinetic energy of the blade model is expressed as,

$$U = U_E + U_R \tag{9}$$

where $\forall$ denotes the velocity vector and V denotes volume.

The general mathematical formulation for finding the governing equations and boundary conditions can be derived using the extended Hamilton's principle,

Table 1 Mesh sensitivity analysis

Mode	2 × 2	4 × 4	6 × 6	8 × 8	10 × 10	12 × 12	Ref. [9]
1	15.462	15.482	15.481	15.479	15.479	15.478	15.842
2	38.907	38.611	38.519	38.503	38.499	38.497	35.121
3	44.476	43.691	43.626	43.612	43.606	43.604	45.298
4	88.443	85.617	85.272	85.195	85.172	85.157	84.975
5	115.646	106.862	104.981	104.6279	104.520	104.482	106.690
6	143.939	110.524	109.642	109.4727	109.427	109.411	108.528

$$\Gamma = \frac{1}{2} \iiint \left(\rho\{\forall\}^T\{\forall\}\right)\mathrm{d}V \tag{10}$$

Finally, the discretisation of the variational of the potential and the kinetic energy using FEM involves eight-noded serendipity elements [15]. Hence, the global matrices are obtained by assembling the elemental mass matrix, elemental strain matrix, and rotational stiffness matrix, and the final equation of motion is obtained as follows:

$$[M]\{\ddot{d}\} + [K_{\text{eff}}]\{d\} = 0 \tag{11}$$

Here, $[K_{\text{eff}}]$ is the sum of the elastic and centrifugal stiffness matrices.

3 Numerical Investigations

3.1 Convergence Analysis

To ascertain the number of elements for carrying out vibrational analysis of FG sandwich blades, mesh sensitivity analysis is carried out for rotating flat FG sandwich blades [9], with aspect ratio = 3, thickness ratio = 20, volume fraction index = 2, and nondimensional rotational speed = 3 (Table 1). Therefore, a 10 × 10 element computational domain is adopted for further analysis to obtain reasonably accurate results unless stated.

3.2 Validation Report

A comparison study is carried out with the work carried out by Lieu et al. [9]. Here, the first six vibration modes of an FG sandwich blade constituting C276 metal alloy and Al_2O_3 materials are compared at different rotating speeds. The comparison results (Table 2) are precise, confirming the present model's robustness.

Table 2 Comparison results for FG sandwich blade for first six modes of vibration and different nondimensional rotational parameter

	$\frac{\Omega}{\omega_1} = 0$		$\frac{\Omega}{\omega_1} = 1$		$\frac{\Omega}{\omega_1} = 2$	
Modes	Present	Ref. [9]	Present	Ref. [9]	Present	Ref. [9]
1	4.5469	4.4743	6.8805	6.9030	11.0586	11.2340
2	25.1653	25.4167	27.9532	28.5235	34.8726	34.0383
3	32.9524	33.0633	33.6181	33.0979	35.5330	35.8589
4	73.3678	73.0937	75.6842	76.2806	80.7033	80.9215
5	88.0051	89.2792	89.2489	92.7844	94.1859	96.5525
6	100.4972	99.5237	101.5491	100.9778	104.5972	103.9269

3.3 *Results and Discussions*

The subsequent results and discussions compare the vibrational frequencies of three different material blades, namely

- Type I blade with SUS304 core and SUS304/Si_3N_4 outer skins maintained at 1100 K.
- Type II blade with Ti6Al4V core and Ti6Al4V/Si_3N_4 outer skins maintained at 1100 K.
- Type III blade with IN718 core and IN718/Si_3N_4 outer skins maintained at 1100 K.

The thermophysical properties (ξ) of the constituent materials obtained from the research articles [8, 16, 17], as given in Table 3, are obtained as the cubic function of temperature.

The geometric and material parameters of the rotating pretwisted curved blades have the blade span $_a = 0.2$ m, aspect ratio $\frac{a}{b} = 3$, thickness ratio $\frac{h}{b} = 10$, layer thickness ratio $\frac{h_m}{h} = 0.5$, span-to-hub ratio $\frac{R_0}{a} = 1$, twist angle $\phi = 30°$, and $l = 2$, radius of curvature.

$Rs = 5a$ and the blade temperature has been maintained constant at 1100 K for the three blades. Here, the frequencies of vibration are defined in terms of nondimensional frequency parameters and rotation in terms of nondimensional rotational parameters, as given in Table 4.

The nondimensional frequency parameters are defined as, $\beta = \omega a^2 \sqrt{\frac{\rho_m h}{D_m}}$, where $D = \frac{Et^3}{12(1-\upsilon^2)}$ is the flexural rigidity per unit length. ρ_m, E_m, and υ_m are the material properties of the metal alloys at an operating temperature of 1100 K. The nondimensional rotational speed is defined as $\frac{\Omega}{\omega_1}$, where ω_1 is the fundamental circular frequency of vibration of the stationary blades.

Table 3 Thermophysical properties of the constituent materials for FG sandwich blades

Property	ξ_0	ξ_{-1}	ξ_1	ξ_2	ξ_3
Silicon nitrate (Si_3N_4)					
E (Pa)	3.4843×10^{11}	0	-3.07×10^{-4}	2.16×10^{-7}	-8.946×10^{-11}
ρ (kg/m^3)	2370	0	0	0	0
Υ	0.24	0	0	0	0
A	5.8723	0	9.095×10^{-4}	0	0
Stainless steel (SUS304)					
E (Pa)	2.0104×10^{11}	0	3.0790×10^{-4}	-6.5340×10^{-7}	0
ρ (kg/m^3)	8166	0	0	0	0
Υ	0.28	0	0	0	0
A	1.2330	0	8.0860×10^{-4}	0	0
Titanium alloy (Ti6Al4V)					
E (Pa)	1.17×10^{11}	0	-6.18313×10^{-5}	-5.36712×10^{-7}	3.3816×10^{-11}
ρ (kg/m^3)	4448	0	-1.4674×10^{-5}	-2.79002×10^{-8}	1.279×10^{-11}
Υ	0.3509	0	-1.46965×10^{-4}	3.20319×10^{-7}	-5.06412×10^{-11}
A	6.346	0	1.54428×10^{-3}	-7.87898×10^{-7}	1.5758×10^{-10}
Nickel alloy (IN718)					
E (Pa)	2.1483×10^{11}	0	-3.58423×10^{-5}	-2.33×10^{-7}	0
ρ (kg/m^3)	8210	0	0	0	0
Υ	0.2382	0	8.39631×10^{-4}	-1.26×10^{-6}	8.39631×10^{-10}
A	7.9107	0	3.033866×10^{-3}	-3.79233×10^{-6}	2.53×10^{-9}

Table 4 List of nondimensional parameters

Nondimensional parameters	Expressions
Nondimensional frequency parameter	$\beta = \omega a^2 \sqrt{\frac{\rho_m h}{D_m}}$ here, $D = \frac{Et^3}{12(1-\upsilon^2)}$
Nondimensional rotational speed	$\frac{\Omega}{\omega_1}$
The ratio for hub radius	$\frac{R_o}{a}$

Comparing the First Six Vibration Modes for the Stationary Blades

Table 5 illustrates the first six vibration modes for the three types of blades in stationary conditions and the parameters defined earlier. The fundamental frequencies for blades in hertz are converted into circular frequencies that are further used to

Table 5 Comparison report of the first six modes of vibration for the three advanced composite blades at stationary conditions

Mode	Frequency (Hz)		
	Type I blade	Type II blade	Type III blade
1	167.67	206.45	172.15
2	863.76	1029.9	892.97
3	1046.1	1234.9	1070.5
4	1525.4	1728.3	1612.7
5	2769.2	3322.3	2849.5
6	3236.3	3801.3	3313.9

calculate the nondimensional rotational speeds for the Campbell diagram. The data in the table also suggests that the Type I and Type III blades have similar frequencies for all the modes of vibration compared to the Type II blades, which have higher frequencies for all modes of vibration. This is owing to the overall material properties of the blades at 1100 K. It should be noted that since the first circular frequencies are different, the final corresponding rotational speeds (Ω), when nondimensional rotational speed $\frac{\Omega}{\omega_1} = 1$ for the three blades, are different.

Comparing the Campbell Diagrams

Figures 3, 4, and 5 illustrate the variation in the nondimensional frequency parameters (β) with the nondimensional rotation speeds (Campbell diagram) for the SUS304 blades at 1100 K. In contrast to Table 1, where frequency values were similar for Type I and Type III blades, the values vary significantly when converted into their corresponding nondimensional forms. This is owing to the constituent metal alloys' material properties used in transforming the frequencies into their nondimensional forms. It is also observed that the β trends (increasing) are similar for all three types of blades, and only the range of the results changes. In stationary blades, the β values for the first and sixth vibration modes vary from 5.722 to 110.454 for the Type I blade, 8.238 to 151.692 for the Type II, and from 4.971 to 95.7052 for the Type III blade. While rotating at $\frac{\Omega}{\omega_1} = 1$, the β values for the first modes and the sixth modes of vibration vary from 17.003 to 118.857 for the Type I blade, from 24.428 to 164.961 for the Type II, and from 14.779 to 103.118 for the Type III blade.

The mode exchange phenomena (two modes come closer and then separate) occur between mode 2 and mode 3 of vibration in all three materials. For Type I, Type II, and Type III, the mode exchange phenomena occur near $\frac{\Omega}{\omega_1} = 0.8$, $\frac{\Omega}{\omega_1} = 0.9$, and $\frac{\Omega}{\omega_1} = 0.7$, respectively.

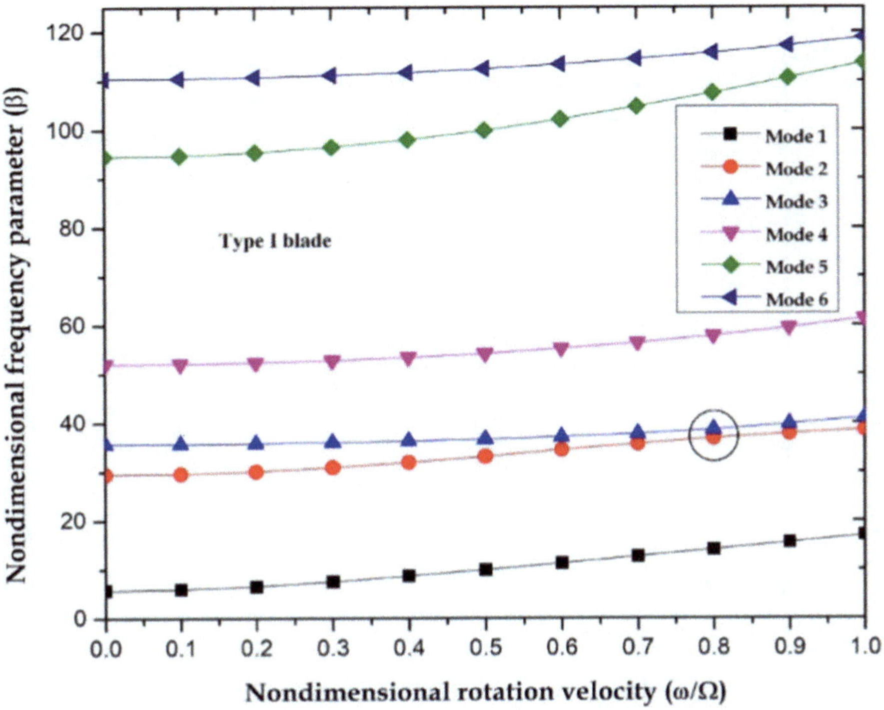

Fig. 3 Campbell diagram for rotating blade with SUS304 as core and SUS304/Si_3N_4 as face-sheets

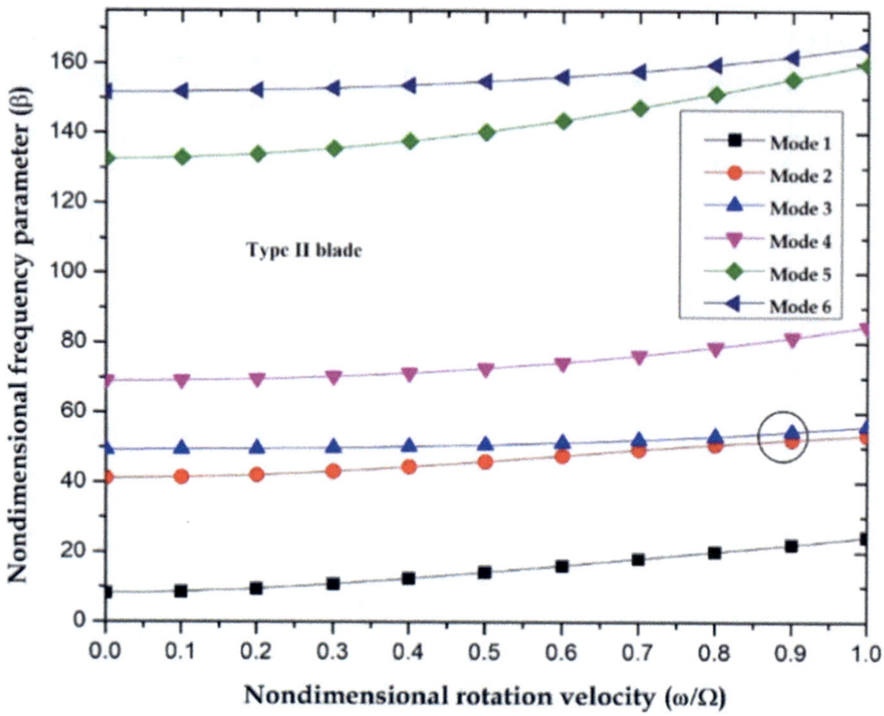

Fig. 4 Campbell diagram for rotating blade with Ti6Al4V as core and Ti6Al4V/Si_3N_4 as face-sheets

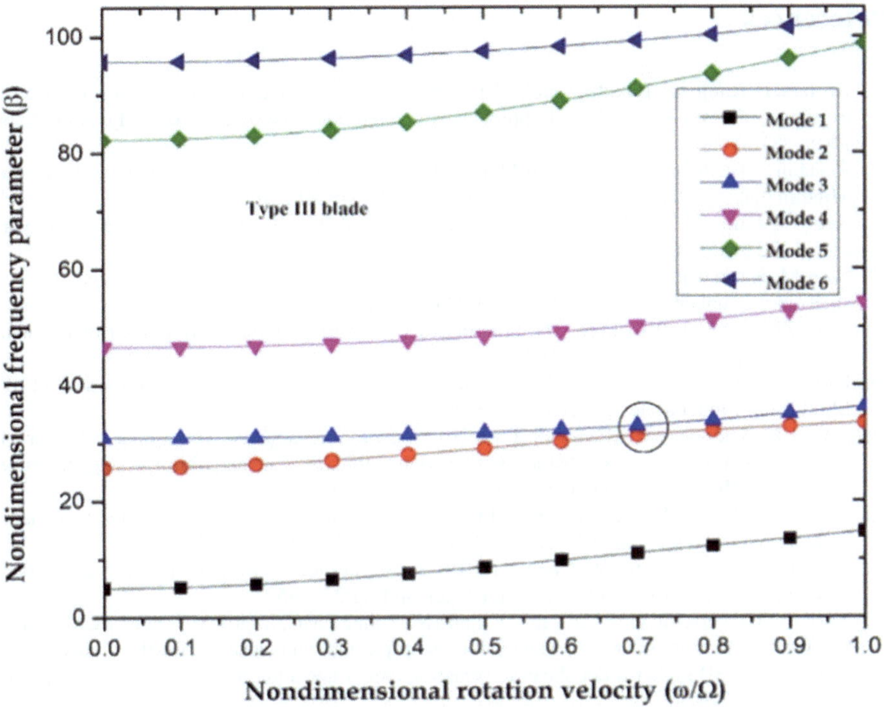

Fig. 5 Campbell diagram for rotating blade with IN718 as core and IN718/Si_3N_4 as face-sheets

4 Conclusion

This paper conducts a comparative vibrational analysis of three different FG sandwich blades made of three different materials labelled as Type I, Type II, and Type III. First, the blade model for the FG sandwich blade is developed using FSDT kinematics, Voigt's micromechanical model. Then, the equations of motion derived via Hamilton's principle and Block Lanczos and FEM approach are used to determine the vibrational frequencies. Finally, the Campbell diagrams of rotating pretwisted curved blades are compared. It is found that although the nature of the Campbell diagrams remains the same, the magnitude and range of the modes of vibration vary significantly. The locations of the mode exchange phenomena between mode 2 and mode 3 vibrations for Type I, Type II, and Type III occur near $\frac{\Omega}{\omega_1} = 0.8$, $\frac{\Omega}{\omega_1} = 0.9$, and $\frac{\Omega}{\omega_1} = 0.7$, respectively.

These results can help the turbine blade developers compare the coated blades made of different materials and further help compare and optimise the blade parameters.

References

1. Rostami H, Ranji AR, Bakhtiari-Nejad F (2018) Vibration characteristics of rotating orthotropic cantilever plates using analytical approaches: a comprehensive parametric study. Arch Appl Mech 88:481–502
2. Rostami H, Bakhtiari-Nejad F, Ranji AR (2019) Vibration of the rotating rectangular orthotropic Mindlin plates with an arbitrary stagger angle. J Vib Control 25(6):1194–1209
3. Kee YJ, Kim JH (2004) Vibration characteristics of initially twisted rotating shell type composite blades. Compos Struct 64(2):151–159
4. Rout M, Karmakar A (2019) Free vibration of rotating pretwisted CNTs-reinforced shallow shells in thermal environment. Mech Adv Mater Struct 26(21):1808–1820
5. Li L, Zhang DG (2016) Free vibration analysis of rotating functionally graded rectangular plates. Compos Struct 136:493–504
6. Parida S, Mohanty SC (2018) Vibration analysis of FG rotating plate using nonlinear-FEM. Multidiscip Model Mater Struct 15(1):26–49
7. Ansari E, Setoodeh AR, Rabczuk T (2020) Isogeometric-stepwise vibrational behavior of rotating functionally graded blades with variable thickness at an arbitrary stagger angle subjected to thermal environment. Compos Struct 244:112281
8. Cao D, Liu B, Yao M, Zhang W (2017) Free vibration analysis of a pre-twisted sandwich blade with thermal barrier coatings layers. Sci China Technol Sci 60:1747–1761
9. Liu LT, Hao YX, Zhang W, Chen J (2018) Free vibration analysis of rotating pretwisted functionally graded sandwich blades. Int J Aerosp Eng:2727452
10. Neves AMA, Ferreira AJM, Carrera E, Cinefra M, Roque CMC, Jorge RMN, Soares CM (2013) Static, free vibration and buckling analysis of isotropic and sandwich functionally graded plates using a quasi-3D higher-order shear deformation theory and a meshless technique. Compos B Eng 44(1):657–674
11. Chen Y, Ye T, Jin G, Li S, Yang C (2021) Vibration analysis of rotating pretwist FG sandwich blades operating in thermal environment. Int J Mech Sci 205:106596
12. Rathore SS, Kar VR, Sanjay (2022) Modal analysis of rotating pre-twisted functionally graded sandwich blade. Mater Today: Proc 56:896–901
13. Rathore SS, Kar VR (2023) Thermoelastic eigenfrequency of pre-twisted FG-sandwich straight/curved blades with rotational effect. Struct Eng Mech 86(4):519–533
14. Reddy JN (2004) Mechanics of laminated composite plates and shells: theory and analysis. CRC Press, Florida
15. Cook RD, Malkus DS, Plesha M (1989) Concepts and applications of finite element analysis. John Wiley Sons, New Jersey
16. Reddy JN, Chin CD (1998) Thermomechanical analysis of functionally graded cylinders and plates. J Therm Stresses 21(6):593–626
17. Lu X, Lin X, Chiumenti M, Cervera M, Hu Y, Ji X, Ma L, Yang H, Huang W (2019) Residual stress and distortion of rectangular and S-shaped Ti-6Al-4V parts by directed energy deposition: modelling and experimental calibration. Addit Manuf 26:166–179

Influence of Ni Interlayer on Mechanical and Microstructural Properties in USW Processed Al–Cu Joints

Tapapriya Mahanta, Soumyajit Das, Bharat Chandra Routara, Abhilas Swain, and Mantra Prasad Satpathy

1 Introduction

In the transportation sector, fossil fuel combustion accounts for approximately twenty four percentage of global carbon dioxide emissions [1]. The majority of this amount is attributed to vehicles powered by internal combustion engines (ICEs) [2]. Due to this alarming condition, carbon emission legislation has been enacted at a global level to prevent the detrimental implications of climate change. In response to these policies, nations have modernized the automobile industry and developed electric vehicles (EVs) in order to reduce carbon footprint [3]. Over the past few years, lithium-ion batteries have become the most widely used power source for a variety of applications from small capacity to large capacity, including electronics devices and EVs [4]. The power of an EV is generated by thousands of interconnected cells, unlike that of a mobile device, such as a cell phone or a laptop, which requires only a few cells. In battery pack manufacturing, the joining of cells and modules is the most critical step that directly impacts the capacity of the battery. The welding joints in EV battery packs usually consist of thin materials that typically range from 0.3 to 1 mm thick, and they are usually welded in the form of laps, fillets, or spots [5]. The electrically conductive metals copper (Cu), aluminum (Al), and nickel (Ni) are commonly employed in automotive lithium-ion batteries as electrodes, current collectors, and bus bars due to their exceptional conductivity and high specific strength. The joining of these dissimilar materials is difficult using traditional fusion welding techniques [6]. Due to their distinct thermo-physical characteristics, such as melting points and thermal conductivities, it is very difficult to establish a complete metallurgical bond between two dissimilar materials [7]. It is also important to note that the presence of intermetallic compounds (IMCs) at the bond interface significantly reduces the

T. Mahanta · S. Das (✉) · B. C. Routara · A. Swain · M. P. Satpathy
School of Mechanical Engineering, KIIT Deemed to be University, Bhubaneswar, Odisha 751024, India
e-mail: soumyajitdasofficial@gmail.com

R. Sahu et al. (eds.), *Advancements in Materials Processing Technology, Volume 1*,
Springer Proceedings in Materials 48, https://doi.org/10.1007/978-981-97-4958-4_28

battery's capacity and structural integrity. The most investigated joining techniques for EV batteries are ultrasonic welding (USW), resistance spot welding (RSW), laser beam welding (LBW), and wire bonding (WB). When compared with other welding techniques, USW has the advantage of requiring less energy and a shorter welding time. In addition, USW is a more appropriate method for joining thin sheets of dissimilar metals with high conductivities.

In the USW method, the joining workpieces are mechanically vibrated at high frequencies to create frictional heat between the surfaces, thereby forming an effective bond. The welding process removes oxides and contamination from the surfaces, resulting in a metallurgical bond that does not require melting, but relies on adhesion and diffusion of the softened metals. In order to achieve high strength joints, it is imperative to control the welding process parameters. Over the past few years, numerous researchers have examined the effects of process parameters on the quality of joints. According to Dhara et al. [8], weld quality can be divided into three categories: under weld, good weld, and over weld. There is an interfacial separation at under weld joints, with no or partial adhesion of the materials. Despite this, it has been observed that over welded joints exhibit circumferential fractures around their joining regions. The highest failure load was observed for good-weld joints with limited interfacial separation and material tearing. In determining the quality of welding, clamping force, welding energy, and oscillation amplitude are of primary importance. It was observed by Zhang et al. [9] that excessive interfacial locking reduced the quality of welding and suppressed the production of heat. With increasing clamping force, there was no monotonic increase in peak tensile shear load. It has been found that welding energy and failure load are related in a parabolic manner as reported by Mohammed et al. [10]. According to Peng et al. [11], Various fracture modes were observed as a result of the difference in welding energy input, which included interfacial failures, button pull-out failures, and transverse through-thickness failures. Liu et al. [12] reported that Al/Cu joints were capable of maintaining their mechanical strength even when the IMC layer's thickness is less than 1 μm. In the case of mixing Al and Cu alloys, a high negative mixing enthalpy allows the formation of Al_2Cu, AlCu, and Al_4Cu_9. In order to eliminate these potential flaws, the USW process involves cold spraying, the surface coating of metal powders, as well as the addition of interlayers between the workpieces. According to Wu et al. [13] the addition of interlayers protects the characteristics of materials by preventing direct contact and mutual diffusion between dissimilar metals. Similarly, an interlayer of Al and Ni foil was studied by Lin et al. [14] for its influence on the weld strength for SS and Ti joint. As a result, the interlayer of Al and Ni had a very positive impact on the formation of a strong bond during the USW.

In this study, Ni interlayers are used to weld 0.5 mm commercial Al sheets and 0.3 mm pure Cu sheets using the USW process. In order to determine the impact of the Ni interlayer on the bond strength, this lap configured dissimilar joints are tested at several weld energy (WE) and different clamping pressure (CP). The bonding mechanism of Al–Ni–Cu joints is analyzed using scanning electron microscopy (SEM) by observing interfacial material mixing, different types of microbonds, and several cross-sectional analyses.

2 Materials and Methods

To conduct this experiment, 0.1 mm Ni foil is applied between 0.5 mm Al sheets and 0.3 mm pure Cu sheets. In Table 1, all materials are listed according to their physical properties. Each specimen is 80 mm long and 20 mm wide and is welded with a 20 mm overlapped position. A lap configuration test specimen is schematically shown in Fig. 1a, and a lateral drive ultrasonic welding system (Telsonic® M4000) runs at a frequency of 20 kHz with a maximum power output of 3 kW. Figure 1b shows the setup for the USW machine. In this experiment, Ni interlayers are used to strengthen Al–Cu joints under different parametric conditions, such as WE and CP. Table 2 provides a summary of the welding parameters employed in this investigation. A Testo 890 thermal image camera measures the temperature at the joint interface during ultrasonic welding. During the tensile shear and T-peel tests, an Instron 3369 machine is used, which is fully computerized and is capable of delivering a load of 50 kN. There is a constant speed of 2 mm/min for the crosshead. Zwick Roell ZHV microhardness testers are used to measure Vickers hardness with a 400 gf indentation load and a dwell time of 15 s. To determine the cross-sectional behavior of Al–Ni–Cu joints, the ZEISS® EVO 18 SEM is used to observe the microstructural profile.

Table 1 Physical properties of the materials

Materials	Ultimate tensile strength (MPa)	Yield strength (MPa)	Thermal conductivity (W/m °C)	Specific heat J/g °C)
AA1060	82.7	28	230	0.900
C11000	250	70	300	0.385
N02200	379	103	60.7	0.453

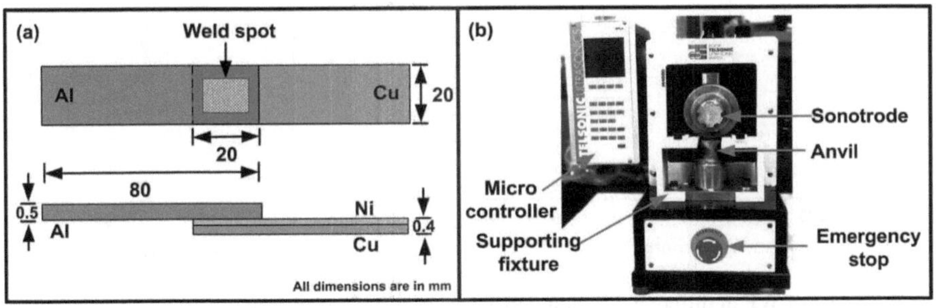

Fig. 1 **a** Schematic illustration of an Al–Cu specimen with a Ni interlayer, and **b** ultrasonic welding machine

Table 2 Weld parameter variations

Factors	Levels				
Clamping pressure (MPa)	0.32	0.34	0.36	...	...
Weld energy (J)	990	1080	1170	1260	1350

3 Results and Discussion

3.1 Tensile Shear and T-peel Strength

The tensile shear and T-peel tests are performed on specimen welds at varying CPs with different WEs. A summary of the results obtained from both tests can be found in Fig. 2a and b. In the graph of the tensile shear load, it is revealed that the maximum load is obtained 1157 N at a CP of 0.34 MPa and a WE of 1170 J. Afterward, the nature of the decrease is observed for further increases in WE. As the WE changed from 1170 to 1350 J, the loads gradually decreased as the Al material exhibited increased softness and flexibility. The layer of Ni in the faying region creates inter-bonding regions instead of brittle IMCs, thereby improving the strength of the weld. Based on the T-peel load test, 288 N of maximum load is achieved when WE is 1170 J and CP is 0.34 MPa. Later stages of the welding process gradually reduced the loads to increase WE, which resulted in a concentration of stress during welding and affected joint strength.

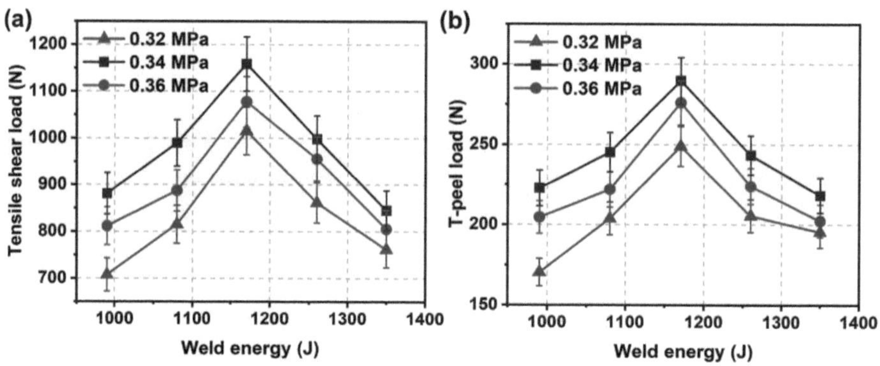

Fig. 2 Effect of WE at different CP on **a** tensile shear and **b** T-peel load

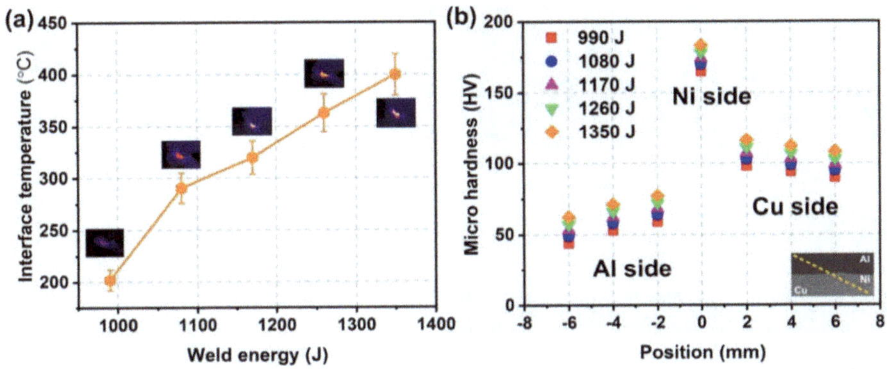

Fig. 3 **a** Temperature profile at 1170 J of WE and 0.34 MPa CP, **b** Vickers microhardness changes

3.2 Interface Temperature Profile and Microhardness Variation

It is essential to note that the temperatures at the interface may have a significant impact on the quality of the weld in a number of ways. There are several possibilities, including plastic deformation, metallurgical reactions, and interatomic diffusion. As the CP increases, the frictional force at the faying surface rises, leading to an upsurge in temperature. A highest temperature of 398.6 °C is recorded at 1350 J of WE. Figure 3a depicts a graph of the weld interface temperature distribution. A severe material deformation resulted in rapid heat dissipation at the interface of the weld joint, causing a gradual increase in temperature at the bonding interface between 1170 and 1350 J. A high WE can also result in excess heat being produced, which can lead to a reduction in the strength of the joint. For a strong bond to be formed between materials, it is necessary to maintain an appropriate weld interface temperature [15].

In order to define the work hardening phenomena, an analysis of microhardness variations in weld cross-sectional areas is conducted. The micro-hardness value is drastically different between the Al and Cu faces. Figure 3b illustrates the distribution of microhardness at the joint contact. The mechanical interlocking between the Al–Ni–Cu joints around the bond interface could be the possible reason for it. Material microhardness profiles are observed to be greater than those of their base metals in all materials. A good reason for this can be found in the severity of the plastic deformation, which results in a finer grain on all levels [16].

3.3 Microstructural Analysis

It is necessary to conduct a metallographic analysis of the weld cross-sectional zone to better understand the deformation inter metallic layer at the joint interface. SEM

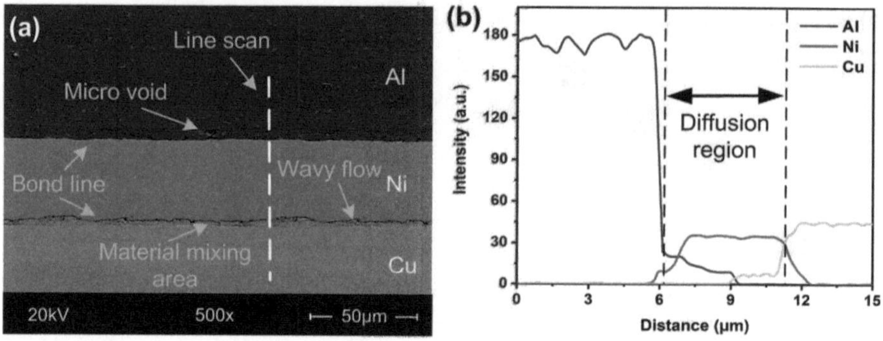

Fig. 4 **a** Cross-sectional image of Al–Ni–Cu joints, and **b** EDS line scan result

images of Al–Ni–Cu welded joints at 1170 J of WE and 0.34 MPa CP are demonstrated in Fig. 4a. Around the weld interface and along the bond line, there are several zones where material mixes and stress concentrations occur. As indicated by wavy flow patterns, the addition of interlayers to Al–Cu joints facilitates metallurgical bonding. A more significant amount of plastic deformation is induced in metals due to the addition of an interlayer. As a result, a strong bond is formed between the sheets of Al–Cu. Figure 4b portrays the EDS line scan results. A sufficient amount of diffusion region is observed around the weld interface in order to form a strong bond. As a result, this factor greatly enhances the strength of the weld [17].

4 Conclusion

This study investigates Al–Cu joints with Ni interlayer under varying CP and WE using USW process. Based on the findings of this study, the following points may be deduced:

I. A greatest tensile shear load of 1157 N is attained at a WE of 1170 J and a CP of 0.34 MPa. Under the aforementioned conditions, the highest load is recorded 288 N on the T-peel tests. However, when the WE is increased to 1170 J, the joint strength decreased significantly.
II. A peak temperature of 398.6 °C is detected at a WE of 1170 J and a CP of 0.34 MPa. In response to a greater energy input, a huge amount of heat is generated at the joint interface. On the Al, Ni, and Cu sides, microhardness values increased after extensive cold work hardening.
III. The cross-sectional analysis indicates no defects at the faying surface. An improvement in weld strength has been observed between Al and Cu sheets after the Ni interlayer was added. A line scan using EDS revealed that the Ni layer has an influence on the mix of materials for bonding.

References

1. IEA (2012) I. CO_2 emissions from fuel combustion highlights. Int Energy Agency 896
2. Hao H, Geng Y, Sarkis J (2016) Carbon footprint of global passenger cars: scenarios through 2050. Energy 101:121–131
3. Eskander SMSU et al (2020) Carbon footprint of global passenger cars: scenarios through 2050. Int Energy Agency 9:120307
4. Wimbadi RW, Djalante R (2020) From decarbonization to low carbon development and transition: a systematic literature review of the conceptualization of moving toward net-zero carbon dioxide emission (1995–2019). J Clean Prod 256:120307
5. Das A, Li D, Williams D, Greenwood D (2018) Joining technologies for automotive battery systems manufacturing. World Electr Veh J 9:22
6. Ni ZL et al (2021) Characteristics of ultrasonic spot welding of Cu/Ni dissimilar metals. Sci Technol Weld Join 26:529–539
7. Feng W et al (2023) Microstructure and texture evolution of aluminum and titanium ultrasonic welded joints. Mater Charact 195:1–7
8. Dhara S, Das A (2020) Impact of ultrasonic welding on multi-layered Al–Cu joint for electric vehicle battery applications: a layer-wise microstructural analysis. Mater Sci Eng A 791:139795
9. Zhang W et al (2018) Corrigendum: effect of ultrasonic spot welding on the mechanical behaviour of NiTi shape memory alloys. Smart Mater Struct 27:85020
10. Mohammed S, Dash SS, Jiang XQ, Li DY, Chen DL (2019) Ultrasonic spot welding of 5182 aluminum alloy: evolution of microstructure and mechanical properties. Mater Sci Eng A 756:417–429
11. Peng H, Chen D, Jiang X (2017) Microstructure and mechanical properties of an ultrasonic spot welded aluminum alloy: the effect of welding energy. Materials (Basel) 10:449
12. Liu J, Cao B (2021) Microstructure characteristics and mechanical properties of the Cu/Al dissimilar joints by electric current assisted ultrasonic welding. J Mater Process Technol 297:117239
13. Wu X, Liu T, Cai W (2015) Microstructure, welding mechanism, and failure of Al/Cu ultrasonic welds. J Manuf Process 20:515–524
14. Lin JY, Nambu S, Koseki T (2020) Evolution of bonding interface during ultrasonic welding between Ni and steels with various microstructure. ISIJ Int 60:330–336
15. Peng H et al (2020) Microstructure and mechanical properties of Mg-to-Al dissimilar welded joints with an Ag interlayer using ultrasonic spot welding. J Magnes Alloy 8:552–563
16. Das S, Satpathy MP, Pattanaik A, Routara BC (2019) Experimental investigation on ultrasonic spot welding of aluminum-cupronickel sheets under different parametric conditions. Mater Manuf Process 34:1689–1700
17. Ni Z-L, Ye F-X (2019) Microstructure and mechanical properties of an ultrasonic spot-welded aluminum-to-aluminum joint: response to interlayer thickness. Materials (Basel) 12:369

Design and Analysis of Cylinder Head Gasket Under Engine Cold Assembly and Cold Start Condition

Viddya S. Patil and Adik M. Takale

1 Introduction

The rubber on the piston head is the most significant passive seal component in a fuel-burning engine. It is an essential component of an engine and has to perform many functions at the same time while the engine is operating. The head sealing needs to preserve a vacuum surrounding the combustion region even at its greatest pressure and temperature. The gasket must be able to seal versus air, cooling liquids, combustion, and engine oil when everything is at its greatest temperature and pressure. The function of this device is to seal off cylinder, fresh water jackets, oil channels, and all external air, liquids, and gases. It is located near the engine's head and casing. Because an engine gasket failure typically causes the entire engine to fail, the area of the sealing that wraps around the cylinder must be sufficiently durable to withstand identical pressures placed on the pistons while also ensuring no leakage of water or combustion gases between the various volumes [1].

The materials utilized or the engine structure, which must be thermally and chemically resistant, cannot harm the engines coolants, oils, or combustion byproducts [2]. The best option for internal combustion engines right now is multi-layer steel head gaskets to be which is developing diesel technology in powerful applications sector. A few factors that have contributed to the seal's popularity include higher dependability, stronger chemical and thermal resistance, increased clamped gasket control, lower bolt force variations, increased degrees of flexibility response, and improved emission control.

The engine gasket's head protects the piston head and all of its operating components. It must be able to withstand the shocks and heat strains imparted by the motor's head and casing. The design of the cylinder head gasket will be determined by the

V. S. Patil (✉) · A. M. Takale
Department of Mechanical Engineering, Sharad Institute of Technology College of Engineering, Yadrav, Kolhapur, Maharashtra 416118, India
e-mail: vidyapatil@sitcoe.org.in

R. Sahu et al. (eds.), *Advancements in Materials Processing Technology, Volume 1*, Springer Proceedings in Materials 48, https://doi.org/10.1007/978-981-97-4958-4_29

type of engine used. According to Srikanth et al. [3] cylinder head may encounter significant stress if the engine exceeds the temperature of and operates outside of its typical operating range, which could lead to head gasket brake-down. Karthikeyan and Sakthivel [4] have done the structural analysis of aluminum cylinder head gasket and results reveal that, when heated to aluminum expands by a factor of two to three comparisons to cast iron. The head may warp due to the difference in rates of expansion during heating that exists intermediate in aluminum head and a C.I. block, as well as additional tension brought on by overheating. In turn, this can result in a loss of restraining power at crucial places and let the head gasket leak.

Contact behavior develops as a result of the engagement between cylinder head and gasket plus interface among cylinder head and the bolt [5]. The FEM is more efficient than the discrete element technique in solving contact problem of surfaces with complex contact surface geometry. The technique was created by Lee et al. [6] to address various contact issues. However, a variety of operating procedures and loading circumstances contribute to warmth being generated inside the engine and its construction. 2020 saw the completion of an investigation by Abik et al. [7] which looked at the flaws of an automotive gasket that had been connected to an engine temperature. To ascertain if the failure will be caused by or have an impact on the combustion temperature, they conducted an investigation. To determine the reason of the gasket's fracture, a finite element method (FEM) must be utilized to evaluate the stress situation under (steady state) temperatures effect overloading. Malik et al. [8] undertook an effort to explore the finite element techniques that analysts are now using and whether their findings agree with those of physical tests. It came to light through comparisons using ABAQUS software and Panasonic put pressure on film that each element type has advantages and disadvantages in terms of real-world correlation, computational cost, and procedural simplicity. Sangle et al. [9] investigated the fretting mechanism depending on fundamental fretting mechanism to analyze the gasket failure through Ansys.

Vijayabaskar et al. [10] performed a thermal monitoring test on a four-stroke engine's cylinder head gasket. This investigation can be used to compare the two gasket components to identify where problems in cylinder head gasket occur. Distortion of the gasket materials is mostly brought on by temperature variations. The numerical simulation of the thermal analysis will be discussed in this paper using the popular commercial FEM program ANSYS.

According to Satish Anande et al. [11], the head of the engine seal is one of the most important components in forming a good seal. That is critical that you choose the proper material for the head of the cylinder seal so that it does not interfere with the assembly of the other engine components. The most often utilized material for cylinder head gaskets is steel for construction. Fukase et al. [12] suggested nonlinear analysis to create conclusions that are more precise. Since we need to provide enough bolt strain for effective sealing, we frequently come across the issue of gasket breakage while tightening the bolts. For the investigation in this work, the multiple-layer cylinder head gasket of a single-cylinder diesel engine was taken into consideration. To minimize bore distortion and obtain the ideal pressure to contact on the combustion chamber gasket, a nonlinear study of the cylinder head gasket is carried out. Modeling was completed using CRE-O 2.0, and ANSYS 15 program was utilized for analysis. As a result, while sealing gaskets, the mechanical issues caused by temperature contact must be taken into account. Zinzuvadia et al. [13] also carried out the FEM analysis of multi-layer cylinder head gasket to find out normal and shear stresses generated in gasket. Prior study on the reactions of heat contact has been conducted by a vast number of researchers. In addition to the element of contact issues, FEM carefully analyzes the behaviors of heat propagation throughout the internal structure of the engine's combustion chamber and the other component. Without accounting for any thermal tensions brought on by water dispersion in the head of the engine, the pre-stressing pressure for the hold-down bolts determines the efficacy of the head seal encapsulation. As a consequence, Lee et al. [6] established a realistic way to enhance the assembly force without exceeding the strength of the material of each component in the engine construction in order to improve the gasket's sealing competence.

This study uses computer modeling for structural analysis with the commercial FEM tool ANSYS. This contact theory-based study investigates the performance of gasket seals as well as the stress/strain response of the piston head seal between assembly loaded circumstances. Study is also conducted on estimating typical gasket pressure, pressures of contact in gaskets for motions between the combustion chamber and gasket, gasket and block, and additionally within gasket layers, for the purpose to get the greatest engine efficiency.

2 Fundamental Theory

2.1 *Contact Theory*

In the mechanical contact finite elements simulation, penalty mechanisms like the Lagrange multiplier approach and the reinforced Lagrangian method are frequently utilized. The ill-conditioning in the penalty mechanisms, however, gets worse as the severity of the penalties rises. The equation system produced by the Lagrange multipliers a position introduces additional unknowns and is not always a positive-definite equation system. Penalty technique and Lagrange multiplier approach are employed in the enhanced Lagrangian method, which gains the benefits of both. Augmented Lagrangian method on contact point region has the following variational weakly form as in Eq. (1)

$$\delta\pi = \int \left(\tau y_N + \frac{\alpha}{2}\gamma^2 N\right)\mathrm{d}r, \tag{1}$$

where y_N is rate at which two touching bodies penetrate one another, is a penalty, and is the Lagrange multiplier. Equation (1) could be modified to take the version of being comparably strong through performing the calculation as

$$M_a + f^{\text{int}} - f^{\text{ext}} + G^T r + P_c d = 0, \tag{2}$$

$$G_v \leq 0, \tag{3}$$

where v denotes the velocity fields of the two bodies, stands for the Lagrangian multiplier, $G^T r$ for the contact force, P_c for overall contact stiffness, and $P_c d$ for contact force (penalty force). Internal, exterior, and inertial forces have been designated by the letters f^{int}, f^{ext}, and M_a, respectively. The contact finite element calculation is controlled by Eq. (2). Inequality restrictions are used in Eq. (3) to describe the touch border between two touch bodies. The key advantage of the modified Lagrange technique for the interaction problem is the fact it makes the penalty methodology more durable and stable while also being a simple method that does not require the development of additional discrete system equations. In finite element analysis, the augment Lagrangian approach is utilized to precisely explain what happens between the combustion head and gasket over many engine operating conditions.

3 Analysis Methods and Finite Element Models

3.1 Models Using Finite Elements

The analysis of the cylinder head with reference to structure is developed in this study using a four-stroke diesel motor with 12 cylinders. Cylinder head gasket assembly model was developed by using Pro/E software at Sharad Institute of Technology College of Engineering Yadrav as shown in Figs. 1, 2, and 3 to evaluate the effectiveness of gasket sealing. Because doing so reduces the number of constraints taken into account during the analysis process and saves on the amount of infinite element counts, only one cylinder's head is taken into consideration in this research for the purposes of witnessing the distribution of normal pressure across the gasket. Material characteristics of each element provided by Greaves Cotton Ltd., Pune used in analysis of structural for this study are listed in Table 1. Cast iron gives the fundamental substance for cylinder head, block, and liner. Carbon steel as (SCM 435) used to make the pre-stressed bolts. Table 2 indicates the details of pressure and deflection values for the MLS gasket body employed in structural analysis (Greaves Cotton Ltd., Pune), the internal portion of the gasket are continuously in touch with a number of parts that possess dissimilar material characteristics.

The following premises are taken into account for structural analysis:

1. Because the 12 cylinder heads are structurally symmetric, just one cylinder head is considered from the complete model to speed up computation and establish a distinct border. This is one of many presumptions made when simulating the manufacture of cylinder heads.

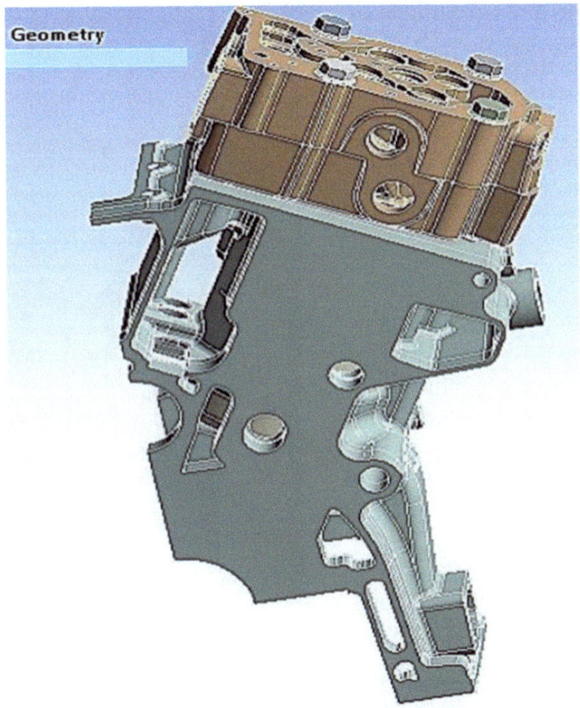

Fig. 1 Assembly of cylinder head, gasket, and block developed in SITCOE, Yadrav

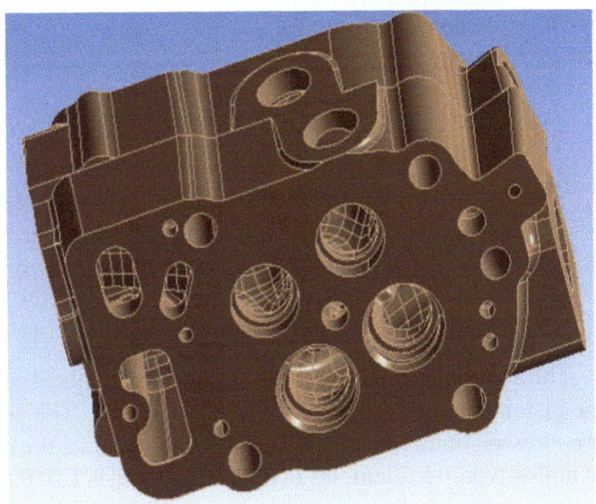

Fig. 2 Cylinder head developed in SITCOE, Yadrav

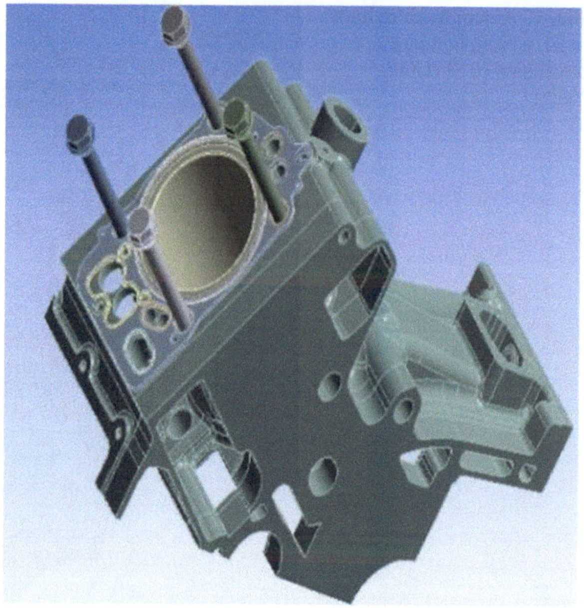

Fig. 3 Assembly of block, liner, gasket, and bolts developed in SITCOE, Yadrav

Table 1 Components list with material attributes for structural analysis provided by Greaves Cotton Ltd., Pune

Component	Material	Young's modulus	Poisson's ratio
Liner	C.I	118	0.29
Bolts	Carbon steel	205	0.29
Head	C.I	118	0.26
Block	C.I	137	0.26

2. Pre-stressing of a bolt necessitates the construction of a partial block for the purpose to evaluate the efficacy of gasket sealing.
3. Complete dimension of the gasket is 2.8 mm, while the inner ring at the center of the combustion area, the general bead, and the half bead in the bolt hole are all taken into account.

The area-to-area touch pair was created by replicating the point of contact between the piston head, gasket, blocks, liner, and studs. The assembled engine fasteners are pre-stressed in direction of given force to simulate the tightening level. Total number of nodes as well as elements in the FEM model is 1,010,514 and 699,901, accordingly. The mesh model of complete structure of cylinder head gasket assembly and cylinder head gasket is developed in ANSYS 16.2 software as shown in Fig. 4.

Table 2 Pressure and deflection values for the MLS gasket body provided by Greaves Cotton Ltd., Pune

Primary bead (Wire ring)		Secondary bead (Full bead)		Half bead		Rubber bead		Other gasket body	
Deflection	Pressure	Deflection	Pressure	Deflection	Pressure	De flection	Pressure	Deflection	Pressure
0	0	0	0	0	0	0	0	0	0
0.025	30	0.003448	0.014125	0.000384	25	0.5	9.107143	0.000384	25
0.05	65	0.100281	0.1507812	0.000768	50	0.7	12.75	0.000768	50
0.075	110	0.200292	0.2034479	0.00134	87.5	1	18.21429	0.00134	87.5
0.1	165.8	0.257948	0.3467917	0.00192	125	1.2	1.2	0.00192	125
0.125	220	0.301313	0.5054375	34.95296	89.59742				

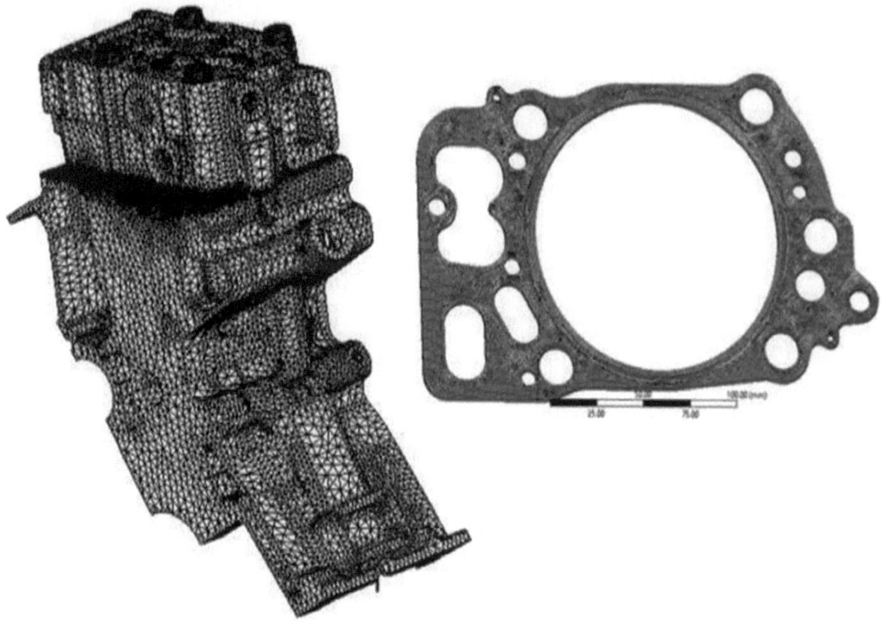

Fig. 4 Mesh model of cylinder head, gasket assembly (left) and only gasket (right) developed in ANSYS 16.2 software

3.2 *Procedure for Structural Analysis*

According to the given loadings, the method of study can be separated into three separate stress steps to demonstrate numerous engine running operations.

1. Assembly and loading
2. Thermal loading
3. Pressure of gas.

From there, only assembly loading and gas pressure have been taken as factors for this work and output of these load step form structural analysis's final results [14].

1. Assembly loading

Pre-stressing pressure of bolt especially is applied to combustion chamber near its strongest point and is crucial in avoiding gas escape from the engine's internal components. In this situation, a determined pre-stressing pressure of 70 KN is given to each bolt to prevent inadequate gasket sealing. To prevent movable body motion, the nodes' removals at the barrier's bottom have been fixed and there is no friction on either side of the block.

2. Gas pressure

The combustion chamber's surface is exposed to gas pressure that results from firing, and the amount of gas pressure varies depending on the cycle's duration. 200 bar pressure of air is added to the load conditions, for numerical simulation of the chilly starting situation for the steady-state analysis.

4 Findings and Conclusions

The location of the normal pressure as exerts itself on the gasket has been used in this study to examine the performance of the gasket sealing. The statistical results for the engine working under different conditions are carefully analyzed by using ANSYS 16.2 software.

4.1 At Cold Assembly Conditions

A head gasket that has been installed between cylinder head and engine block is gently compressed upon tightening the head bolts, enabling the gasket's soft facing material to conform to the slight irregularities on the head and block deck surfaces. This enables the gasket to become a cold seal, minimizing the leakage of coolant until it started. The gasket sealing of the engine is believed to be completely free of any additional external loading when it is in a static state. Thus, in the previous example the main source of loading is pre-stressing of the bolt. Additionally, varied types of an engine design and stroke quantity determine the amount that bolt is pre-stressed. The parametric technique for pre-stressing bolts is used as an end product. The findings unambiguously show that the cable ring location is the location where the gasket experiences its maximum contact pressure and minimum normal pressure. The results of analysis (ANSYS 16.2 software) in Fig. 5 show the conclusions of total gasket deformations in assembly conditions and Fig. 6 provides normal gasket pressure on the connection. It should be emphasized that the degree to which gaskets seal depends mostly on the amount of preconditioning force applied during cold assembly. While Fig. 7 shows the findings of a parametric study of the preconditioning force of bolts. In this case, the findings show that the low force of contact on the seal exceeds safety factor of design demand as well as occurs at the same position. It should be noted that as contact pressure reduces from 158.47 to 123.41 and preconditioning force of the bolt lowers from 70,000 to 40,000 N, the gasket sealing performance relies on how much pre-stressing force the bolts undergo throughout cold assembly settings.

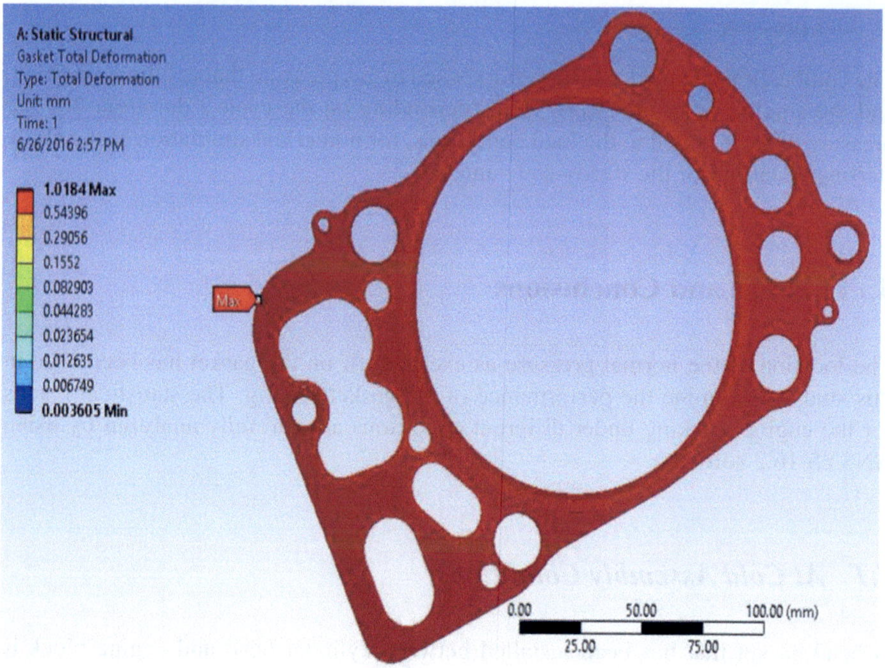

Fig. 5 Gasket total deformations at assembly condition (ANSYS 16.2 software)

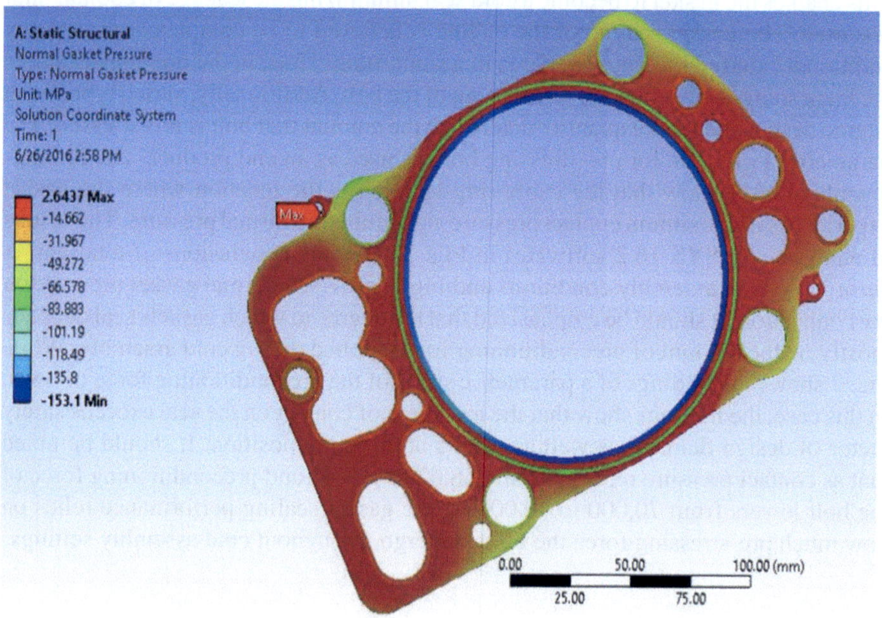

Fig. 6 Normal gasket pressures at assembly condition force intensities (ANSYS 16.2 software)

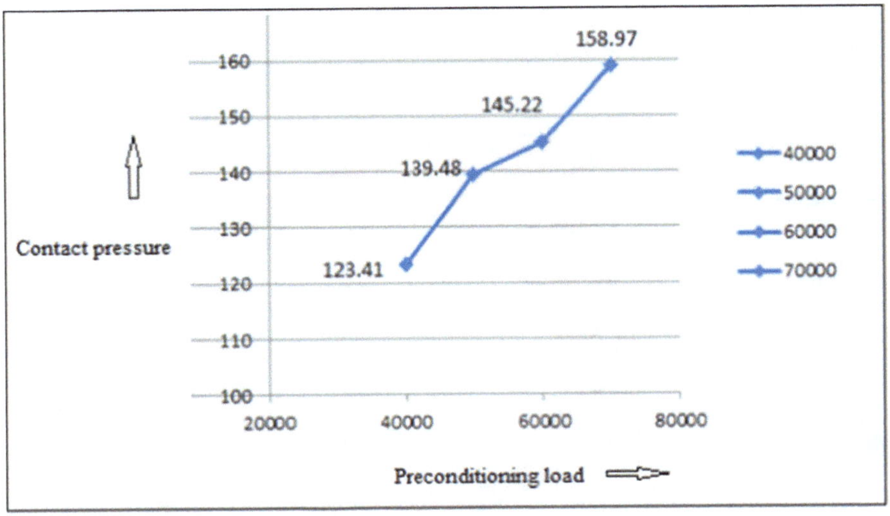

Fig. 7 Gasket's weakest contact pressure at various preconditioning (ANSYS 16.2 software)

4.2 At Cold Start Condition

The loading requires both the assembled pressure and the gas pressure to be set, emulating operational stage a cold start. Similar to cold assembling, the position of the least effective gasket sealing and the dispersion of ordinary pressure on the seal's projecting surface are also present. When the active vector of the gas flow resists the pre-stressing power of the nuts and bolts, as contrasted to when the assembly is cold, the gasket's sealing ability is greatly diminished. As seen in Fig. 6, the inner ring is evident where the lowest normal pressure on the seal occurs, and its value remains greater than the assembly condition.

Figure 9 shows the findings from the investigation of normal gasket pressures under initial cold circumstances, whereas Fig. 8 shows the results of total gasket deformations under cold beginning circumstances. Figure 10 depicts the results of parametric examination for fastener pre-stressing force. The current data suggests that the least pressure applied on the gasket occurs at the same spot and meets the protective value of the design criterion. It should be mentioned that the efficiency of gasket encapsulation is dependent on the degree of bolt preconditioning force during cold construction instances, as the contact pressure reduces from 158.47 to 123.41 when the bolt pre-stressing power decreases from 70,000 to 40,000 N.

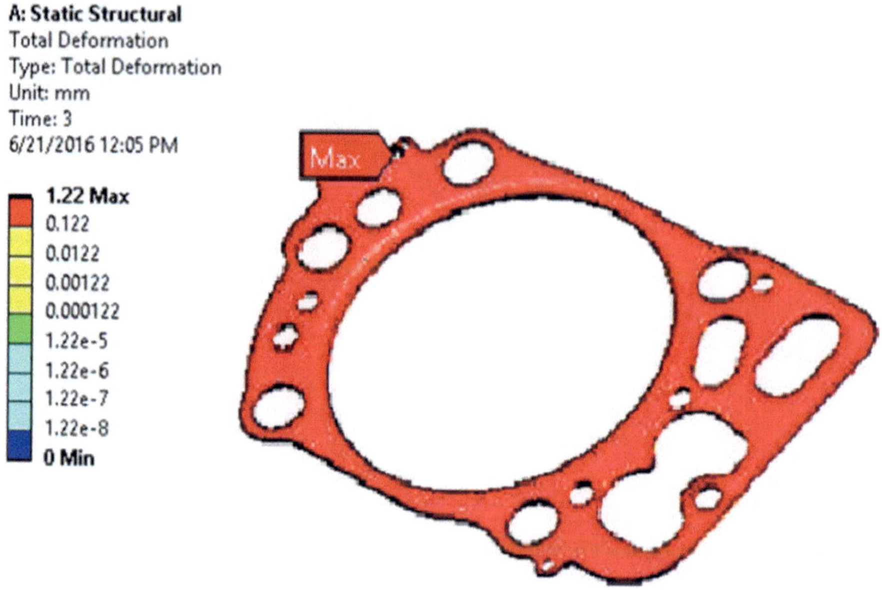

Fig. 8 Total gasket deformations at cold start condition (ANSYS 16.2 software)

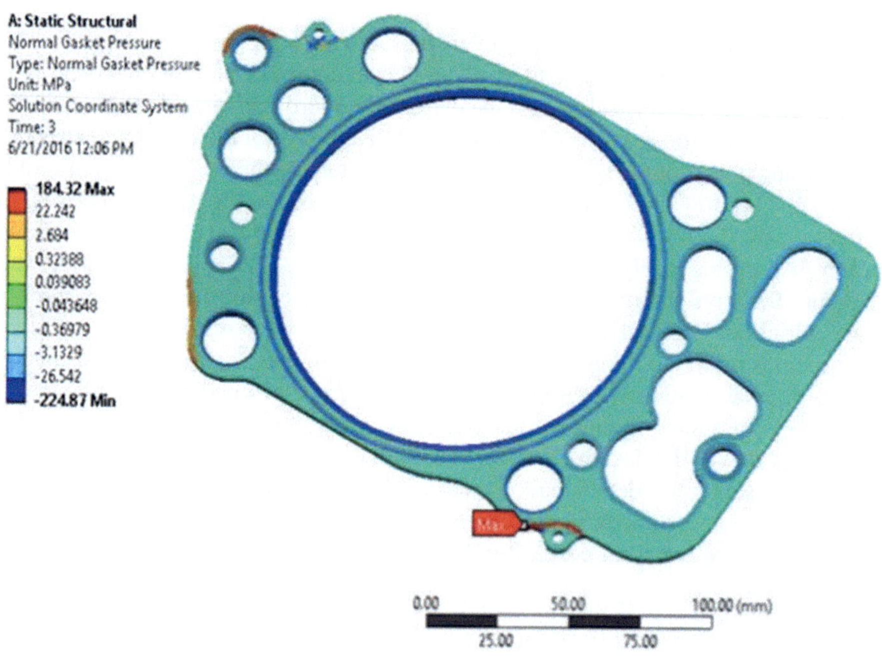

Fig. 9 Normal gasket pressures at cold start condition (ANSYS 16.2 software)

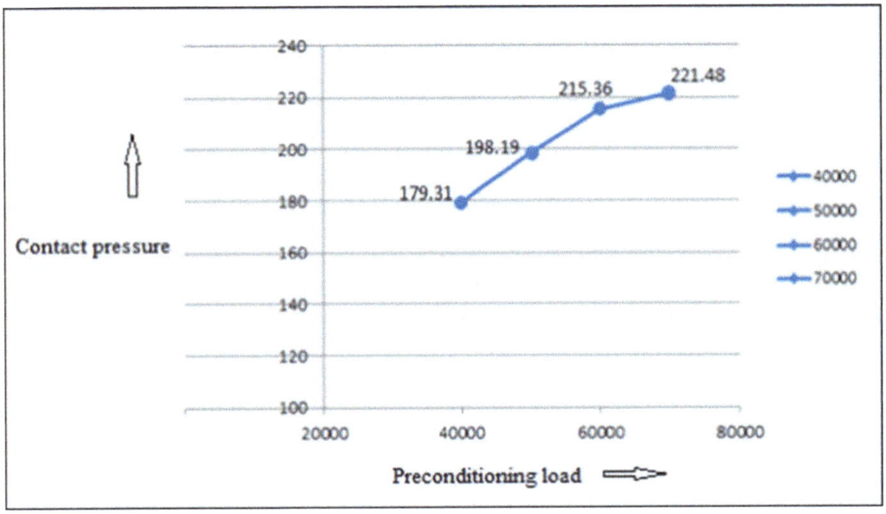

Fig. 10 Smallest contact pressure on seal at different values of the preconditioning force (ANSYS 16.2 software)

5 Conclusion

With application on numerical simulation and finite element analysis, in this study, a structural examination of a cylinder head gasket will be undertaken under assembly and initial cold conditions. At the end, the ability of the gasket sealing is the main area that is affected by the preconditioning force of the fasteners, which generates the external loading. The part sizes of gasket and deck faces have an impact on the resolution of the findings; decreasing the dimensions of elements boosts resolution at the expense of computational cost. As a result, an effective technique for improving the gasket's sealing capacity was presented, and the position of the lowest pressure in contact that had been determined on it was discovered. In the early stages of design, the construction of the gasket in the area with the poorest sealing can be concurrently enhanced.

References

1. Popielas F, Chen C, Obermaier S (2000) CAE approach for multi-layer-steel cylinder head gasket. SAE technical paper series
2. Raub JH (1992) Structural analysis of diesel engine cylinder head gasket joints. SAE Paper 921725
3. Srikanth M, Balakrishnan BM (2015) Cylinder head gasket analysis to improve its thermal characteristics using advanced fem tool. Int J Mach Constr Eng 2(1)
4. Karthikeyan S, Sakthivel P (2015) Design and structural analysis of aluminum cylinder head gasket under engine operation. Int J Eng Dev Res 3(2)

5. Jorwekar PP, Birari YV, Nadgouda MM (2006) Cylinder head gasket contact pressure simulation for a hermetic compressor. In: International compressor engineering conference at Purdue, C071, pp 1–5
6. Lee C-C, Chiang K-N, Chen W-K, Chen R-S (2005) Design and analysis of gasket sealing of cylinder head under engine operation conditions. Finite Element Anal Des 41(11):1160–1174
7. Abik R, Anto Jeberson A, Ashif Ibrahim K, Balamurugan, Prabaharan R (2020) Selection and optimization of gasket materials. Int J Inno Res Sci Eng Technol 9(2)
8. Malik S, Rajesh K, Vijayaganapathy D Dr (2018) Head gasket finite element model correlation. Int J Pure Appl Math 120(6):3965–3974
9. Sangle D, Ozarde A, McNay G (2019) Cylinder head gasket fretting simulation for high horse power engine. SAE Int J Adv Curr Pract Mobility 1(1):133–136
10. Vijayabaskar L (2016) Thermal analysis of cylinder head gasket using Ansys. Imp J Interdiscip Res 2(12)
11. Anande S, Chaudhari A, Rodge MK Dr (2016) Modification in design of cylinder head gasket for reduction in bore distortion and to achieve optimum contact pressure. Int J Mech Eng Technol (IJMET) 7(5)
12. Fukase S, Takahashi T, Kimura J (1999) Experimental study of static and dynamic behavior of the cylinder head gasket in a turbocharged diesel engine with intercooler. SAE technical paper series
13. Zinzuvadia BK, Kanani JB (2015) Design and analysis cylinder head gasket of oil engine. Int J Adv Eng Res Dev 2(5)
14. Vinod Kumar D, Udaya Kiran C Dr, Vijaya Kumar Y (2013) Design and analysis of gasket sealing of cylinder head under engine operation conditions. Int J Inno Eng Technol 2(3)

Grain Refinement by Ultrasonic Treatment (UT) in Al–Al_3Zr Composite

Vishesh Agrawal, Manoj Khobragade, Ashish Thakur, and Y. Y. Mahajan

1 Introduction

The desire to reduce weight and a goal of high efficiency and performance in the material are the driving forces behind the usage of metal matrix composites in the automotive, agricultural, aerospace, and mining sectors. Aluminum metal matrix composites (AMMCs) with high specific stiffness and high strength may be utilized in long-term applications where weight reduction is a key factor, such as robots, high-speed machinery, rotating shafts, automobile engine, brake components and many more applications [1]. A kind of composite materials known as in-situ composites is one in which the reinforcement is produced by reaction within the matrix during manufacturing. Researchers frequently use the in-situ method of composite synthesis because it has a number of advantages over traditional stir casting, including fine particle size, a clean interface, good wettability of the reinforcement with the matrix, and uniform distribution of reinforcement in comparison with other processes. In addition, in-situ processing through casting route is a cost-effective approach for mass production when compared with other techniques like powder metallurgy and spray forming [2].

According to research on the microstructure of as-cast Al–Mg–Cu composites reinforced by in-situ nano-sized Al_2O_3 dispersoids, the size, continuity, and sphericity of intermetallic particles (Al_2CuMg) formed during cooling and solidification of the composite as well as the size and sphericity of α-Al dendrites were all significantly influenced by the ultrasonic treatment of the melt prior to casting [3]. Under ultrasonication, a higher solute content seems to promote grain refining efficiency. Enhanced heterogeneous nucleation under cavitation appears to be the source of grain refinement during ultrasonication [4].

V. Agrawal (✉) · M. Khobragade · A. Thakur · Y. Y. Mahajan
Visvesvaraya National Institute of Technology, Nagpur, Maharashtra 440010, India
e-mail: visheshvnit@gmail.com

R. Sahu et al. (eds.), *Advancements in Materials Processing Technology, Volume 1*,
Springer Proceedings in Materials 48, https://doi.org/10.1007/978-981-97-4958-4_30

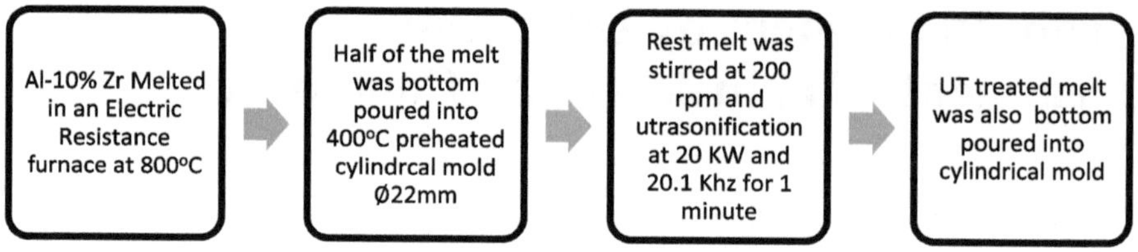

Fig. 1 Schematic of casting

For Al–Al_3Zr systems, several investigations have been conducted [5–7]. One such investigation is the use of powder metallurgy to create an Al_3Zr-reinforced Al matrix composite via in-situ synthesis in vacuum. Because of the Al_3Zr particles dispersion within the matrix of aluminum, the strength of composite is increased [8]. However, a thorough investigation of how UT affects the Al–Al_3Zr system microstructural characteristics and the microstructure's refinement is still ongoing [9, 10]. The current work examines the impact of UT on the evolution of the Al–Al_3Zr system microstructure in terms of grain size.

2 Experimental Procedure

The goal is to cast Al–Al_3Zr composite with UT and NUT, to compare their microstructural properties. An Al-10% Zr composite was melted in an electric resistance furnace at 800 °C. With the help of stirrer made of zircon-coated SS, mixing was done at the rate of 200 rpm. The density of Al_3Zr particle is higher than the density of aluminum. So to avoid the settling of Al_3Zr particle in the bottom during casting of Al–Al_3Zr composite, stirrer is used [11]. For casting of the melt, cylindrical mold of 22 mm diameter and 170 mm height was used. This mold was preheated at 400 °C before pouring molten metal. The cylindrical mold was preheated to avoid the thermal shock [12]. Half of the melt was bottom poured into the mold. The remaining melt undergoes for the ultrasonic treatment via Ti6Al4V sonotrode for the time duration of 1 min. Further, this UT-treated melt was casted in the same dimensional mold.

The schematic of experimentation is shown in the Fig. 1.

3 Results

3.1 Microstructural Analysis

For metallographic observation, a section of the sample 10 mm × 10 mm × 10 mm was cut and the surface was prepared by metallographic practice. The samples were etched using Keller's Reagent to reveal its microstructure. The prepared surface

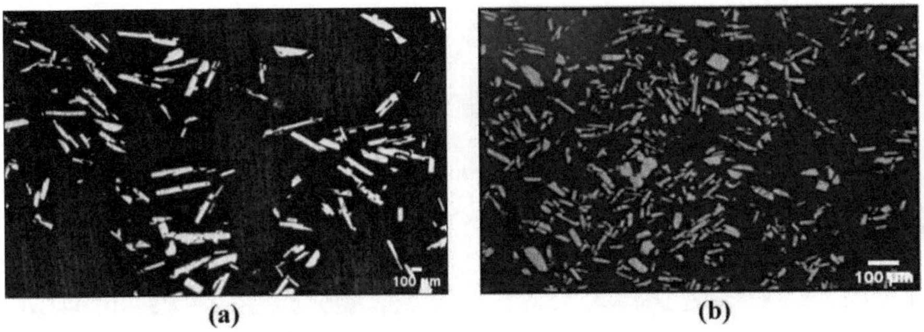

Fig. 2 Optical microstructure of **a** NUT composite, **b** UT composite

was observed in a Bright field optical microscope (Make-Zeiss). The observed microstructure is shown in Fig. 2.

4 Discussion and Analysis

The following section discuses about the results obtained from experiments and their related mechanisms. In optical microstructure, Fig. 2b, more second phase particles of rectangular shape were observed in UT-treated Al–Al_3Zr composite then compared to the NUT Al–Al_3Zr composite in Fig. 2a. Also, more particle rupture was observed in UT when compared with NUT composite.

4.1 Particle Size Analysis

The software Image J was used to calculate the particle size distribution and area fraction using thresholding method. Figure 3 shows the processed image of microstructure.

For NUT composite Fig. 3a, the average particle size was 355.07 µm and for UT-treated Al–Al_3Zr composite Fig. 3b, the particle size was 292.24 µm. So, in UT-treated Al–Al_3Zr composite, we observed that particle size is decreased. Particle size is decreased by 17.7%, as shown in Fig. 4.

$$\text{Particle Size Reduction} = \frac{355.07 - 292.24}{355.07} * 100 = 17.7\%.$$

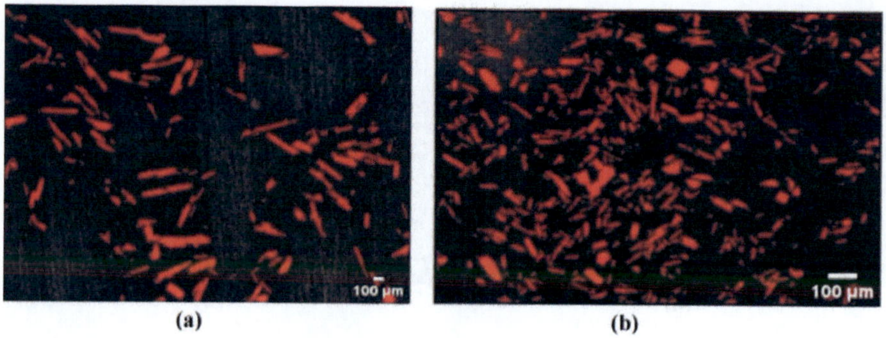

Fig. 3 Thresholded image of **a** NUT composite and **b** UT composite

Fig. 4 Graph of mean particle size of UT and NUT composite (obtained by image analyzing software)

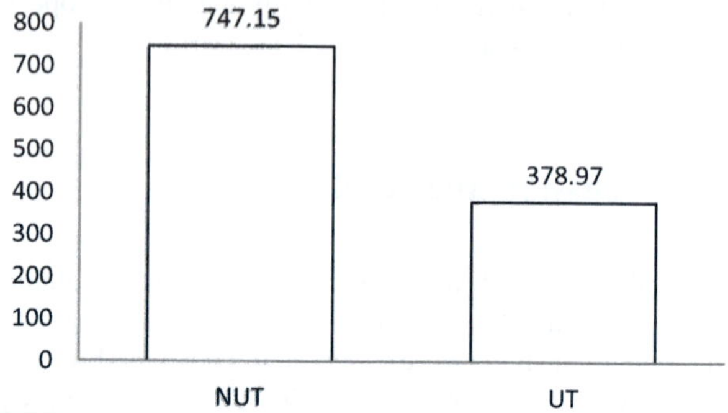

4.2 Particle Breaking Mechanism

This reduction in particle size can be attributed to the ultrasonification applied during casting. Ultrasonic treatment relies on acoustic cavitation (the formation and implosion of cavitation bubbles) in the melt caused by high-frequency (ultrasonic) vibrations with amplitudes greater than a particular threshold (Blake threshold).

Acoustic streaming and cavitation are produced by an ultrasonic radiator (sonotrode). Acoustic streaming refers to the flow of a sound wave through a melt, as well as the stirring and mixing of the melt. Cavitation is the formation, expansion, and implosion of a bubble caused by the pressure of sound waves in the melt [13].

After an acoustic cavitation bubble forms, it pulsates and eventually collapses (Fig. 4c), resulting in impact pressures that can locally exceed 400 MPa for a short period of time [14]. This pressure results in the fragmentation of Al_3Zr particles in the melt. The schematic of cavitation is shown in Fig. 5.

Moreover, Al_3Zr particles acts as a good grain refiner [15]. The fragmented Al_3Zr particles represent high-potency nucleants, they may act as nucleation points for Al grains. So with UT, nucleating sites increase which is a kinetic aspect. Thus, more nucleation sites are available for the aluminum matrix.

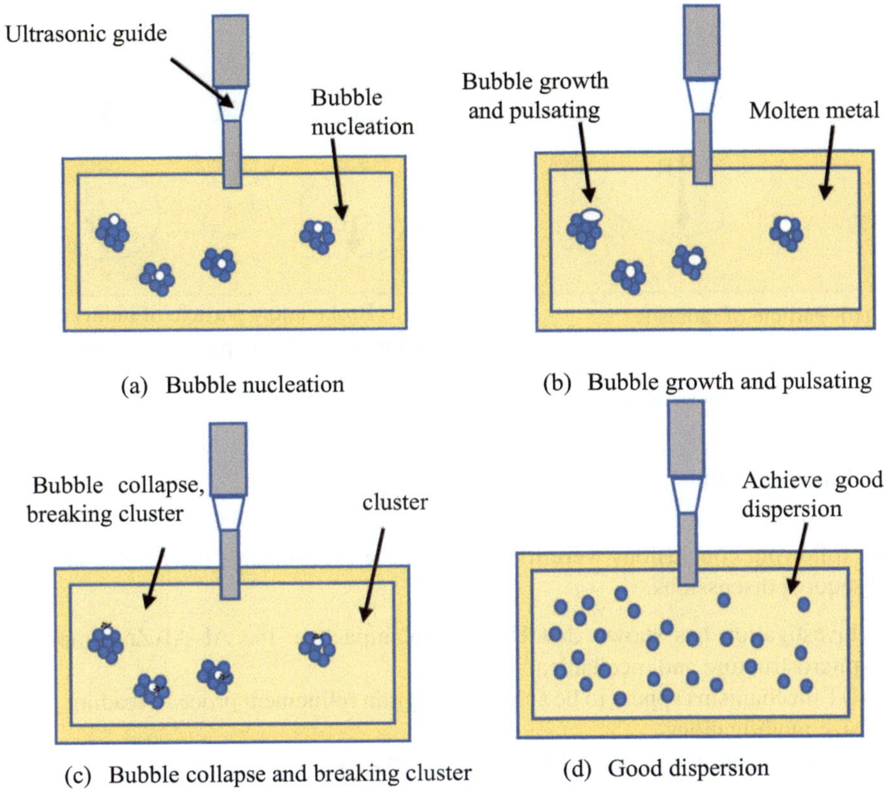

Fig. 5 Schematic of ultrasonic treatment

4.3 *Volume Fraction Analysis*

For NUT composite Fig. 3a, the mean volume fraction of Al_3Zr particles was 4.61% and for UT-treated Al–Al_3Zr composite Fig. 3b, the volume fraction was calculated to be 10.42% by averaging area fraction of multiple images.

ΔG_r (critical excess free energy) is related to the radius of the nucleating sites. After UT, ΔG_r reduces, which gives us idea about the reduction of critical radius of particle and increase in nucleating sites [16]. It may also be said that formation of compounds reaction becomes more favorable after ultrasonic treatment UT, due to which more Al_3Zr particles were formed. Due to these reasons, an increase in area fraction of Al_3Zr after UT is observed. This is shown in Fig. 6. The refined and multiplied primary Al_3Zr particles are the primary source of grain refinement during the UT process [17].

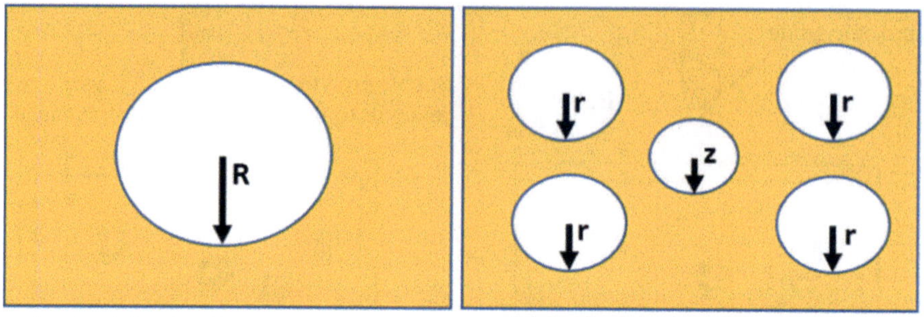

(b) Particle of radius R

(b) Broke into 4 particle of radius r and formation of extra particle of radius z.

Fig. 6 Schematic showing increase in area fraction

5 Conclusion

The following conclusions were formed from the experiments, the results, and the subsequent discussions.

1. Investigation has shown that UT has an impact on the Al–Al_3Zr composite microstructure and mechanical properties.
2. UT mechanisms appear to be active in the grain refinement process, leading grain size modifications.
3. Area fraction was increased from 4.61 to 10.42%.
4. Grain refinement was accomplished by reducing particle size from 355.07 to 292.24 μm.

References

1. Mavhungu ST, Akinlabi ET, Onitiri MA, Varachia FM (2017) Aluminum matrix composites for industrial use: advances and trends. Procedia Manuf 7:178–182. https://doi.org/10.1016/j.promfg.2016.12.045
2. Pramod SL, Bakshi SR, Murty BS (2015) Aluminum-based cast in situ composites: a review. J Mater Eng Perform 24(6):2185–2207. https://doi.org/10.1007/s11665-015-1424-2
3. Chen C, Guo Z, Luo J, Wang W, Hao J (2016) Effects of ultrasonic treatment on microstructure and properties of Al-based composites reinforced by in situ Al_2O_3 nanoparticles. High Temp Mater Processes (London) 35(2):169–175. https://doi.org/10.1515/htmp-2014-0150
4. Kotadia HR, Qian M, Das A (2018) Solidification of aluminium alloys under ultrasonication: an overview. Trans Indian Inst Met 71(11):2681–2686. https://doi.org/10.1007/s12666-018-1446-1
5. Zhang W, Xu K, Long W, Zhou X (2022) Microstructure and compressive properties of porous 2024Al-Al_3Zr composites. Metals (Basel) 12(12):2017. https://doi.org/10.3390/met12122017
6. Xu D, Long W, Zhou X (2020) Microstructure and corrosion resistance of Al_3(Zr, Ti)/Al composite prepared by powder metallurgy. Adv Compos Lett 29:2633366X2090616. https://doi.org/10.1177/2633366X20906160

7. Srinivasarao B, Suryanarayana C, Oh-Ishi K, Hono K (2009) Microstructure and mechanical properties of Al–Zr nanocomposite materials. Mater Sci Eng, A 518(1–2):100–107. https://doi.org/10.1016/j.msea.2009.04.032
8. Pal A, Verma A, Kandpal BC, Saxsena S (2015) Stir casting of metal matrix composites—a review. Int J Comp Math Sci
9. Li C, Wu S, Lü S, Li J, Liu L, Xia L (2021) Effects of excessive Zr content and ultrasonic treatment on microstructure and mechanical properties of Al-Zn-Mg-Cu alloy. Metals (Basel) 11(4):632. https://doi.org/10.3390/met11040632
10. Gupta R, Daniel BSS (2021) Strengthening mechanisms in Al_3Zr-reinforced aluminum composite prepared by ultrasonic assisted casting. J Mater Eng Perform 30(4):2504–2513. https://doi.org/10.1007/s11665-021-05558-x
11. Oladijo OP, Awe SA, Akinlabi ET, Phiri RR, Collieus LL, Phuti RE (2021) High-temperature properties of metal matrix composites. In: Encyclopedia of materials: composites. Elsevier, pp 360–374. https://doi.org/10.1016/B978-0-12-819724-0.00096-3
12. Danilov P, Kudryashova O, Khmeleva M (2018) Crystallization of metals in the ultrasonic field. MATEC Web Conf 243:00006. https://doi.org/10.1051/matecconf/201824300006
13. Lebon GSB, Tzanakis I, Pericleous K, Eskin D (2018) Experimental and numerical investigation of acoustic pressures in different liquids. UltrasonSonochem 42:411–421. https://doi.org/10.1016/j.ultsonch.2017.12.002
14. Wang F, Qiu D, Liu Z, Taylor J, Easton M, Zhang M (2014) Crystallographic study of Al_3Zr and Al3Nb as grain refiners for Al alloys. Trans Nonferr Metals Soc China 24(7):2034–2040. https://doi.org/10.1016/S1003-6326(14)63309-4
15. Kim H, Suslick K (2018) The effects of ultrasound on crystals: sonocrystallization and sonofragmentation. Crystals (Basel) 8(7):280. https://doi.org/10.3390/cryst8070280
16. Umeda J, Nishimura N, Fujii H, Jia L, Kondoh K (2021) In-situ formed Al_3Zr compounds reinforced Al composites and tribological application. Crystals (Basel) 11(3):227. https://doi.org/10.3390/cryst11030227
17. Subroto T et al (2020) Structure refinement upon ultrasonic melt treatment in a DC casting launder. JOM 72(11):4071–4081. https://doi.org/10.1007/s11837-020-04269-3

An Expansion of Ferrate Treatment Technology Domain for Se(IV) and Se(VI) Removal from Aqueous Medium

Saida Shaik, Soni Jha, Sanchita Chakravarty, Priya Ranjan Mishra, and Rina Sahu

1 Introduction

Selenium (Se) has a substantial role as a micronutrient in animal and humans at low concentrations. Selenoproteins plays a crucial role in the process of DNA construction, thyroid hormone synthesis, antioxidant defense, fertility, and reproduction in the animal and human bodies. However, consumption of excess selenium (> 1.4 mg/L) would create severe risk and harms animal and human health. According to World Health Organization (WHO), intake of selenium concentrations over 400 μg per day may lead to health issues such as gastrointestinal and neurological problems, respiratory distress, hair and nail loss or brittleness, and fatigue, to name some. Selenium exists in various states named as elemental (Se^0), selenide (Se^{2-}), selenite ($SeO_3{}^{2-}$), and selenate ($SeO_4{}^{2-}$). Among all oxidant forms, selenate $[Se^{VI}O_4]^{2-}$ and selenite $[Se^{IV}O_3]^{2-}$ oxyanions are more toxic. Water is the primary source of selenium intake in humans and animals. Selenium mobilization in water may take place through natural processes such as weathering of plant life, soils, bedrocks, and anthropogenic activities such as mining, smelting, and refining of selenium ores and concentrates [1, 2]. Therefore, the removal of Se(VI) and Se(IV) from water bodies has aroused great concern all over the world.

S. Shaik (✉)
Metallurgical and Materials Engineering Department, Indian Institute of Technology Kharagpur, Kharagpur, West Bengal 721302, India
e-mail: shahid.saida@gmail.com

S. Jha · S. Chakravarty
Analytical Chemistry Group, MER Division, CSIR-National Metallurgical Laboratory, Jamshedpur 831007, India

P. R. Mishra (✉) · R. Sahu
Metallurgical and Materials Engineering Department, National Institute of Technology Jamshedpur, Jamshedpur 831014, India
e-mail: mishranit90@gmail.com

R. Sahu et al. (eds.), *Advancements in Materials Processing Technology, Volume 1*, Springer Proceedings in Materials 48, https://doi.org/10.1007/978-981-97-4958-4_31

There are several technologies based on physical adsorption, complexation, heterogeneous photocatalysis, and chemical oxidation available for the removal of both inorganic and organic effluents from water bodies. Among all, the use of the chemical oxidation technique is renowned as the famous and cost-effective method for water cleaning. Some of the popular oxidants used for the cleaning of both organic and inorganic wastes from water sources are; chlorine, hypochlorite, hydrogen peroxide, ozone, sulfur dioxide, and ammonium bisulfate. But, the use of these oxidants generates intermediate toxics, and they are inefficient for the complete removal of effluents. In recent years, the use of eco-friendly oxidants based on the ferrate ligand complexes and ferrate salts are promising for the organic and metallic effluents removal from water. Ferrates (IV, V, VI) have high oxidation properties in both acid and basic medium, the oxidation–reduction potential of ferrates in the acid medium is 2.20 V, and in the basic medium, it is 0.72 V. In a single dose, ferrate (Fe(IV)) can simultaneously remove toxic metals, drugs, pesticides, and industrial toxins from water sources. It can also aid in inactivating disinfects and human pathogens like viruses, spores, bacteria, and protozoa. Another advantage of using ferrates is that it forms an environmentally friendly byproduct in the form of insoluble iron (III) hydroxide [3–5].

Ferrate's utilization and efficiency in removing the organic and different metallic effluents like Cu, Ni, Ag, As, etc., from the water bodies has been documented in several publications for decades [6–11]. There are few papers available on selenium removal from water bodies by using ferrate technology. Still, there is a significant gap regarding the zeta potential and size of the precipitates formed during the selenium removal process. The objective of this paper is to discuss the reaction mechanism, kinetics of selenium removal, zeta potential variations of ferrate precipitates, and size dependency on the pH of the water systems.

2 Materials and Methods

2.1 Materials Used

Analytical grade sodium selenate anhydrous (Na_2SeO_4) and sodium selenite (Na_2SeO_3) were used as the sources for selenium (VI) and selenium (IV), respectively, in water samples. Analytical grade ferric nitrate and potassium hypochlorite are used for the synthesis of ferrates. Analytical grade pure HCl and NaOH were diluted to 1 M using deionized water for the pH adjustment experiments. These chemicals were used as such without any prior purification. NIST traceable certified reference material standards solutions procured from Thermo Fisher Scientific were used in the calibration and determination of selenium concentration. For the preparation of all the solutions distilled Milli-Q Water with conductivity less than 18 $M\Omega.cm^{-1}$ is used.

2.2 Synthesis of Potassium Ferrate

There are several methods based on wet chemical oxidation methods, electrochemical processes, and thermal route methods available for the synthesis of potassium ferrate. Generally, for large-quantity productions, thermal base routes are used for potassium ferrate synthesis. But, the purity of the final potassium ferrate produces is very low (30–50 wt% potassium ferrate) in the thermal process. Wet chemical oxidation and electrochemical methods are used for the laboratory scale experiments, in which over 98 wt% pure potassium ferrate is obtained as the final product. In the present work, the wet chemical method was used for the synthesis of potassium ferrate using ferric nitrate and potassium hypochlorite as precursors [3]. To check the purity of the final product, all the material was dissolved in chilled water and filtered with 0.45 μm membrane filter paper to separate the colloidal ferric hydroxide and oxides. The results show that the purity of potassium ferrate in the final product obtained was 97.8 wt%.

2.3 Analysis and Characterization

The pH value of the solutions of Se(IV) and Se(VI) was measured by the use of 794 Basic Titrino with 728stirrer (Make: Metrohm). The equipment has the facility to measure pH value as well as the temperature of the solution for temperature correction ± 2 °C. The concentration of Se(IV)and Se(VI) was measured by the use of ICP-OES (Thermo Scientific iCAP 7600 DUO ICP-OES) and AAS-HG (Varian 240 with VP 100). The concentration of ferrate solution used in the adsorption study was analyzed by volumetric titration method [12]. Total iron concentration in the control sample and treated sample solutions was also determined by ICP-OES.

Malvern Nano-Zeta Sizer (ZS) is used to study size and zeta potential of the floc formed on the reduction of ferrate ion to ferric hydroxide by photon correlation spectroscopy (PCS) alternatively known as dynamic light scattering (DLS) or quasi-elastic light scattering (QELS). PCS is a technique that is able to deal with the measurement of particle size distribution in the nano-size region. PCS is applied to suspension formed after the reduction of ferrate ion to ferric hydroxide in aqueous medium at different pH values.

2.4 Experimental Design

In the present study, three types of solution samples were synthesized which were; (i) blank; the sample formed by the addition of ferrate in deionized water, (ii) sample formed after addition of ferrate to synthesized selenium-containing water, and (iii) control sample of water containing synthesized selenium without the addition of ferrate. At first, we studied the influence of pH on the efficiency of Se(IV) and Se(VI) ions removal. The pH value of the solutions was adjusted by 1 M solution of hydrochloric acid with the help of 794 Basic Titrino. In all the experiments, 100 ml of two different synthetic Se(IV) and Se(VI)-induced water samples was used with the overall concentration of Se(IV) and Se(VI) having 0.5 mg/L. Potassium ferrate solution with 13.5 mg/L of total Fe content (freshly prepared in chilled deionized Milli-Q water and filtered) was added to the sample while stirring, and pH of the solution was maintained at the required quantity. After 30 min solutions were filtered with 0.45 micronNylon6, membrane filter paper. After that, the filtered water samples were sent for ICP-OES analysis to analyze the percentage of selenium removal. Later on, the effect of potassium ferrate concentration variation on selenium removal was carried out in the range of 0.5 to 12 mg/L as Fe. This experiment was carried out at pH 3.5 $\pm$ 0.4 and at 0.5 mg/L concentration of Se(IV) and Se(VI) each in separate samples. For each study, 100 ml of synthesized selenium-containing water is taken, and the required amount of potassium ferrate was added to the solution. After 30 min reaction time, the samples were collected using filtration and sent to ICP-OES analysis as mentioned in pH variation analysis. Zeta potential and particle size analysis were done by using DLS method to know the mechanism of selenium ions removal and changes in oxidation states of ferrate in water samples. The amount of potassium ferrate was adjusted to the final concentration of Fe equal to 12 g/L. For each experiment, 10 ml of filtered precipitate iron oxide was taken for this study, and a small amount of HCl was added along with continuous stirring. The reproducibility was confirmed in triplicate experimental runs. Finally, to investigate the adsorption kinetics of selenium ions on ferrates batch type, adsorption experiments were carried out in 100 ml synthesized solutions. Adsorption isotherm on ferrate was studied for six different concentrations of Se(IV) and Se(VI) in the range of 0.1–2.0 mg/L. In these experiments, all initial pH values were below 7.0, ensuring that all added ferrate(VI) ~ 12 mg/L was decomposed during the oxidation step. After 60 min of contact time, the samples were filtered and sent for the chemical analysis. All the experiments were performed at room temperature and constant stirring speed 250 rpm.

3 Result and Discussion

3.1 Effect of pH Variation on Selenium(IV) and Selenium(VI) Removal

The effect of initial pH value (2.5–8) on the adsorption of Se(IV) and Se(VI) in the presence of potassium ferrate was given in Fig. 1. It shows that the percentage of both Se(IV) and Se(VI) ions removal favored at low pH values. Almost complete Se(IV) content was precipitated out at the pH values 3.5. While, complete removal of Se(VI) form requires even further low pH values, which is noted as 2.6. A sharp increase in both selenium metallic ions removal form water sources observed at pH values below 5. An increase in pH values over 5 almost shows zero removal of both Se(IV) and Se(VI). Figure 2 represents the speciation of both Se(IV) and Se(VI) with the variation in pH values. It also shows that, at pH above 9 both Se(IV) and Se(VI) exhibit the SeO_3^{-2} and SeO_4^{-2}, which are readily soluble in water and hard to remove. In the pH range 3 to 9, Se(IV) ions exit in both ionic SeO_3^{-2} and complex $HSeO_3^{-1}$ form. As the pH lowers (pH < 3) hydroxyl complexes like H_2SeO_3 and $HSeO_3^{-1}$ are the most stable phases for Se(IV) ion in aqueous solutions. These complexes have low electronegativity compared to ionic forms, which makes them harder to precipitate on the surface of the iron oxide. In the case of Se(VI) ions, some amount of SeO_4^{-2} ionic forms exists even at pH below 1, which indicates that Se(VI) form, it shows less complexation and high electronegativity which make it very easy to adsorb on positive surface charge iron oxides. Low adsorption efficiencies of both Se(VI) and Se(IV) are obtained even though both SeO_3^{-2} and SeO_4^{-2} are stable species at high pH values. This is due to the accumulation of OH^- ions (at pH > 5, as shown in Fig. 2) around the iron oxides at high pH values iron oxides which reduces the surface charge of the iron oxide [10, 13, 14]. In general, anion (Se(IV) and Se(VI) or SeO_3^{-2} and SeO_4^{-2}) adsorption is favored by the high electropositivity on the iron oxide surface. In an acidic environment, iron oxide is surrounded by H^+ ions which aid in increasing the positive surface charge of iron oxides.

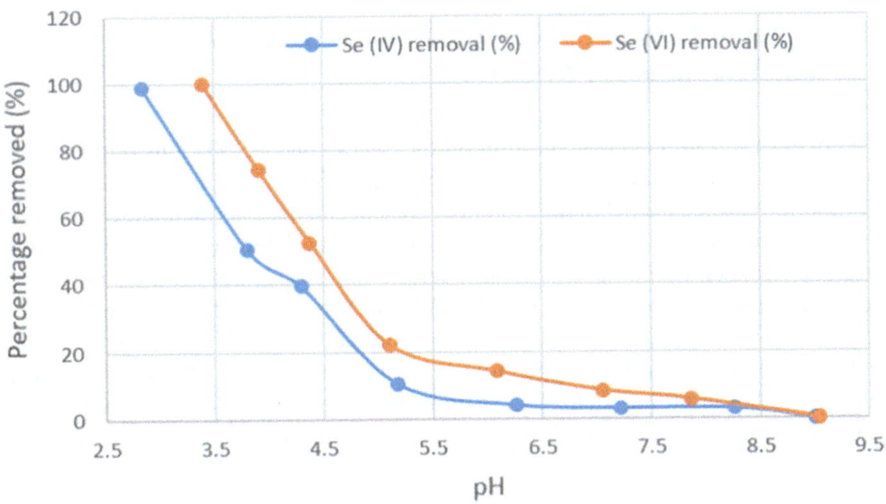

Fig. 1 pH variation study on Se(IV) and Se(VI) removal

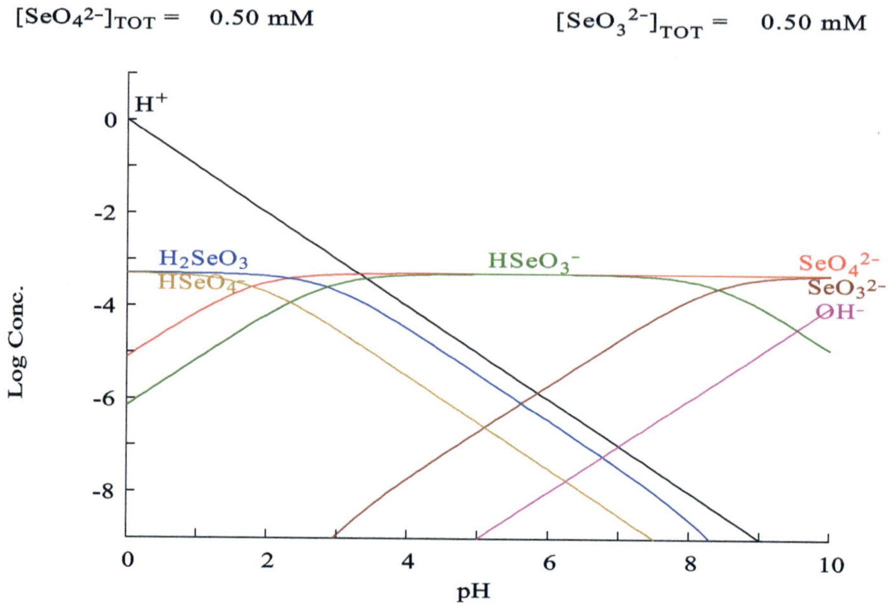

Fig. 2 Species distribution of Se(IV) and Se(VI) ions with the variation in pH

3.2 *Effect of Ferrate Concentration on Selenium(IV) and Selenium(VI) Removal*

The effect of ferrate ion concentration on the Se(IV) and Se(VI) ions removal was investigated in the ferrate ion concentration range 0.5 to 12 mg/L Fe. Based on the pH variation results, the pH of the system in present calculations maintained constant as 3.5. Figure 3 shows the efficiency of Se(IV) and Se(VI) removal as a function of ferrate ion concentration (as Fe). At low ferrate ion dosage Se(VI) ion shows high precipitation tendency compared with Se(IV) ion. While, at high concentrations (Fe > 5) Se(IV) ion has high removal efficiency compared to Se(VI) ions. This is may be due to the high release of oxygen ions at high concentrations of ferric ions which helps in oxidation of Se(IV) to Se(VI) form. However, both Se(VI) and Se(IV) get completely removed at about 12 mg/L ferrate concentration.

3.3 *Effect of pH on the Variation in Zeta Potential and Precipitate Size of Iron Oxides*

Figure 4 demonstrates the zeta potential and size changes in iron oxide precipitate formed with variation in pH. From the figure, it is evident that surface charge of iron oxides decreases with an increase in pH values, as mentioned earlier in Sect. 3.1. The

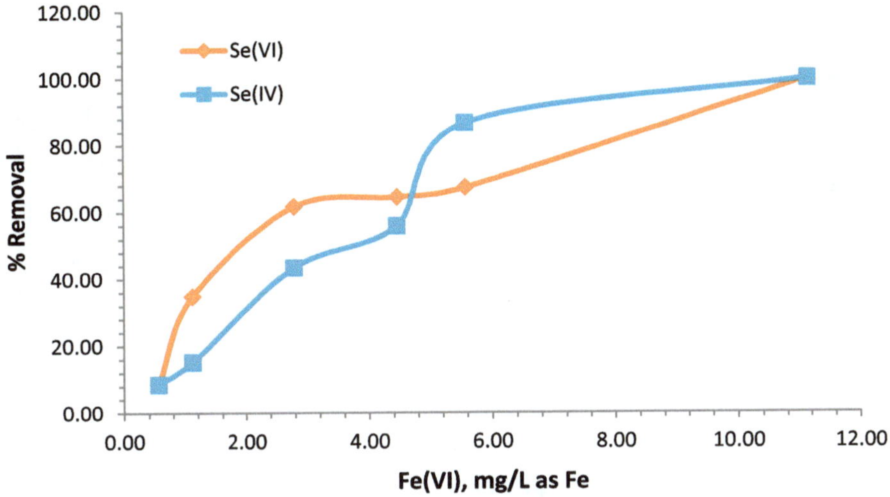

Fig. 3 Effect of ferrate concentration on removal efficiency of Se(IV) and Se(VI)

surface charge of adsorbent is neutral or zero at pH equals to 5.5, which is indicated by the point of zero charge (PZC). Increase in pH values is greater than 5.5, the surface of adsorbent induced with a negative charge, which creates the repulsion toward surrounding selenium anions (SeO_3^{-2} and SeO_4^{-2}). In contrast, at low pH (< 5) surface of adsorbent has a positive charge which creates the attraction forces to surrounding selenium anions. Figure 4 also shows that the average particle size of adsorbents formed increases with a reduction in pH values. An increase in the size of the particles may be attributed to the adsorption of selenium ions on the surface of iron oxide precipitates.

3.4 Selenium Adsorption Isotherms

The nature of the interaction between adsorbent and adsorbate is described by using the adsorption isotherm. An investigation of the adsorption isotherm on ferrate has been studied and reported with conventional adsorption isotherms studies. Batch adsorption experiments were carried out with pH value near 3.5, keeping the ferrate concentration constant. The experimental data were fit into the linear forms of Langmuir, Freundlich [15–17] isotherm equations, which were labeled in Eqs. (1) and (2). The correlation coefficients obtained from the fits are used to describe the appropriate adsorption mechanism of selenium on

$$\frac{1}{q_\mathrm{e}} = \frac{1}{k_\mathrm{L} q_\mathrm{m}} . \frac{1}{c_e} + \frac{1}{q_\mathrm{m}}, \tag{1}$$

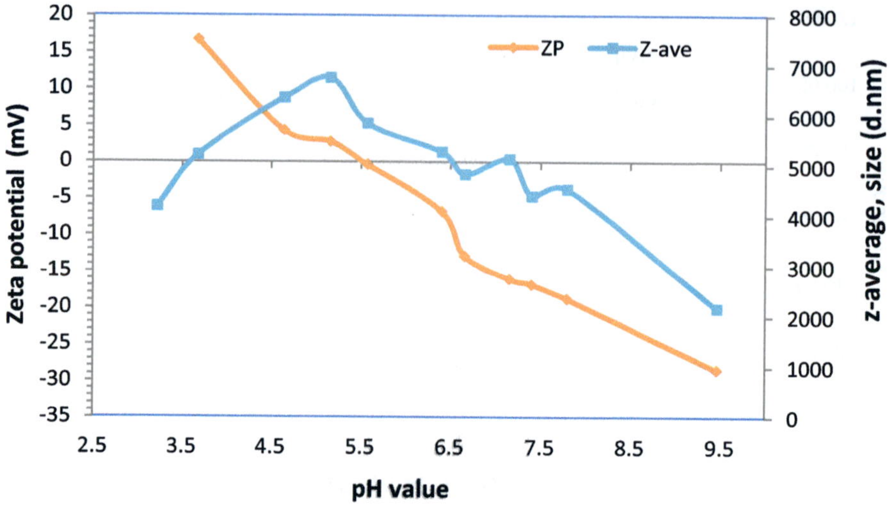

Fig. 4 Effect of pH on the zeta potential (ZP) and average size (Z-ave) of the potassium ferrate suspension

$$\ln q_e = \ln k_F + \frac{1}{n} . \ln c_e, \tag{2}$$

where C_e represents the initial Se(IV) or Se(V) concentration (mg/L), q_e indicates the equilibrium selenium ion concentration after the addition of adsorbate (mg/g), q_m is the maximum adsorption capacity mg/g, k_L is the Langmuir constant (L/mg), k_F (mg/g)/(mg/L)$^{1/n}$) and n are the Freundlich constants. Figure 5 and Table 1 indicate the adsorption isotherm data of selenite (Se(IV)) and selenate (Se(VI)) fitted using the Langmuir and Freundlich isotherm equations. It shows that the adsorption data of both selenite and selenate fits well in the Langmuir isotherm model equation with better correlations coefficients (R^2) values 0.91 and 0.98, respectively. The maximum adsorption capacity (q_m) for selenite and selenate on ferrate ion obtained from the Langmuir equation was noted as 35.21 μg g^{-1} and 250 μg g^{-1}, respectively. Whereas the rate constant (k_L) values for selenite and selenate removal were noted as 0.0252 min^{-1} and 0.07 min^{-1}, respectively.

3.5 *Mechanism of Selenium Ions Removal*

Potassium ferrate has high oxidation potential in both acidic and basic medium, and it acts as a powerful oxidizing agent for many of the inorganic as well as organic compounds. During the oxidizing process, Fe(VI) loses electrons and forms Fe(III) ion which is a good coagulant and adsorbent. Figure 6 indicates the E-pH diagram of iron and water systems. It shows that at high potentials ($E_{SHE} > 2$) $FeO_4{}^{-2}$ ion is

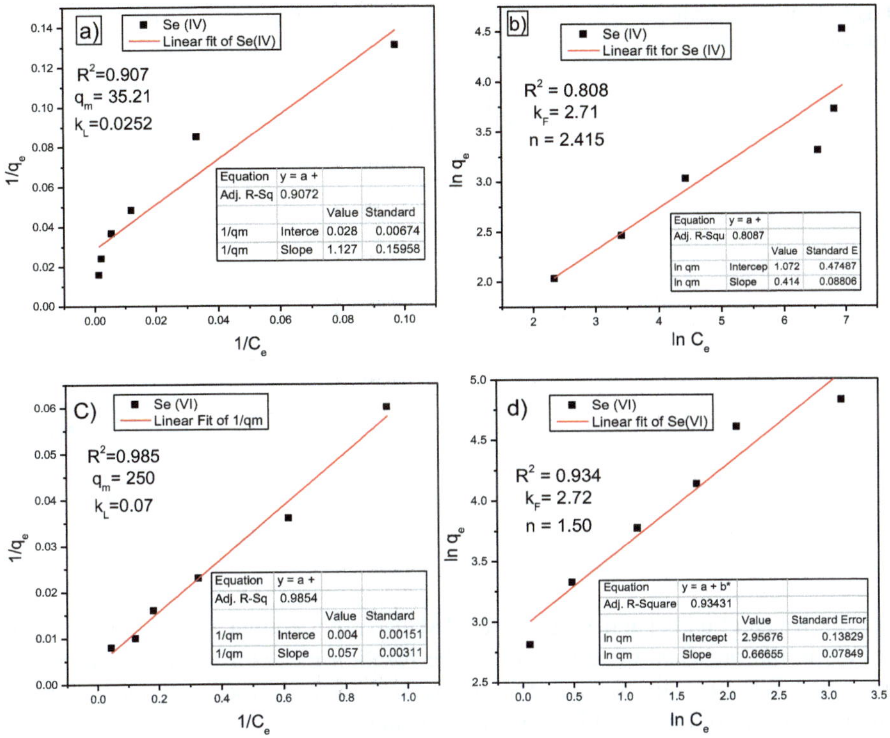

Fig. 5 Adsorption isotherms, **a** Langmuir isotherm for selenite adsorption, **b** Freundlich isotherm for selenite adsorption, **c** Langmuir isotherm for selenate adsorption, **d** Freundlich isotherm for selenate adsorption

Table 1 Parameters of Langmuir and Freundlich model for selenite and selenate adsorption onto ferrate

Selenium species	Langmuir model			Freundlich model		
	q_m (μg g^{-1})	k_L	R^2	n	k_F	R^2
Se(IV)	35.21	0.025	0.907	2.4	2.71	0.808
Se(VI)	250	0.07	0.985	1.5	2.72	0.934

the stable phase for Fe. At low potentials ($0.7 < E_{SHE} < 2.2$) and pH less than 0.5, Fe^{3+} species are the stable phases, and Fe_2O_3 are equilibrium species that existed in normal operating conditions. In water, ferrate ion releases oxygen ion and forms iron(III)oxide. The reactions followed during the ferrate oxidation are labeled in Eqs. (1) and (2). Figure 7 indicates the E-pH diagram of selenium in the water system. It shows that at high potentials ($E_{SHE} > 1$) Se(VI) ionic forms (SeO_4^{-2} and $HSeO_4^{-2}$) are most stable. By comparing both iron and selenium E-pH diagrams, it is evident that in the presence of FeO_4^{-2} (at $E_{SHE} > 2$), all the selenium exists in the form of

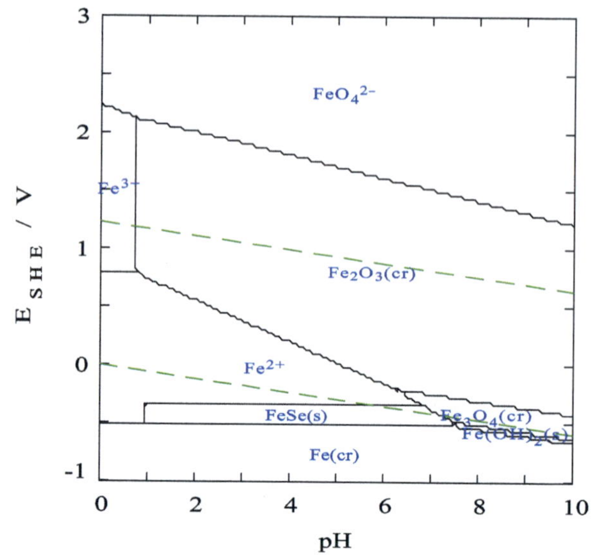

Fig. 6 E-pH diagram for iron and water system

Se(VI). Xu and Fu [14] and Viktor et al. [18] also reported that Se(IV) oxidized into Se(VI) in the presence of ferrate ions. The reaction mechanism of this process is described in Eq. (3). In the final step, the SeO_4^{2-} ions adsorbed onto the positively charged surface of Fe_2O_3 and formed precipitates.

$$FeO_4^{2-} + 8H^+ + 3e^- = Fe^{3+} + 4H_2O \tag{3}$$

$$2FeO_4^{2-} + 10H^+ + 6e^- = Fe_2O_3 + 5H_2O \tag{4}$$

$$2FeO_4^{2-} + 3SeO_3^{2-} = Fe_2O_3 + 3SeO_4^{2-} + 2O^{-2} \tag{5}$$

3.6 *Removal Efficiency in Real Wastewater Sample*

The actual wastewater sample collected from a copper processing plant is studied. Preliminary studies indicate the removal efficiency in wastewater is tricky due to the presence of other metal ions in wastewater. In the presence of other anions and cations removal efficiency of Se(IV) and Se(VI) decreased, it may be due to the simultaneous oxidation and adsorption of other anions and cations present in the sample. Increasing the Fe(VI) dose increases removal efficiency, which is ascertained from Table 2. It shows that an increase in dosage enhances the selenium removal in contaminated water. Wang et al. [19] also noted that the presence of contaminants like Ca, Mg, Cu, etc., reduce the selenium adsorption on ferrate ions.

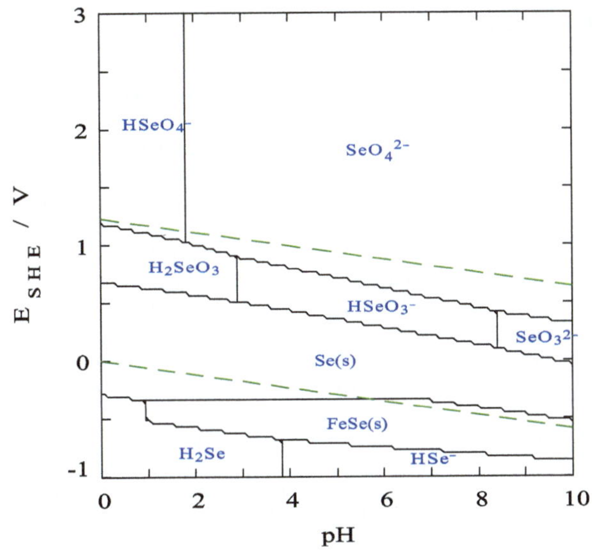

Fig. 7 E-pH diagram of selenium and H_2O system

Table 2 Removal percentage of selenium in the real wastewater sample

Dosage Fe(VI), mg/L	Removal Se_{total}(%)
17.2	15.90
30.5	40.82
68.8	85.56

4 Conclusion

In present work, the species distribution mechanism of selenite and selenate ions based on the pH and Eh has been investigated. The pH and ferrate ion dosage have a significant impact on the selenium ions removal and surface properties of iron oxide. An increase in pH value reduces the adsorption efficiency of both selenium ions (Se(IV) and Se(VI)). This is due to the accumulation of OH^- ions around the precipitated iron oxide, which creates the repulsive forces on negatively charged selenium ions. An increase in ferrate ion concentration enhanced both selenium ions removal efficiency. The adsorption isotherms show that both selenate and selenite follow the Langmuir isotherm model during the removal process. Thus, the adsorption results suggest the homogeneous nature of adsorption surfaces or sites on iron oxides. Maximum selenate and selenite adsorption on given ferrate ion dosage (12 g/L Fe) are noted as 35.21 $\mu g\ g^{-1}$ and 250 $\mu g\ g^{-1}$, respectively.

References

1. Lemly AD (2004) Aquatic selenium pollution is a global environmental safety issue. Ecotoxicol Environ Saf 59:44–56
2. Navarro-Alarcon M, Cabrera-Vique C (2008) Selenium in food and the human body: a review. Sci Total Environ 400(1–3):115–141
3. Li C, Li XZ, Graham N (2005) A study of the preparation and reactivity of potassium ferrate. Chemosphere 61:537–543
4. Maity S, Chakravarty S, Bhattacharjeea S, Roy BC (2005) A study on arsenic adsorption on polymetallic sea nodule in aqueous medium. Water Res 39:2579–2590
5. Murmann RK, Robinson PR (1974) Experiments utilizing FeO_4^{2-} for purifying water. Water Res 8:543–547
6. Yin Z, Xu S, Liu S, Xu S, Li J, Zhang Y (2020) A novel magnetic biochar prepared by K_2FeO_4-promoted oxidative pyrolysis of pomelo peel for adsorption of hexavalent chromium. Biores Technol 300:122680
7. Filip J, Yngard RA, Siskova K, Marusak Z, Ettler V, Sajdl P, Sharma VK, Zboril R (2011) Mechanisms and efficiency of the simultaneous removal of metals and cyanides by using ferrate (VI): crucial roles of nanocrystalline iron (III) oxyhydroxides and metal carbonates. Chem–Eur J 17(36):10097–10105
8. Zhang H, Zheng L, Li Z, Pi K, Deng Y (2020) One-step Ferrate (VI) treatment as a core process for alternative drinking water treatment. Chemosphere 242:125134
9. Prucek R, Tuček J, Kolařík J, Filip J, Marušák Z, Sharma VK, Zbořil R (2013) Ferrate (VI)-induced arsenite and arsenate removal by in situ structural incorporation into magnetic iron (III) oxide nanoparticles. Environ Sci Technol 47(7):3283–3292
10. Sharma VK, Zboril R, Varma RS (2015) Ferrates: greener oxidants with multimodal action in water treatment technologies. Acc Chem Res 48(2):182–191
11. Yates BJ, Zboril R, Sharma VK (2014) Engineering aspects of ferrate in water and wastewater treatment–a review. J Environ Sci Health Part A 49(14):1603–1614
12. Schreyer JM, Thompson GW, Ockerman LT (1950) Oxidation of chromium (III) with potassium ferrate (VI). Anal Chem 22(11):1426–1427
13. Ghernaout D, Naceur MW (2011) Ferrate (VI): in situ generation and water treatment–a review. Desalin Water Treat 30(1–3):319–332
14. Xu L, Fu F (2020) Se (IV) oxidation by ferrate (VI) and subsequent in-situ removal of selenium species with the reduction products of ferrate (VI): performance and mechanism. J Environ Sci Health Part A:1–9
15. Altın O, Özbelge HÖ, Doğu T (1998) Use of general purpose adsorption isotherms for heavy metal–clay mineral interactions. J Colloid Interface Sci 198(1):130–140
16. Foo KY, Hameed BH (2010) Insights into the modeling of adsorption isotherm systems. Chem Eng J 156(1):2–10
17. Crini G (2006) Non-conventional low-cost adsorbents for dye removal: a review. Biores Technol 97(9):1061–1085
18. Viktor Z, Wang L, Ma J (2020) Promotional effect of Mn (II)/K_2FeO_4 applying onto Se (IV) removal. J Hazard Mater 384:121264
19. Wang J, Li J, Xie L, Liu Q, Zeng H (2020) Understanding the interaction mechanism between elemental selenium and ferric hydroxide in wastewater treatment. Industr Eng Chem Res. https://doi.org/10.1021/acs.iecr.0c00533

MIX
Papier aus verantwortungsvollen Quellen
Paper from responsible sources
FSC® C105338

If you have any concerns about our products,
you can contact us on
ProductSafety@springernature.com

In case Publisher is established outside the EU,
the EU authorized representative is:
Springer Nature Customer Service Center GmbH
Europaplatz 3, 69115 Heidelberg, Germany

Printed by Libri Plureos GmbH
in Hamburg, Germany